Dying for Victorian Medicine

Also by Elizabeth T. Hurren

PROTESTING ABOUT PAUPERISM: Poverty, Politics and Poor Relief in Late Victorian England

Dying for Victorian Medicine

English Anatomy and its Trade in the Dead Poor, c. 1834–1929

Elizabeth T. Hurren
Reader in the History of Medicine, Oxford Brookes University, UK

© Elizabeth T. Hurren 2012
Softcover reprint of the hardcover 1st edition 2012 978-0-230-21966-3
All rights reserved. No reproduction, copy or transmission of this publication may be made without written permission.

No portion of this publication may be reproduced, copied or transmitted save with written permission or in accordance with the provisions of the Copyright, Designs and Patents Act 1988, or under the terms of any licence permitting limited copying issued by the Copyright Licensing Agency, Saffron House, 6–10 Kirby Street, London EC1N 8TS.

Any person who does any unauthorized act in relation to this publication may be liable to criminal prosecution and civil claims for damages.

The author has asserted her right to be identified as the author of this work in accordance with the Copyright, Designs and Patents Act 1988.

First published 2012 by
PALGRAVE MACMILLAN

Palgrave Macmillan in the UK is an imprint of Macmillan Publishers Limited, registered in England, company number 785998, of Houndmills, Basingstoke, Hampshire RG21 6XS.

Palgrave Macmillan in the US is a division of St Martin's Press LLC, 175 Fifth Avenue, New York, NY 10010.

Palgrave Macmillan is the global academic imprint of the above companies and has companies and representatives throughout the world.

Palgrave® and Macmillan® are registered trademarks in the United States, the United Kingdom, Europe and other countries.

ISBN 978-1-349-30515-5   ISBN 978-0-230-35565-1 (eBook)
DOI 10.1057/9780230355651

This book is printed on paper suitable for recycling and made from fully managed and sustained forest sources. Logging, pulping and manufacturing processes are expected to conform to the environmental regulations of the country of origin.

A catalogue record for this book is available from the British Library.

A catalog record for this book is available from the Library of Congress.

10 9 8 7 6 5 4 3 2 1
21 20 19 18 17 16 15 14 13 12

Pro Vitis Praeteritis Et Futuris

This book is dedicated to
Professor Steven Andrew King
&
Master Thomas Philip de Banke

Contents

List of Tables	viii
List of Figures	ix
List of Illustrations	x
List of Abbreviations	xi
Acknowledgements	xii
Preface	xv

Part I   An Historical Landscape

1	Chalk on the Coffin: Re-Reading the Anatomy Act of 1832	3
2	Restoring the Face of the Corpse: Victorian Death and Dying	41
3	A Dissection Room Drama: English Medical Education	74

Part II   An English Anatomy Trade

4	Dealing in the Dispossessed Poor: St. Bartholomew's Hospital	119
5	Pauper Corpses: Cambridge and Its Provincial Trade	175
6	Balancing the Books: The Business of Anatomy at Oxford University	218
7	'Better a third of a loaf than no bread': Manchester's Human Material	264
	Conclusion	303

Notes	312
Select Bibliography	356
Index	373

Tables

4.1	Overview of the Body Trade by London Location, 1832–72 & 1885–1930	145
4.2	Infirmaries and Workhouses that Sold Bodies to St. Bartholomew's Hospital, 1832–72, 1885–1930	146
4.3	Bodies Bought from London Voluntary Hospitals, 1832–72, 1885–1930	153
4.4	Street Deals Sold to St. Bartholomew's Hospital, 1832–72, 1885–1930	155
4.5	Bodies Bought from Prisons, Asylums and Mental Health Hospitals, 1832–72, 1885–1930	157
4.6	The Average Number of Bodies Dissected, Ratios of Male: Females Corpses by Decade at St. Bartholomew's Hospital, 1832–72, 1885–1930	161
4.7	The Disease Categories of Corpses Dissected at St. Bartholomew's Hospital, 1832–72, 1885–1930	166
5.1	Mr Rosebrook Morris the Surgeon's Death Register: Deaths Due to a Lack of Medical Relief in the Brixworth Union, c. 1836–1844	180
5.2	Geography of Suppliers to Cambridge Anatomical School, 1855–1920	210
6.1	Travelling Anatomists: The Railway Journeys made for Public Engagement Reasons by Professor Alexander Macalister and Professor Arthur Thomson in Provincial England, c. 1880–1890	232
6.2	Body Parts & Whole Cadavers Purchased for Dissection by Professor Arthur Thomson on behalf of Oxford University Anatomy School, 1885–94	234
6.3	Pauper Bodies Sold to Oxford Anatomy School, 1895–1929	237
6.4	Arthur Thomson's List of Medical Students and their Career Destinations after Oxford in the late-Victorian Era	243
6.5	City-Coroner's Cases in Oxford, 1877–1888	252
6.6	Oxford Anatomy School's Expenditure on Human Material, 1885–95	256

Figures

4.1	The Total Number of Bodies of the Poor Bought by St. Bartholomew's Hospital for Dissection, c. 1832–1929	132
4.2	Number of Bodies per Decade Acquired by Anatomists at St. Bartholomew's Hospital, c. 1832–1929	142
4.3	Number of Male and Female Bodies Sold to St. Bartholomew's Hospital, 1832 to 1929	160
4.4	Bodies of Males and Females Acquired for Dissection per Decade at St. Bartholomew's Hospital, 1832 to 1929	162
4.5	Age Range of Bodies Sold to St. Bartholomew's Hospital where n = 5062 bodies, 1832 to 1929	164
5.1	The Number of Bodies Sold for Dissection to Cambridge University, 1855 to 1920	191
5.2	The Age Profile of the Bodies Sold to Cambridge Anatomy School, 1855–1920	200
6.1	The Number of Bodies Sold to Oxford Anatomy School, 1885–1929	253
7.1	Anatomy Inspectorate Returns from Manchester Anatomists on Poor Law Purchases, 1834–1871	290
7.2	Number of Pauper Bodies from Crumpsall Workhouse, May 1883–May 1884, where n = 194	293
7.3	Bodies Sold from Crumpsall Workhouse to Manchester Medical School, 1860–1890	298

Illustrations

1.1 ©Shropshire Record Office on behalf of Shropshire Local History Society, poster styled 'Anatomists – To be Sold by Auction (Without Reserve) on Tuesday the 8th of November 1836 at Starvation Hall' — 26

1.2 ©Northampton Library Local Studies Room, poster styled 'The Brixworth Stakes, The Event will run on the Poor Law Course, Brixworth, on the 29th inst… – NB No horsemen allowed on the course [the poor cannot vote]' — 27

3.1 ©Wellcome Trust Image Collection, Slide Number, L0013321, 'The dissection of a young beautiful woman directed by J. CH. G. Lucas (1814–1885) in order to determine the ideal female proportions', chalk drawing by J. H. Hasselhorst, (1864) — 107

3.2 ©Wellcome Trust Image Collection, Slide Number, L0002687, Victorian Photograph of 'The Interior of the Department of Anatomy at Cambridge University, 1888' — 110

3.3 ©Wellcome Trust Image Collection, Slide Number, L0013441, Victorian Photograph of 'The interior of Edinburgh University dissection room, 1889' — 112

3.4 ©Wellcome Trust Image Collection, Slide Number, L0014980, Victorian Photograph of 'The Interior of the Dissecting Room, Medical School, Newcastle-Upon-Tyne, 1897' — 112

3.5 ©Wellcome Trust Image Collection, Slide Number, L0039195, Victorian Photograph of 'The Interior of a dissecting room: five students and/or teachers dissect a corpse at University College, London' — 113

4.1 Locations of Bodies coming in from the larger Infirmaries and Workhouses, purchased by St. Bartholomew's hospital, 1832–1930 — 159

5.1 ©Cambridge Library Local Studies Room, Cambridge cartoon, anti-anatomy, 1858 — 194

Abbreviations

BL	British Library
Bodl. Lib.	Bodleian Library
COS	Charity Organisation Society
CRO	Cambridgeshire Record Office
DMOP	District Medical Officer for the Poor
DWP	Department of Work and Pensions
JRL	John Radcliffe Library Manchester
LGB	Local Government Board
LMA	London Metropolitan Archives
LRO	Leicestershire Record Office
MLSR	Manchester Local Studies Room
MRDRA	Medical Relief Disqualification (Removal) Act 1885
MRO	Greater Manchester Record Office
NPLAA	New Poor Law Amendment Act, referred to as New Poor Law
NRO	Northamptonshire Record Office
OB	Old Bailey Court Records
OHA	Oxfordshire Health Archives
ORO	Oxfordshire Record Office
PLB	Poor Law Board
St BHA	St Bartholomew's Hospital Archive
TNA	The National Archives

Acknowledgements

This book would not have been feasible without the generous support of the Wellcome Trust in London. I am grateful for a Research Leave Award granted in 2009–10. It gave me the time to write up archive material that was collected from numerous sources over the past ten years. In particular, I would like to express my sincere thanks to Dr Tony Woods. His enthusiasm for the historical work of the medical humanities has been crucial. He has never flinched in his commitment to help realise innovative research. Likewise, Dr Nils Fietje and Liz Shaw have been generous with their time and funding efforts on my behalf. Dr Chris Stock and Tom Zeissen have been very supportive of the public engagement work undertaken on this and other research projects in which it has been a pleasure to represent the Wellcome Trust. I am also grateful to Professor Diana Woodhouse, Pro-Vice Chancellor for Research at Oxford Brookes University, for sponsoring a matching semester of research leave. It follows that any errors or discrepancies are mine and do not reflect the support of others in academic life.

This book is the product of a long research journey through countless dusty archives. I am grateful to many archivists. Special mention is made of St Bartholomew's hospital trust, the Bodleian Library, Oxford Record Office, Oxfordshire Health Archives, Cambridge University Library, Cambridge Record Office, the British Library, Manchester Local Studies, Manchester Record Office, the John Radcliffe Library, the National Archives, Northamptonshire Record Office, Leicestershire Record Office, and all those repositories cited in the footnotes that held Poor Law records.

The aim in this book has first and foremost been a human one, set out in the Preface. For this reason I am grateful to those scholars who pioneered work in this field and who by identifying the gaps in historical thinking left the various doors open to future research opportunities. In particular, I would like to express my sincere admiration and thanks (in alphabetical order) to some I know well and others at a distance: Joanna Innes, Thomas Laqueur, Ruth Richardson, Michael Sappol, Julie-Marie Strange, and Keir Waddington. I owe a debt of gratitude to Keir for alerting me to the St Bartholomew's dissection registers at the 'Locating the Victorians' conference in 1999. I hope that the case study in Chapter 4 has repaid finally his generosity. Michael too shared his innovative research and gave generously of his excellent scholarship and time on a visit to England in 2007–2008. Joanna asked the most difficult questions over a drink in Oxford and helped sharpen my focus. Julie-Marie has always expressed her enthusiasm for new ways of thinking about the poorest and their death experiences in her admirable work.

A special mention should also be made in respect to Dr Yolande Erraso who acted as my research assistant while waiting to begin a post-doctoral study of the history of cancer at Oxford Brookes University. Dr Ina Scherder has shared her recent research on Irish anatomy, as has Dr Fiona Hutton on medical education for her thesis study on mid-Victorian attitudes which I co-supervised in 2007. I would also notably like to express my thanks to Professor Detlev Mühlberger with whom I often discussed research over a glass of red wine. He was very supportive of the scholarship that underpins this book before his untimely death in 2008. He is still missed by colleagues who valued his decency and determination never to defend the indefensible.

I am also grateful to the Leverhulme Trust which granted me a Study Abroad Fellowship in 2008–2009 to investigate the influence of Paris anatomy on London medical education in the nineteenth century. In turn, the EHESS in Paris hosted research papers. Professor Paul André Rosenthal (now at Sciences Po) and Professor Patrice Bordelais were very supportive of the new research. I appreciated the long talks over their generous lunches that improved my theoretical thinking and French conversation. The experience of living and researching in Paris has strengthened both the breadth and depth of background research, which, though reduced to a footnote in places, nonetheless informs the intellectual underpinnings of this book.

I have been very fortunate also in having a wide circle of friends outside academic life. As one of the world's worst cooks, I want to express my thanks to those who have fed me on a regular basis. There have been too many to name (but *you know who you are*) who have saved me from the lonely baked potato. I would, however, like to make a special mention of two in particular. Glynis Chapman makes not only the best soup but also the sort of trifle that I thought only my Irish grannie could whip up. Thank you for the delicious home-cooked food, loving care, and for being a kindred spirit. Clare de Banke brought a gift into my world on the day that my godson, Thomas Philip de Banke, was born. Thank you for laughter, fig rolls, and fun-filled days finger-painting. It has been a long journey from finals to maternity room but I would not have missed a moment.

xiv *Acknowledgements*

I have also been fortunate to have made friends in academic life with people whose work I admire and who retain that rare quality, a sense of balance in their personal lives. Professor Peter King started out as my PhD supervisor. Over the last ten years his friendship has been loyal and support unflinching. Though he votes one way, and I the other, we both share a commitment to social justice that is beyond petty politics and has inspired this book. Professor Keith Snell published my first academic article. Down the years, I have admired his commitment and first-rate research scholarship in English local history. His friendship, laughter over good food, and interest in art are certainly at the heart of this book. Professor Thomas Sokoll has not only pioneered scholarship on pauper letters but also shared with me his love of classical music and jazz. It has been a pleasure to read the work of someone who believes in the importance of renewing friendships.

Others I see less often than I would like. For acting as hosts on weekends away, I would like to thank sincerely: Professor Andreas Gestrich and his wife Kate Tranter in Trier and London; Professor Silvie Hahn and her husband Stan in Salzburg; Marie Laywine in Dorset for her inspiring paintings; Professor Paul André Rosenthal, Dr Manuela Martini, and their children, Zelda and Marta, in Paris; my aunt Kathleen and mother, Maureen, in Northern Ireland, who waft gentle breezes of loving support across the Irish Sea; and not least, Bruce Balmer, who walks me sword side in Oxford and Italy.

Finally, this book is dedicated to Professor Steven Andrew King and Master Thomas Philip de Banke. Its Latin inscription contains a world of meaning that I know they will recognise both today and in the years ahead – Pro Vitis Praeteritis et Futuris.

Preface

It is a snowy February in 2009. A café has just opened in a fashionable part of Oxford. Sitting down to drink a latte, a lady next to me rushes out to catch the bus. Her hands full of shopping bags, she leaves behind a thumbed copy of a popular daily newspaper. On the front page the headline announces that a famous actress in a television soap has miscarried. Before long, I become aware that the headline has caught the eye of a woman working in the café. Slowly she is wiping the vacated table beside me. I notice that she is middle-aged, with a pleasant, open face and a local Oxford accent.

The lady leans over and enquires whether I would like to read the paper. Politely, I decline. She lingers by the table. Folding the newspaper in her hands, the lady gazes at the headline. Catching her eye, I smile back. Perhaps it was a look of empathy. Maybe she simply needed some human contact. I sense she wants to talk and remark that it is a sad story. She takes this as a cue to open up the conversation. But I am still surprised that she chooses to share with me, in a public place, one of the most intimate experiences of her life. The headline has triggered a private pain. Later I suspect that the lady has not voiced it in nearly 30 years.

> I had a baby, just like that. He was so beautiful. He died at six months. I had to give birth. They gave me a choice. I wanted to see him you see. I got to hold him in my arms. He was so beautiful. He had all his little legs and arms. Only his nails were missing from his fingers. They come later. He was a beautiful boy. After, I wasn't able to have any children. They took everything away. He really was beautiful. It really helped me just to be able to hold him in my arms. I held on to him for a long time. He was beautiful my baby boy. I remember everything about the feeling of him in my arms. He was so beautiful.

As her words fade out, I am suddenly no longer the historian of medicine drinking a quiet café latte. I let slip under the table an academic article that I have been reading on the history of death and dying. It seems immaterial in the face of such a painful memory. Through this lady, the past has leapt up at me from a newspaper. I listen very carefully, aware of the lady's need to speak without interruption. I know it instinctively. It is what makes me human and I hope a decent historian.

The irony is not lost on me that the stories at the centre of this book, which you are now reading, have come back to life. The encounter also reminds me that some emotional experiences are universal. They are intrinsic to human nature. Across time the medicine and science differ but the human dilemmas in death endure. To this statement I can add the footnote, caveat, and historical opinion. But here at the start of this journey of words that would detract from the human encounter.

The bare medical facts are that the café lady had a baby. She carried it in her womb, until it died at six months. The doctor worked with her. He helped her give birth. She wanted to hold her baby. The procedure was very painful. Later she had a hysterectomy. Holding the child helped her trauma. She never forgot. But this human experience also betrays a much deeper historical truth, that history is a wasteland of forgotten lives. It is peopled by untold stories and private memories scattered inside ourselves, the archives of our most private thoughts. Often they are inarticulate, sometimes they are very subjective, and frequently they have been edited by family myth, poor memories, and careless custodians. Like this lady's story, over and over again, the pain is too great to record on paper. This stranger in a café encounter is a keen reminder that despite our human imperfections the central narrative of the history of medicine still speaks to us about what we share, rather than what sets us apart. It is our common humanity that we need to talk about much more in a biomedical age.

As I leave behind the lady in the coffee shop, I know that this book will be richer for the window on the past that she opened for me. And that I should pay very careful attention – not just to what she said – but to what she edited out. The story stopped. She left me with an image of her beautiful son. No name. No funeral details. No sense of burial or a grave. To ask would have been insensitive and intrusive. In my heart, I know that there could have been another unspoken history behind her silence.

I had been researching how poor women living in Oxford sometimes sold their stillbirths for dissection. The high fee – up to a year's wage – even in the midst of such grief and pain – would have been the pragmatic option for someone living in poverty. The anatomy sale would have been covert, transacted at night, generally by an intermediary supplying medical schools. Trainee doctors studying medicine would have dissected and dismembered the human remains. When the dissection was finished the anatomists would have buried what was left. Seldom would the supplier have been invited to the service but the lady would have saved the cost of an expensive funeral. It is these stories that are at the heart of this book – the men, women, children, infants, and stillbirths sold for dissection. Sitting in a fashionable Oxford café, it is easy to be dismissive about this reality. We can be sceptical that grieving women could ever have contemplated, much less actually sold their dead children for a fee and the price of a funeral. Riding a morning bus across the city-centre, I recall my Irish grannie's homespun wisdom. *Judge not*, she'd say, *until you have walked a mile in another person's shoes*. I know that the academic life of Oxford in 2009 bears no relation to the world-without-welfare that so many experienced in the nineteenth century.

The lady's buried history reminds me that my task here is to find the stories that were lost. The central aim of the chapters that follow is to try to touch again the unspoken memories. I am struck by the fact that it is where those lives connect – in the historical space between dissected body and burial in a pauper grave – that this book can make a difference. I know by experience that the history of anatomy is like a dry stone wall. In its crevices are poor law records, medical school archives, parish graveyard entries, hospital admissions, infirmary discharge registers, coroners' reports, and pauper letters. This literature of scattered memories, momentarily touching transient lives, provides clues about the sorts of stories that have been missed. I am not the first to navigate this historical course. There is much that I am building upon. I hope therefore that the new material here is a tribute to past innovation, for this book does contain an original historical contribution. In the archives connections exist provided researchers think laterally about the nature of the Victorian information state. If the dissection record was destroyed, then the burial entry for the body when it went into the earth will have survived. In Britain, bureaucracy has been betraying its secret medical history since the Anatomy Act (1832, 2&3. William IV c. 75) decreed that all dissection subjects must have a basic Christian burial in consecrated ground. Every dissection burial after the Births, Deaths and Marriages Registration Act (1836, 7 & &, William IV c. 86) has a documented story to tell. The lady's narrative in the café ended before the funeral. In this book the fact of legal burial denotes a new beginning. It is a chance to rediscover the paper trail of a lost anatomical past.

Anatomists destroyed most of the straightforward evidence about their work. But we can work backwards from the point of death into the archives until research touches the lives of those compelled by poverty to give the gift of their body to medical science. This book's central aim is to dignify each name with an address, a place of birth, a sex, age, family connections, place of belonging, sometimes a physical description, long after dissected remains were interred and marginal lives were filed and forgotten. What you are about to read does not simply provide the first detailed facts and figures on the economy of supply in dead bodies that underpinned the history of anatomy and medical education from Victorian times. It seeks to make new human connections about our collective medical past. To lift out of a sea of bureaucracy the sorts of life stories that should not have been lost. For, make no mistake, aspects of this secret anatomy trade are still flourishing on the internet now.

And these people misplaced by history – their names scrubbed out, their life stories forgotten, their medical gift seldom recognised by posterity – are not so far removed from where you are sitting reading this book.

Sometimes, in fact, they are just a coffee cup apart.

Part I
An Historical Landscape

1
Chalk on the Coffin: Re-Reading the Anatomy Act of 1832

The year is 1858. Another mid-Victorian winter has been dark and dreary. Londoners watch for the first sign of spring. By early February change is in the frosty air.[1] The mercury level in a thermometer on the wall of Camden Square churchyard rises sharply to 41.2° Fahrenheit.[2] The sun's rays warm pavements stained by muddy feet. People once more congregate on the busy streets and thoroughfares. Newspapers report the first flowering of snowdrops and crocuses in the capital's parks.[3] Yet their fragrant beauty cannot quench the stench of pollution. Across London foul drains and cess pits stink with raw sewage. Breathing in the strong scent of rotten eggs, fashionable ladies hold silk scarves to their mouths. John Welsh, the meteorologist, abandons the weather observatory at Kew gardens.[4] London's poor air quality has aggravated his breathing difficulties and on the advice of his doctor he rents temporary lodgings at Falmouth until the stagnant smells are blown away by the March winds. Most of London's poor do not have the luxury of leaving.[5] In the alleyways, lodging houses, and cramped housing, the shafts of early spring sunlight never seem to penetrate a narrow world. The spectre of hunger shapes the daily grind.[6] Neighbours gossip about rising food bills, more hungry mouths to feed, and having to pay rent money to avoid debtors' prison. As bad as things are, worse is to come. News has yet to reach the poorest that new legislation is about to exaggerate their pauperism by the summer of the Great Stink of London.[7]

In February, Parliament is sitting to pass the Medical Act (1858, Victoria, c. 90).[8] It will become law on 2 August 1858.[9] For those living just above the threshold of absolute poverty this 'Act to regulate the Qualifications of Practitioners in Medicine and Surgery' will have profound consequences for their customary rights in death.[10] It will increase the demand for dead bodies by the medical profession. The legislation decrees that future doctors must now study for two years in human anatomy at a designated medical school before being licensed to enter general practice.[11] A medical register and licensing qualifications will be overseen by the newly formed General Medical Council. It will police the boundaries of professional medicine. More doctors mean more bodies. The business of anatomy is about to burgeon.

In the early spring of 1858, the first stirrings of this new reality are coming to court at the Old Bailey across the Thames from Parliament. A trade in dead bodies will form a crucial part of the evidence. As events in the court case unfold (see next section) the general public will learn in the popular press that a profitable trade in the dead has recently become law. Supply networks are being established across London.[12] Financial incentives are paid to a wide variety of body dealers, since 'trade wasn't limited by county or national boundaries, only by the pace at which bodies rotted'.[13] The medical profession has become reliant on these supply chains. The timing of the Old Bailey case is therefore very embarrassing. Most welfare agencies are publicity-shy about being exposed as anatomy suppliers in London and the provinces. Officials are privately concerned about the level of resentment in poorer communities.[14] To increase circulation rates, eager newspaper editors are more determined to expose the link between the exploitation of the poor and the professionalisation of medicine.[15] The Old Bailey case is therefore scrutinised by the press, politicians, the medical fraternity, and the general reading public. This intense media attention concerns those anxious to lobby Parliament to pass the Medical Act without revision in the early spring of 1858. Before looking at that controversy in more detail, it is important to briefly set the scene of the body trade in the mid-nineteenth century.

There were three key pieces of new legislation that facilitated the expansion of the medical profession and its body trade. These were, in order of importance, the Anatomy Act of 1832 (William 2&3, c. 75), the Poor Law Amendment Act of 1834 (William c. 44, 251, 313), and the Medical Act (1858, Victoria c. 90) introduced above. The first statute permitted the poorest to be dissected; the second decreed that workhouses should hand over any abandoned corpses to be studied; and the third gave anatomists the official power to buy human material generated by a body trade to teach medical students. The cumulative impact of the new legislation was to penalise the most vulnerable members of Victorian society in death for what was seen at the time as the crime of poverty. By the time the Medical Act was introduced into Parliament it was widely accepted that the teaching of human anatomy in dissection theatres, and detailed morbid anatomy on the dissection table, was an essential feature of a more professional form of medical training. From the 1820s most English medical students trained abroad in Paris, or at the larger voluntary teaching hospitals in Edinburgh or London. This meant that in Britain anatomists (before the Anatomy Act) worked on criminals hanged for murder whose death sentence also included dissection, an additional physical punishment reflecting the gravity of homicide. Generally a dissection was staged at Surgeons' Hall in London, or bodies were displayed in provincial anatomy theatres like those at Cambridge and Oxford universities.[16] The problem with this supply mechanism was that criminal bodies were always in short supply. Historically the murder rate lagged behind the demand for corpses, and that trend gave rise to a black market trade in the dead. Medical students needed more bodies but their supply mechanisms were often immoral.

It is important to appreciate that at the time there was no publicly funded system of medical education. Students paid basic course fees and were charged 'extras' to dissect cadavers and body parts. In England, that human material became a valuable commodity and anatomists paid high prices for corpses. Meanwhile the poor tried to pool their meagre resources to bury their dead for cultural and religious reasons. A pauper funeral avoided the social stigma of having to sell a loved one's remains to anatomists. These cultural sensibilities tended to exaggerate body shortages because anatomists tried to buy the dead poor in larger numbers to make up any shortfall in criminal cases. Body dealing thus became very profitable, often attracting unscrupulous characters. These profiteers were called resurrectionists, nicknamed 'sack men'. They exhumed the dead from shallow graves in overcrowded cemeteries. Soon stealing corpses for profit became the norm across London. In extreme cases those notorious activities also gave rise to a new criminal offence, murdering for a quick supply deal.[17] The crime of 'Burking' involved killing the lonely, usually by targeting those who bought a space in a shared bed in anonymous lodging houses each night. Sometimes dispossessed street-sleepers simply disappeared and turned up dead on the dissection table. There was a lot of anxiety, imagined and real, about dying and the fate of the corpse. In poorer communities, these unpalatable facts of life were the origin of the tradition of 'a wake'. The bereaved would stay 'awake' for three days after the death of a friend or loved one to watch the body being interred securely in the grave. If they did not, then a 'sack man' could exhume the fresh corpse for dissection. Many cadavers were stolen and traded at the back doors of private anatomy schools at night. This secret trade was exposed after a series of moral panics about 'Burking' in the 1830s, catalysts that legalised the dissection of the poor.

In 1832, the Anatomy Act was passed to stop the crime of body-snatching. The bereaved rich and middle classes objected to a property crime that transformed their dead relatives into a medical commodity on the dissection table. Politicians decided instead that the poorest, whose families that could not afford a pauper funeral, should be given six weeks to find enough money to bury their dead. Those who failed to raise enough funds from funeral raffles held in local public houses, or by begging in the street, were compelled by the new statute to hand over pauper cadavers for dissection. Now those who died in destitution fell into a new legal category, the 'unclaimed'. What this meant was that they died in absolute poverty and their bereaved relatives, if penniless, had no civil rights. Few contemporaries doubted that the Anatomy Act exaggerated the experience of being poor. Finding enough money to give a loved one a pauper funeral was a major financial crisis for the majority of the poorest families in Victorian England.[18]

Health-care providers argued that the destitute should be dissected to repay their welfare debt to Victorian society. In the human anatomy theatre they would be improving medical education in the national interest. The Anatomy Act therefore inflamed cultural and religious tensions regarding a lack of respect for the dead. It worsened community relations in the poorest neighbourhoods. Riots in places like Cambridge and York threatened the social order. Yet the body trade never abated despite the public outcry. In 1858, once the medical profession began to expand, the pressure for pauper bodies was intensified. It is curious that despite the widespread resentment that the new legislation created, its business practices and profit margins remain obscure in the historical literature. In this book that economy of supply will be exposed for the first time. Historical detective work will show how and in what ways medical education came to rely on its body trade in whole cadavers and body parts. The finer detail of the business of anatomy was pivotal to the success of the Medical Act.

Throughout the chapters that follow, record linkage will piece together the sorts of human stories that led to the dissection table and an anatomical burial. Those cases will be firmly rooted in a rich range of new source material extracted from infirmary and poor law records, coroner's reports, and anatomy school archives, as well as letters written by the poor objecting to dissection. Although dissection registers were often later destroyed, alternative parish burial records, for instance, contain details of anatomical burials. It is worth emphasising that the paperwork of the Victorian information state was extensive and today betrays the business of anatomy in ways that historians have overlooked. This means that a wealth of detail has been reconstructed as part of the meticulous preparation for the case studies chosen for this book. The intention is not to overwhelm the reader with that detail but instead to celebrate the fragments of pauper lives that can be recovered and which merit closer historical reading. It is a central contention of this book that the dissected body has a narrative that can be revisited by working back from the point of death into the archives. This will reveal the true reality of the Victorian underworld. It was peopled by a crowd with individual stories to tell about their broken lives. Those paupers cannot be seen unless this new type of medical and welfare history concentrates on a secret body trade, its role in medical education, and the human face of so much abject poverty. The overall aim is to produce an accessible and lively account of the lost property of long forgotten pauper stories, to bring a dead history back to life. For this compelling reason it is now time to revisit *Rex versus Feist* at the Old Bailey in 1858. This case symbolises so many hidden and secretive dealings in the dead, emblematic of the different types of discourses that people this book. So, instead of starting with a more traditional re-reading of the Anatomy Act and a review of its standard historical literature (both of which come later), we are first going to concentrate on a notorious case that gave voice to a trade in the dead poor and the unscrupulous dealings on the eve of the Medical Act in London and provincial England.

The case of *Rex versus Feist*

Rex versus Feist opened on 22 May 1858 at the Old Bailey.[19] Albert Feist was the master of St. Mary workhouse in Newington, South London. He had been indicted for selling the corpse of Mary Whitehead to Guy's hospital for dissection. Feist was charged with delaying the burial of her corpse, selling it on for dissection, and then burying it some days later without her family's knowledge. The Whitehead family were meanwhile present at a false funeral staged by the master. They buried an aged unknown person. The master had falsified the burial rites to disguise the fact of Mary Whitehead's dissection. Her corpse meanwhile had been dissected down to its extremities. Anatomists had severed the trunk, removing the limbs and brain. There was little left to bury. This brutal reality was not a casual mistake or a bureaucratic slip. Feist was also charged with '63 other COUNTS [sic]', in which he 'unlawfully did fail and omit to provide for the decent interment of the said body, and did for lucre and gain to himself, deliver the said body to a certain hospital for the purpose of dissection'. It is alleged that each time he staged a false funeral and profited from the body sale, contravening the terms of the Anatomy Act since it outlawed trading corpses for profit.[20] Before examining the detailed evidence, the jury was given a summary of local poverty conditions. These put the case for the prosecution in context and were relevant to the discussion that followed.

St. Mary's Newington was situated in the diocese of Southwark, next to Lambeth Palace on the south bank of the Thames.[21] A workhouse had been built here in 1734.[22] In a parliamentary report of 1777, it was recorded that there was accommodation for up to 200 inmates. After the New Poor Law Amendment Act (hereafter called the New Poor Law) was passed in 1834, 18 guardians of the poor were elected to oversee the administration of the workhouse and its inmates. Just under 45,000 residents were eligible for poor relief provided they were destitute or in dire need of medical care.[23] The New Poor Law appeared strict but guardians had the discretion to interpret its provisions leniently. They tended to apply the statute according to their financial interests. If the local economy was vibrant then poor relief was generous. In recession benefits outside the workhouse were withdrawn.

In St. Mary's Newington, poor law officials were exacting but they also made sensible calculations.[24] Paupers first had to sell all their belongings to receive public assistance. The relieving officer then issued a ticket to permit them to enter the workhouse. This ensured that welfare was each claimant's last resort. Once inside the workhouse pauper inmates had to wear the badge of poverty by putting on a pauper uniform. Men, women, and children were filtered into separate wards. They worked for their keep – the men in stone yards, the women doing ropemaking, washing, mending, and cleaning. In reality most residents only entered the workhouse when in dire need. Even in a location like St. Mary's Newington, guardians gave a small dole – known as outdoor relief – to supplement low wages if the claimant had fallen on hard times. A small payment in kind often made up any shortfall in clothing, food, rent, or work materials. The baker, butcher, shoemaker, and so on, provided aid on presentation of a Poor Law ticket. It was a harsh but practical welfare system.

By 1858 there were around 50,000 paupers per annum claiming welfare in London.[25] These were unevenly spread across the capital. In the wealthier parishes of Paddington, St. George's Hanover Square, St. James Westminster, and the City of London unions, around 5504 paupers were supported. Collectively they cost ratepayers £3,030,300. This was a burdensome bill but the contributions were collected from well-off property owners. In the East End unions, some 44,005 paupers – eight times as many – drew on the same amount of rates, £3,155,613 per annum. Such ratepayers had much lower income levels and so the financial pressure was intense.[26] The Common Metropolitan Act (1867) was passed to rectify this situation by making all ratepayers in the capital share the financial responsibility for London paupers. In the meantime the system of rate collection and welfare funding was unfair. It meant that there were a lot of disparities in pauper's weekly allowances – 7s 3d for a pauper in the West End compared to just 2s 6d in the East End. Those who needed it most were paid the least.

Against this backdrop, Newington parish decided to build a new workhouse.[27] Ratepayers calculated that they could not afford to pay 'in kind' pauper allowances anymore (like rent money, bread, meat, alcohol for pain relief, and shoes to walk to work). The number of the poorest was increasing daily throughout the 1850s. Guardians commissioned instead a larger workhouse. It was a purpose-built building on the south side of Westmoreland Road in the parish of Walworth next to St. Mary's Newington. They planned to offer the destitute workhouse care and hoped to persuade others to practice self-help. In 1858, both workhouse sites were being used until the building works were finished. The master, Alfred Feist, was in charge of the transition from the old to the new premises. He had worked in this capacity for two years by the time his case came before the Old Bailey on 22 February 1858. Few suspected that he was also running a profitable body trade from the premises.

Edward Corder Durmer was the first witness to be called in *Rex versus Feist*. He confirmed that he was 'one of the churchwardens and guardians of Newington parish'. When it had been brought to his attention that Alfred Feist was trading the dead bodies of the poor for profit 'he [the master] was discharged in consequence of these discoveries'. Despite the allegations, Durmer testified that Feist's

> conduct, as master, was exemplary, as far as my observation went, until I heard this accusation – I have had the opportunity of knowing the inmates intimately, and they speak generally well of him, those who are well conducted; those, who are otherwise, would of course not like his discipline'.[28]

In the opinion of Durmer 'there was no disregard of the feelings of the poor under his care'. The boards of guardians had been very pleased with Feist's work. They 'twice' recommended to the central Poor Law Board (hereafter PLB) that the master's salary was raised at its annual review to reward 'his good conduct'. Nonetheless, Mr. Robinson QC, acting for the prosecution, was anxious to press Durmer about the Dead House in the workhouse.

Most workhouses had a Dead House.[29] The solid square brick structure with a flagstone floor stood in the stone yard at the back of the main building. Dead paupers were transferred there to be washed, dressed, and laid out for burial. A central culvert was cut into the stonework. Here leaking body fluids drained from decaying corpses. Rigor mortis set in quickly and so bodies had to be prepared for burial within a day of death. Their orifices were cleaned, legs and arms tied to keep the corpse together, the body washed and dressed in a backless shroud tied around the torso.[30] They were then placed in an elm coffin because the wood was more water-resistant than oak and this meant that body fluids did not leak out before the funeral. Often cheap 'deal' – pinewood – was used to save costs at a quick burial. Generally a nurse, a taskmaster, and an undertaker regulated this daily work. Grieving relatives were permitted to visit the Dead House by appointment to view their loved one's remains. Those who could afford to carried a woollen shroud or black cloth to dress the corpse. Most were anxious to see the name of the deceased chalked on the coffin to make sure that they were burying the correct body. It was also important to check that the coffin had been nailed down with brass or pewter tacks, to witness that it would go securely into the earth.[31] This tradition was very important both before and after the Anatomy Act. It prevented dissection, illegal or otherwise. If the pauper had died from a contagious disease, then the Dead House was closed to visitors to prevent contagion. The Old Bailey recorder noted that in St. Mary's Newington 'the number of paupers in the workhouse averaged about 550' in 1857 to 1858.[32] Around 30 per cent died per annum. The Dead House was always an important source of body supply for nearby medical schools.

A statement was then read out to the court, witnessed by Harry Burrell Franall Esq.[33] He was the Inspector for the PLB responsible for the regulation of workhouses in south London. The previous summer he had been called in to investigate the Dead House arrangements in St. Mary's Newington after local complaints. The Old Bailey judge decreed that Franall's previous report was admissible as evidence. Franall alleged that Feist confirmed to him: 'I keep one key of the Dead House and there is another kept by the porter.' This was a normal arrangement to keep the pauper bodies secure, away from prying eyes. Either Feist or the porter opened the Dead House door to 'Mr. Hogg', the undertaker, who was contracted to bury the dead by the board of guardians. Robert Hogg worked at 89 St. George's Street, in Southwark.[34] He went to the workhouse 'two or three times a week' to take away the dead for burial.[35] Hogg was normally discreet about his funeral contracts but not in court. He was given immunity from prosecution in return for his full cooperation. His pre-trial statement provided evidence of a long list of people involved in the body trade who were summoned to the Old Bailey.[36] Their testimony was damaging to Feist's defence. All the witnesses claimed that Feist had been running a body trade from St. Mary's Newington workhouse.

The first count to be investigated was the case of Mary Whitehead, aged 75. She died of chronic bronchitis on Friday, 30 June 1857, in the workhouse. Her daughter, Louisa Mixer of No. 30 Pitt Street, St. George's Road, testified that she 'received intelligence' of her mother's death on 'the day that she died'; she 'saw her about half past 2 o'clock, or from 2 to 3 o'clock in the afternoon – I was told that she died about 10 o'clock in the morning'. Louisa wanted to ensure that her mother was buried with dignity. She explained, 'I went into the Dead House with the nurse to put a night dress on my mother as she lay in the coffin'. She needed closure, to physically see, hold one last time, and touch her dead mother. Louisa enquired when the pauper funeral would be and asked the workhouse master whether 'more than two could follow' behind the funeral procession. He said, 'Yes, by payment of 1s. 6d. each'. The fee had to be paid on the day of the funeral to the undertaker, Mr. Hogg.

On the Monday morning, Louisa Mixer returned to witness her mother's burial. She asked and was allowed to see the body a second time in the Dead House. Her sister accompanied her this time. The master 'told us not to go too near – after we had looked at the body – I remained there, perhaps five minutes'. He then said, '*Now ladies that will do* [sic]'. The sisters left the Dead House together, they waited another half hour, and then the master called for '*The friends of Mary Whitehead* [sic]'. Louisa confirmed that she and her sister saw 'a coach and hearse, we got into the hearse – I did not see the coffin put in the hearse – I followed the body believing it to be my mother's – it went to the Victoria Cemetery, and I saw the coffin put in the ground'. The body was placed in a pit grave in Victoria Park Cemetery in Hackney.[37] She did not know that the pauper grave belonged to the anatomy school of Guy's hospital.

About five months after the funeral, Mr. Burgess (senior workhouse clerk) alerted Louisa Mixer to the fact that she had not buried her mother. On auditing the books, he found that the body of Mary Whitehead had instead been sold to Guy's hospital for dissection. Another corpse had been substituted and a false funeral staged. Alfred Feist had personally signed the removal order authorising Mr. Hogg, the undertaker, to take it for dissection and he swapped the corpses. The facts were exposed because William Bull, living at 24 Trafalgar Street, Newington, went to the police. He told them that 'he had been [master's] clerk to Mr. Feist at the workhouse, and it was his duty to enter the deaths of all the paupers into a book and that book, Mr. Hogg, the undertaker was in the habit of examining'. He was not happy about what was happening on the funeral days and alerted Mr. Burgess (senior clerk) to the fraud. In court, Alfred Poland, principal teacher in anatomy at Guy's hospital, likewise confirmed, 'that [Mary Whitehead] was dissected, and afterwards sent back, in order that the remains might be buried'.

The prosecution called Mr. Hogg, the undertaker, to the witness stand to try to uncover how the body substitution system operated. He explained that in Mary Whitehead's case, 'I have no recollection whether there were three bodies in the hearse or two, but there must have been one that was brought from the hospital for internment, and that was buried in the place of Whitehead'. He clarified that 'the body that Mrs. Mixer had seen was taken to Guy's hospital; it was by the direction of Mr. Feist that this was done'. Hogg was cross-examined. The defence asked him to explain how it was possible to make the substitution so easily, the implication being that the master was unaware of what Hogg was doing. He denied this and stated that 'there was no name' on Mary Whitehead's coffin: 'the names are only put on in chalk by the paupers when they bring them down stairs' to the Dead House from the sick wards. It was easy to change an identity and simply swap one set of human remains for another. Chalk on the coffin symbolised the transient personal histories of paupers often filed and forgotten in Poor Law accounts.

Emma Greenland, 'widow of Charles Greenland', resident of Proctor Street, Newington, was then called to give evidence. Her testimony helped explain to the jury the practical workings of the Dead House. In early May 1857, Emma's husband lost his job due to ill health. The Greenland family calculated that their best option was for Charles to enter the workhouse to get proper medical care. Emma stayed at home to look after their children. She found part-time work – washing, mending, and child care – to make their meagre ends meet. On the '15th May

1857', Charles Greenland 'became an inmate of Newington workhouse'. Emma testified that, 'I was not in the habit of going to the workhouse to see him'. She was not heartless and had 'applied to see him, but was not allowed to do so'. The master and mistress both refused her entry. She could not get much time off work during proper visiting hours. The 'first time' that she met the master formally was 'on the same night that her husband died'. She called at the workhouse around 'dinner time, between 1 and 3 o'clock' but was turned away 'as it was not visiting day'. When she returned in the evening, Feist told her that her husband had 'been dead about a quarter of an hour'. The master said, 'You'd better not see him now – you'd better come next morning at 9 o'clock'.

Emma Greenland returned the next day with her brother-in-law, John Greenland. A nurse showed them into the Dead House where her husband's body was displayed. Emma Greenland testified that Feist asked whether 'we had made arrangements about the burial?' The family had little money. Emma was too distraught to answer. John Greenland spoke for her. Emma told the court that John said to the master 'that it was out of my [Emma's] power to bury him'. Feist replied, 'Very well, the parish will bury him'. They enquired about the timing of the funeral, anxious to walk behind the parish coffin to the grave. The master referred them to Mr. Hogg, the undertaker, who took charge of the burial arrangements. Emma Greenland also 'applied for the clothes he had on'. Emma recounted that the master 'said if I had his clothes I must have his body, as the clothes he had on came to the parish'. She explained to the jury that 'he went in with a very good clothes, [which] were borrowed...and he was only there [in the workhouse] five days'. Emma needed the clothes because her dead husband did not own them. They were bought from a pawn shop and she had to repay the debt. The master refused. She confirmed that she did not attend the funeral: 'I had my children ill, and could not attend'. Besides, she alleged, on the day after her husband's death, 'his wife and child had been kept standing there all day, and were not admitted, to view the body'. She had been given just one brief glimpse of her husband's remains in the Dead House the morning after his death, which she judged hard.

Emma's brother-in-law, John Greenland, promised her he would walk behind the procession of the dead. John checked that the paupers who assisted in the Dead House had done their duty: 'the name of Greenland was written in chalk on the coffin'. He asked to see the body before burial. Again Feist kept putting him off: 'I was told to go into the waiting room, and there I sat, I should say nearly half an hour until my patience was exhausted and I went into the yard – I did not see the body that morning'. Instead he saw 'three or four coffins, lying on the ground, and a hearse and a coach – I afterwards came out and followed a body, which I supposed to be my brother's – I saw the coffin buried'. Later it came to light that Charles Greenland had been dissected and dismembered at Guy's hospital. Three witnesses confirmed the facts. Mr. Hogg (the undertaker), Mr. Mark Shattock (Guy's accountant), and Mr. Alfred Poland (principal anatomy teacher) told the jury that Charles Greenland's grieving family were unaware of the supply deal. They were also kept ignorant of the fact that accepting a pauper funeral meant agreeing to dissection and dismemberment. At the funeral, Greenland's corpse had been substituted for another body. There was little left to bury.

These were very serious allegations. The evidence seemed damning. As the pauper witnesses gave their respective testimonies, common complaints were voiced.[38] Feist discouraged grieving relatives from seeing the physical remains of their loved ones. There was usually a delay of up to a week before the human remains were interred by Mr. Hogg the undertaker. In the meantime the corpses had been sold to Guy's hospital. When relatives pressed to see the body, they were shown a coffin with the lid already nailed down. They assumed, incorrectly, that they were burying their loved one with dignity. A younger body was often substituted for an older corpse. The master instructed the Dead House staff to shuffle the human remains. His actions kept the covert anatomy trade away from prying pauper eyes. This detailed court testimony exposed a body business that seldom came to the public's attention.[39] Further evidence uncovered the operation of the anatomy trade and just how vulnerable paupers were to exploitation in death.

Next Mary Thompson was sworn in at the witness stand. She stated: 'I am the wife of George Thompson, who died in the workhouse at Newington – he died on 10th March, 1857'. Mary confirmed that she had been in the workhouse with him. They 'had been there two years last November'. Mary's duties were to 'wash and clean about the house'. She knew the master well and sought permission to see her husband's body on the day he died. Mary reported that, 'he said I should, and I did'. On the morning of the pauper funeral she was given time off to attend the burial. Feist asked her 'whether there was a black wrapper or a gown to put on me – nothing was procured for me'. So she walked in her pauper uniform. She confirmed that, 'I saw the body of my husband that morning in the Dead House'. Mr. Hogg, the undertaker, and the taskmaster in the Dead House had been present but not Feist. She lingered because 'I wished to remain until the coffin was nailed down'.

But Hogg and the taskmaster 'hurried me out of the Dead House'. She explained that, 'I followed the remains of my husband, but I am given to understand I did not'. She never spoke to Feist again about the incident. An audit and subsequent police investigation brought to light that her husband's corpse had been substituted too. Like the others, it had been sold for dissection to Guy's hospital.

The Matthews sisters supported these pauper testimonies. Rebecca Matthews of Pleasant Row, East Lane, Walworth, testified that she was the sister of 'Phoebe Clarke who died in the workhouse at Newington, on 21 February, 1857'. Rebecca Matthews and her middle sister, Sarah Crutchley, both visited Phoebe Clarke just before she died. They came back to the workhouse the next day, a Sunday morning, to view the body. Feist agreed to their visit and accompanied them to the Dead House door. On this occasion he stayed outside in the garden leaving the Dead House taskmaster, a man called 'Percy', to show them the corpse. The following Wednesday, the day of the funeral, Rebecca and Sarah returned to walk behind Phoebe's procession. Rebecca did not want to look again at the body in its open coffin. Sarah agreed to make sure that the arrangements were proper. She told the court, 'I was stronger than my sister, she was very poorly, and she could not go – I went in the second time because I felt anxious to see her once more before she was screwed down'. But she never got a second glimpse of the corpse. She recounted, 'A man was there apparently knocking the nails into the coffin'. Soon, 'they got into the vehicle (a hearse) to go off to the funeral'. Neither sister noticed Mr. Feist. They assumed they had buried their sister and that was the end of the matter. Again Mr. Alfred Poland, the principal teacher in anatomy at Guy's hospital, confirmed that Phoebe Clarke had been dissected. The reality was that this young woman was a valuable teaching tool.

The Old Bailey testimony was long, involved, macabre, and enthralling. It lifted the curtain on the dark underworld of a body business that was supposed to have been outlawed by the Anatomy Act. Bodies were not being resurrected from graves at night as in the eighteenth century. Instead they never reached the earth intact. Most were traded behind the closed doors of a London workhouse. The judge wanted to know about the financial incentives involved. These shocked the jury. John Gannon, the medical officer of Newington workhouse, confirmed that the body trade was lucrative. His job was to sign the medical certificates with the cause of death to release the bodies for sale. He claimed he had not benefited financially. He was aware that 'it would be desirable that all cases should go to the school [of anatomy at Guy's], but I had nothing to do with it'. Gannon alleged that the master kept the business to himself. Feist did admit in his witness statement to the PLB inspector that he received 'a gratuity' from Guy's hospital for being a regular body supplier. But, he insisted, this was not a sale fee, which would have been illegal. He always called it 'a gratuity', 'a grace and favour payment', 'not a profit'. This pedantry about the legal wording kept the master just the right side of the Anatomy Act. He told the court that his financial incentive was 'generally given a little after Christmas – the last gratuity I got amounted to 20*l*'. When pressed about whether he was paid in cash for the bodies, Feist replied: *'Never, by God'* [sic]. The fact of being paid by cheque was an important legal defence. The profit margin was a sleight-of-hand financial technicality. Twenty guineas was a huge sum of money, a year's wage for some Poor Law officials. The judge demanded more details.

Mr. Mark Shattock, Guy's hospital accountant, gave evidence that: 'I have from time to time paid Mr. Feist money – in 1856 I paid him 19*l*. 10*s*., and in 1857, I paid him 26*l*'. He confirmed that, 'I paid him that as a gratuity for his trouble ... – it approaches to somewhere about 10*s*. for each body'. The accused was charged with 64 counts of deception, but this evidence showed that he had been paid for about 82 bodies during his two-year employment. The judge pressed the accountant about the numbers in the supply chain at Guy's hospital. Shattock stated that he did not purchase the bodies personally. The master dealt with the porter at the anatomy school in the hospital. Shattock processed the paperwork and receipts. He had done so since 1849 and as far as he was aware Feist had simply taken over the supply deal in 1856 from the previous master. Shattock thought that the gratuity was reasonable. It was remuneration for the number of certificates that had to be completed when a body was passed over. These he explained were time-consuming. A master's Poor Law salary did not cover the administrative costs and so the hospital judged it prudent to pay a gratuity from petty cash. The choreography of dissection cases and their certification was complex and important to set in context.

There was a certificate of death; a certificate of body removal; a certificate of notification to the hospital; a certificate of undertaking for removal; a certificate of receipt by the anatomists; a notification to the Anatomy Inspectorate that a body had arrived; a return certificate saying that the dissection was complete; a notification of removal for undertaking; a certificate of collection; a certificate to request a Christian burial; a certificate for the funeral; and, finally, a certificate of interment.[40] Twelve certificates betrayed the 'bureaucrat's bad dream' that became the Anatomy Act.[41] The twist in this anatomical tale was that the workhouse master admitted that, 'I destroy all such receipts at the expiration of every quarter'.[42] In the majority of cases it was form-filling for the wastepaper basket. Occasionally in a local archive the entire certificate run for a body survives, usually pinned to a file because it was a controversial case.[43] Yet destroying the certificates did not eradicate evidence elsewhere about the supply chain in the Victorian information state.[44] A wealth of detailed information survives, often cross-referenced for audit purposes to life histories created by asylums, infirmaries, and workhouses. These facilitate an historical reconstruction of the trade that lies at the heart of this book.[45] The Old Bailey case is just one example of alternative historical routes into the business of anatomy. Yet, as we shall see in section three of this chapter (the literature review), it is one the historical community has seldom explored.

At the close of the Old Bailey case, Mr. Justice Wightman invited Mr. Robinson QC (prosecution) and Mr. Matthews (defence counsel) to make their final remarks to the jury. Counsel on both sides first asked to approach the bench because the case had sparked legal controversy. The problem was that, as the defence pointed out, the Anatomy Act was 'permissive and not of a prohibitory character'. Under sections '2, 7, and 9' of the legislation the master was empowered with the authority of guardians in respect 'of the lawful possession of such bodies' that died in his care. He had the legal discretion to dispose of any human remains in a way that he deemed to be appropriate. Moreover, under section 7 of the act, unless 'any known relative of the deceased had requested the body to be interred without examination' explicitly, then the master had acted in a legal manner within the permissive terms of the legislation. The prosecution countered that the master had not advertised the fact that paupers could object to dissection. He had not arranged for posters to be set or someone to explain their implications in the workhouse waiting room, a common complaint at the time and one explored in the next section in more detail when we re-read the Anatomy Act. Nor had he told the bereaved that a pauper funeral meant anatomy first and burial later. The evidence had shown that grieving relatives were not given the legal option to stop the corpse from being examined by a medical practitioner for the purpose of dissection. It was self-evident, claimed Mr. Robinson, that in the case of Mary Whitehead her 'friends...had expressed a desire that the body should be buried'.

The master 'would be entrusted with the body for internment' and no more. In other words, the paupers had made their feelings known in the Dead House and been ignored for profit. They had wanted to view the body, sometimes dress the corpse, and witness the coffin lid being nailed down. All checked that the chalk on the coffin bore their loved one's name.

The finer details of legal argument were debated by counsel. Mr. Justice Wightman in his summing up to the jury said that he was 'disposed to consider, that the mere selling of a body for purposes of dissection, might be an offence at common law [an act of trespass for instance], independent of the statute' of the Anatomy Act. He directed that the jury must decide whether the master traded the bodies for 'lucre and reward'. This would be a 'misdemeanour'. The jury retired to consider their verdict; they found the defendant guilty. The foreman of the jury stated that:

> The defendant caused to be delivered the dead bodies of Mary Whitehead and others, delayed the burial of them for an unreasonable length of time, in order that they should be dissected in the mean time; that he did so for the purpose of lucre and gain; and that he caused the appearance of the funerals of the parties to be gone through, with a view to prevent the relatives, in many of the cases, requesting the body to be interred without being subjected to anatomical examination; and that the persons who appeared as relatives of the deceased, might have required the body of the deceased, in any of the cases, to be interred without anatomical examination. Upon this finding, THE COURT, ordered a verdict to be entered, of GUILTY on the 5th, 6th, 7th and 8th Counts – Judgement Reserved...

The defendant was admitted to bail, to appear and receive judgement on 10th May next.[46]

It was noted that Mr. Hogg, the undertaker, had been given immunity from prosecution. The jury expressed 'their regret' about this plea-bargaining, since it prevented the undertaker from 'standing by the side of the defendant'. They thought that Hogg was pivotal to the business of anatomy. The body trade could not have operated without his cooperation. He was complicit in the case. And importantly, as we shall see in Chapter 4, this is not the last time that we will encounter Mr. Hogg's undertaking services. He transferred his body business from Guy's to St. Bartholomew's hospital, a revelation that historians have missed in the archives. National and provincial newspapers meanwhile soon reported the case's outcome in full.⁴⁷ The *Bury and Norwich Post*, for instance, carried a typical write-up in its 2 March 1858 issue:

> On Wednesday last, Alfred Feist, late master of Newington workhouse was charged with the disposal of bodies of paupers for the purpose of dissection for profit, the remains of the dissected bodies being substituted for other deceased paupers so the mourners instead of following the remains of their own relatives to the grave followed some other person. Guilty but no sentence passed on sureties to appear and receive – judgement is a question of the law.⁴⁸

What editors wanted to know was how representative was the case? Would the master be prosecuted or protected by the Poor Law? Was this body trade rife in places where workhouses and anatomy schools stood side by side? These were not unreasonable questions to pose under the circumstances. The case was certainly very embarrassing for the medical fraternity in the capital given that their new education standards were bound to exacerbate body supply problems. It was this political context that afforded Alfred Feist legal protection.

The case went for judicial review on 10 May 1858. Alfred Feist was acquitted on appeal. There were too many people involved in the trade inside the New Poor Law. Central government judged that it was not in their best interests to secure a prosecution. At the Old Bailey, *Rex versus Feist* (1858) created legal precedent.⁴⁹ It established in case law that the principle of consent was not a bereaved pauper's civil right in common law, under the terms of the Anatomy Act, or according to the clauses of the New Poor Law. Many paupers, as we shall see in the rest of this book, got caught out by the legal technicalities. It meant that the chalk on the coffin could be rubbed out. The corpse could be swapped. False funerals were staged. In the space between death and burial, pauper bodies became the staple subjects of a nineteenth-century medical education. The case of *Rex versus Feist* was not a one-off scandal. The chapters in this book emphasise that it was emblematic. Elsewhere the timing of the body trade had subtle differences, but its core features were replicated. Newspaper reporters sensationalised the Old Bailey case. They did not need to do so. The real story was far more sinister. The business of anatomy was in fact disturbingly routine. In the capital, throughout the provinces, and even overseas, the body trade of anatomists was commonplace and thriving by the Victorian era.⁵⁰ Only a re-reading of the Anatomy Act sets this pauper reality in context.

A re-reading of the Anatomy Act

Ruth Richardson was the first person to call for a re-reading of the Anatomy Act.[51] The next section engages with her historical contribution. In the meantime, here it is important to explore her insightful comment about the need to read the Act properly. The best place to start is not in fact 1832 but the recent committee hearings of the Human Tissue Act (2004, Elizabeth II, c. 30). On 5 February 2004 the Standing Committee G – Human Tissue Bill – was convened to debate the various clauses of the new legislation before the House of Commons.[52] Its *Hansard* proceedings were printed online. They are informative about civil-service speak and the difference between rhetoric and reality. Dr. Richard Taylor (Wyre Forest MP) questioned why clause 30, subsection 4 of the Human Tissue Bill, concerning restrictions on transplants involving a live donor, was unintelligible. He commented:

> The Under-Secretary talked about darkness descending upon him. Well, every time I look at the wording of the clause, darkness descends on me. I am sure that the Ministers were not responsible for its wording, so I can be very rude about it. The word "regulations" is used four times; the phrase "regulations may provide" is used twice; and phrases such as "subject to" and "in accordance with" are also included. Even a child at school would know that the same words and phrases should not be used time and again.[53]

Dr Evan Harris (Oxford, West and Abingdon MP) insisted: 'Before the hon. Gentleman sits down, is it not worth reading out subsection (4); otherwise readers of *Hansard* might not be aware what it is he objects to?' The clause was then read out:

> Regulations under subsection (3) shall include provision for the decisions of the Authority in relation to matters which fall to be decided by it under the regulations to the subject, in such circumstances as the regulations may provide, to reconsideration in accordance with such procedures as the regulations may provide.[54]

Dr. Andrew Murrison (Westbury, Conservative MP) remarked that it was obvious that 'the clause is completely impenetrable'. Dr. Stephen Ladyman, (Parliamentary Under-Secretary of Health at the time) took a civil-service view. He replied that it was 'difficult, certainly, but not impenetrable'. To a packed committee he announced:

Biology is complicated, which is why the brightest people become biologists. People who are not quite bright enough to be biologists but are still quite bright become lawyers, and they will help us to unravel the clause. The Bill is...a 'highway code'...of best guidelines...and people should not get too worked up about the fact that some lines and paragraphs in the Bill were more understandable to lawyers than the rest of us.[55]

In the world of parliamentary bureaucracy the historical echoes are often deafening. An equivalent mentality lay behind the drafting of the Anatomy Act in 1832.

Historical accounts seldom re-appraise the wording of the Anatomy Act by a simple reading out loud exercise. This opening chapter proposes to do so in order to try to get closer to the reception of the terms of the legislation in the ears of the poorest.[56] A summary of the Anatomy Act, like the New Poor Law, was supposed to be pinned to workhouse walls, church doors, and places where the poorest might congregate. The theory at the time was that if the labouring poor had knowledge of their predicament then they would see the folly of applying for welfare assistance.[57] Paupers had the legal right to protest in person before guardians of the poor, in print to local newspapers, and by petition to the PLB, and its successor, the Local Government Board (hereafter LGB), if their claims for a pauper burial were refused.[58] But they first had to prove entitlement and to do this they needed to read a copy of the Anatomy Act in person.[59] Any pauper who did so soon realised that he or she was confronted with a document awash with specious statements. How would it then have looked on the workhouse wall and how much of the text would the poorest have been able to read before walking away?

Richardson sets in context that 'the final text of the Act contained twenty-one clauses. Of these, sixteen were concerned with administration and interpretation, and a further three served to clarify the law concerning the legality of bequest and dissection.'[60] There was also a 'custody clause' – whereby if a person died nameless, friendless, and intestate (the case for many illiterate paupers), then workhouse masters and guardians of the poor became the legal executors of their human remains with the discretion to dispose of their bodies for dissection. There was a final clause 'that abolished the use of dissection for murder' because the crime of poverty was now to be punished.[61] Throughout it was filled with ill-defined references to 'the unclaimed', 'claiming', 'claim', 'the subject', and so on. The Anatomy Act script, like the New Poor Law, was in essence a deterrent. It was designed to evade any request for accurate information. Importantly, it is the document's tone that merits still greater historical attention. The rhythm of the sentence construction, as well as slippery definitions and terms, must have daunted the destitute who first heard it disseminated to them by a third party speaking out loud.

Few historical accounts of the Anatomy Act point out that illiteracy kept alive a strong oral culture among the poorest in Victorian society.[62] The way that the pauper spokesperson read out the notice would have shaped local sentiments. Standard texts sometimes neglect to appreciate the fact that these speaking performances would have been rehearsed in crowded public houses and local markets, sifted by gossip, rumour, and casual chit-chat. In other words, historians still need to reconnect to the *performance* of the legislation to really appreciate the Anatomy Act's impact.[63] Starting with the first section of the statute and reading it to an audience is an instructive experience. Even at this historical distance, the legislation betrays how aloof its official language must have seemed to those on the margins of Victorian society. In whatever dialect, regional accent, or timbre one speaks, across each line of text there is one clear message, namely that in death the poorest are to be demeaned by dissection. To really appreciate the cultural impact an analysis is necessary of the way that the specific phrasing is structured. There is an historical irony about the form of the statute's wording.

The script of the Anatomy Act is an expression of a confluence of competing interests – the medical profession on one side, the poorest on the opposite, and welfare agencies in the middle.[64] Yet this observation can distract the modern reader from an unexpected opening observation. The text of the legislation has a very traditional format. Although it is written in civil-service speak – a standard legal language – the first sentence is 14 lines long. This structure of expression would have been very familiar to the poorest. The tradition of the sonnet – a 14 line poem – written as an expression of love, often in memory of the dead, is well-documented in Europe since the sixteenth century.[65] It is very unlikely that civil servants intended to draft the Anatomy Act like a sonnet. But nonetheless that is how the poorest could have read it or listened to its opening refrain being spoken to them. At this juncture in the discussion it is essential to pause and read out the first 14 lines in the illustration overleaf to appreciate the point being made here, in the way that it was necessary during the recent Human Tissue Bill.

A Bill (as Amended on the second re-commitment) For Regulating Schools of Anatomy dated 8th May 1858, House of Commons Parliamentary Papers

(1) **WHEREAS** a Knowledge and the Causes and Nature of
(2) sundry Diseases which affect the Body, and of the best
(3) Methods of treating and curing such Diseases, and of healing and
(4) repairing divers Wounds and Injuries to which the Human Frame is
(5) liable, cannot be acquired without the aid of Anatomical Examination:
(6) **And whereas** the legal supply of Human Bodies for such Anatomical
(7) Examination is insufficient fully to provide the means of such Knowledge:
(8) **And whereas**, in order further to supply Human Bodies for such purposes,
(9) divers great and grievous Crimes have been committed, and lately
(10) Murder, for the single object of selling such purposes the Bodies of
(11) the Persons so murdered: [*] **And whereas** therefore it is highly expedient
(12) to give Protection, under certain Regulations, to the Study and Practice
(13) of Anatomy, and to prevent as far as may be such great and grievous
(14) Crimes and Murder as aforesaid; **BE it therefore Enacted**[66]
[*] **volta**

When the legislation was printed the layout of the words on the page reiterated the sonnet form.[67] It not only read, but looked, like a sonnet. The formatting of the text in black-and-white print on the page also matched cheap ballad material.[68] Many English sonnets would have been advertised this way among the poorest in the past.[69] Looking in more detail at the Anatomy Act's opening and the brief history of the sonnet form expands on this observation.

In the Petrarchan sonnet, introduced into England from Italy, there are two distinct phases in the poem – an opening eight lines (octave), then a volta (known as the turn in an argument), and the closing six lines (sestet).[70] First Sir Thomas Wyatt and then Henry Howard, Earl of Surrey, re-shaped the English style of this Petrarchan format. They used iambic pentameter, or blank verse, because ten stresses on five key syllables in each line made the spoken word very lyrical in English. The sonnet was then a soliloquy in praise of love or lost love after death. The later Shakespearean sonnet varied this basic format again by dividing its 14 lines into 3 quatrains (of 4 lines each).[71] In this case the volta (or change of mood) usually comes at lines 8–9 and the poem ends with an epigram, a final couplet containing the moral of the sonnet.

Turning to the Anatomy Act, the tradition of the sonnet format is slightly varied again. Its metrical pattern is akin to political sonnets in the seventeenth and eighteenth centuries penned by Milton and Wordsworth respectively.[72] The octave in this instance has 10 ½ lines, instead of the more traditional eight. Of these, there are 5 opening lines (the medical argument), then 2 lines (a teaching justification), and 3 ½ lines (outlining the legal imperative). The volta then comes at line eleven. This means that the sestet has been compressed. The effect is to give greater emphasis to the political epigram in the final 3½ lines. It stresses the disempowerment of the poorest in 'the Study and Practice of Anatomy' (capitalised for further emphasis).[73] To achieve this accent the word 'whereas' is repeated at the start of lines 1, 6, 8, and the middle of line 11 (just after the volta), giving structure to the message of the blank verse. In spoken parlance, the overall effect is to make the Anatomy Act's opening sonnet echo Milton's radical political style. This observation is loaded with irony since the Anatomy Act was disempowering the destitute. So what was unintended is remarkably familiar in its oral and visual *performance*.

It is important not to overstate this connection. We need to quantify how many paupers read it in this way, as well as under which specific circumstances. Did guardians always post up the new notice or was it hidden in the clerk's desk drawer? Was just a summary posted up or the entire Act? Did the literate poor hear its opening refrain and quickly walk away. How many were simply resigned to another Poor Law notice? These unanswered questions should not distract historians from the possibility that the tradition of the sonnet was well known in pauper circles. It had a very flexible 14-line structure and there could have been a strong historical connection where it was recognised. Indeed, evidence survives in agricultural paraphernalia in Shropshire and theatre bills in Northamptonshire that suggests the poor may have made a similar connection in their political literature.[74] Posters objecting to the Anatomy Act were often styled like local auctions, ballads, and even horse races (see illustrations 1.1 and 1.2). Anything that was familiar became the blueprint for political protest. The work of Jon Lawrence on the 'politics of place' supports this observation that historians should not underestimate the ability of the poor to protest, and the creative ways in which they expressed their political outlook.[75] Marc Brodie too has recently observed that the poor's culture of the 'theatre and the street' is often misconstrued; historians 'need to look at other forms of evidence' embedded in their oral cultures 'to give us clues about...the alternative understanding they could create'.[76] Re-reading the statute in its original layout supports the general historical position that legislators had no understanding whatsoever about the impact of the statute on the poorest; not the content, nor the way it was structured, or how the written word was heard. All fear starts with a familiar reference point. It generally has some basis in reality. This is what makes it so mendacious. At a reading of the Anatomy Act there was a genuine sense that in a world-without-welfare there was no choice in death for the bereaved but to accept dissection.

Returning to the reading exercise, the next 33 lines of the statute's prose have no full stop.[77] Of these, the first 19 lines are punctuated by commas but not a single key pause. There is no colon or semi-colon. It is a breathless read; one long, ambling sentence structure that is dense with the terms of its legal authority. It literally starts at the top and works its way down to the bottom of society. The pauper that *performed* its power structures was left in no doubt that they came a long way down the list of professional priorities. It is an extensive list of medical expertise which was then authorised:

> Any Member or Fellow of any College of Physicians or Surgeons, or any Graduate or Licentiate in Medicine, or any Person lawfully qualified to practice Medicine in any part of the United Kingdom, or any Professor, Teacher or Student of Anatomy.[78]

The medical fraternity just needed two Justices of the Peace to certify that each medical person can 'carry on the Practice of Anatomy'. These certificates were countersigned at a 'County, City, [or] Borough level'. In other words, it drilled down a demanding list until it reached the pauper place of belonging beneath the weight of legislation. If any pauper paused to question the Bill's terms of authority, a new salaried Inspectorate of Anatomy now stood in everyone's way. They reported to, and were regulated by, the Home Department (later the Home Office). Its daily workload

26 *Dying for Victorian Medicine*

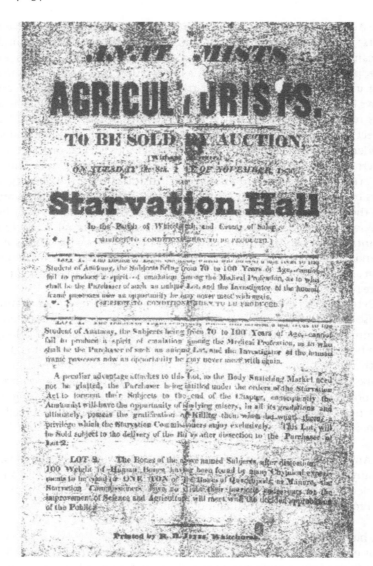

Illustration 1.1 ©Shropshire Record Office, Anatomy Act poster styled 'Anatomists – Agriculturists – To be Sold by Auction (Without Reserve) on Tuesday 8th November 1836 at Starvation Hall in the parish of Whitchurch in the County of Salop [Shropshire]'. I am grateful to Mr Paul Anderton who sent me a copy of this fragile pauper poster in 2009. The poster was found when someone was cleaning out house contents and passed to a local museum who decided not to display it but place it in Poor Law archives.

> This event will be run on the Poor Law Course, Brixworth, on the 29th inst., and we understand the following are entered :
>
> **THE BRIXWORTH COLT.**—This is a well bred little horse, who has always gone straight and in good form. They say he is likely to be a stayer, and although he met with a serious accident, we understand he is heavily backed to win.
>
> **SIDESMAN.**—This is rather a weedy animal by "Clericus," evidently not quite up to weight. He will probably go to the post, but is more suitable to carry a *lady quietly*, than get across country.
>
> **PREVARICATOR.**—Although descended from a good old sire, he runs cunning, and has lately been thrown out of training. He is, however, again entered, but his party have no confidence in him ; he's more fit for a *selling* race.
>
> **POST MORTEM.**—This is a well bred horse by "Medicus," but is aged and obliged to run in blinkers. Although a good stud horse, he has not pace enough for the present race.
>
> **RUFUS.**—This horse's pedigree is not to be found in the Racing Calendar. He is stable companion to Prevaricator, and his trainer must have entered him to fill the race. He's in too good company to win.
>
> N.B.—No horsemen allowed on the course.

Illustration 1.2 ©Northampton Library Local Studies Room, poster styled 'The Brixworth Stakes, The Event will run on the Poor Law Course Brixworth, on the 29th inst… – NB No horsemen allowed on the course [the poor cannot vote]'.

would remain confidential in view of public sensibilities. The door of the anatomy theatre was being kept closed to pauper scrutiny. Overall then, in the opening 47 lines of the Anatomy Act – containing 20 clauses – we read an exhaustive account.[79] In practical terms, the first 14 lines are written in a familiar form once associated with the tradition of the sonnet celebrating the dead or radicalising the politics of protest but now associated with disempowering the poorest. The next 33 lines are very difficult to say out loud in an oral culture. The terms of their authority – parliamentary, medical, and legal – are extensive. Anyone reading them to a pauper audience has to engage with specious statements embedded in repeated legal phrases and little punctuation. The *performance* gives voice to vulnerability. Key words in the text substantiate this point.

The next obvious but important observation is that the word dissection was not defined or described in the statute. The term used instead was 'Anatomical Examination'.[80] This remained a moot point until the recent Human Tissue Bill rectified the obscure wording. In 2004, clause 55 (1) stated explicitly under the subsection 'General Interpretation' that: *'anatomical examination* means examination by dissection for anatomical purposes'. It elaborated in clause 55 (2) that:

> In this Act, references to the carrying out of an anatomical examination are to the carrying out of an examination by dissection for anatomical purposes of the body of a deceased person, and where parts of the body of a deceased person are separated in the course of such an examination, include the carrying out of an examination of the parts for those purposes.[81]

Never once in the original legislation was the cutting of the body into parts either discussed or explained clearly to contemporaries. There was always a sense that something was being withheld but it was never made clear what that was and of course this only increased suspicion on the part of the poor that they were being exploited.

Indeed throughout the original Anatomy Act the term 'Anatomical Examination' appeared 13 times in the whole document. How would the poor have understood this phrasing at the time? One potential reading of the term 'examination' is that the body is touched, probed, inspected, and viewed, but not necessarily cut into parts: not dismembered, not reduced to a trunk, not just the extremities left intact. The *Oxford English Dictionary* in the mid-nineteenth century defined 'examination' as a 'careful and accurate inspection, investigation, or inquiry, taking evidence in search of the nature and qualities of substances...from the Latin *examinatio*'.[82] There was also an interesting cross-reference in medical dictionaries of the time to 'examination...pertaining also to a medical assessment, a diagnosis, or consultation'.[83] It is moreover noteworthy that guidelines issued to Victorian coroners recommended that most bodies should be 'examined' or 'viewed' but not necessarily cut up to establish the cause of death.[84] Typically this entailed looking at the external appearance of the body, for instance, in drowning cases. Post-mortem examination was less common because it was too expensive and thus resented by ratepayers who paid coroners' bills. In other words, even an educated person of middle income and a working medical knowledge might get the wrong impression about what the phrasing 'anatomical examination' actually meant in physical terms. Small wonder the illiterate poor were sometimes befuddled.

A related observation is that the term 'Anatomical Examination' was often foreshortened to simply 'Examination' throughout the Anatomy Act transcript. In clauses 15–25, for instance, 'Examination' was used four times in ten lines, with just one reference to anatomy nearby on the page.[85] The only time that 'dissected' appeared in the text was under clause 30 that repealed the act of dissection for criminals convicted of murder; these were executed in former times and sentenced to anatomical study. The intention was to suggest that the crime of being 'dissected' became distant on the written page when in fact more people had come into its ambit. 'Anatomical Examination' was therefore a choice phrase in the Anatomy Act. Complete dissection actually meant extensive dismemberment: severing the trunk, separating the limbs, using all body 'material', injecting it with preservation fluid, cutting the cadaver with a lancet, sometimes for up to two years in each teaching cycle. This was the true reality by the 1890s. There was not much left physically when the anatomist had finished with the corpse. In this respect, the original wording of the Anatomy Act was calculating, inaccurate, and misleading. The impoverished, as we shall see in Chapters 4 through 7, were often confused by its sleight-of-hand use of creative language to disguise their dissection.

Other key phrases that would have mattered to the poor also appear throughout the legislation and merit close textual reading. Of these, the reference to a 'decent Coffin or Shell' and to 'be decently interred' stand out.[86] In the nineteenth century the yardstick of 'decency' in death tended to be shaped by three local factors.[87] There were first long-held death customs to consider on a regional basis. These tended to be calculated according to financial priorities confronting guardians of the poor. There was then secondly the question of the good of public order. If guardians were parsimonious it was important not to disturb local social relations by being too harsh. In agricultural areas it was the poor that harvested the crops; in towns and cities they produced the goods in local factories. Business depended on economic stability and death customs were an important litmus test of 'class' relations. The third issue to take into consideration was the capacity of the rate base when under financial pressure to fund burial rites. Pauper funerals were sometimes a lower priority when unemployment was high. Standards and their cultural terms of decency were always a delicate balancing act between paupers' needs and guardians' pockets. One way to test local sentiment is to research the physical burial space set aside to inter the dead. Chapter 2 will examine Victorian notions of death and dying. Here how much space the poor got in death once dissection was finished is analysed. Those physical parameters illustrate how standards of 'decency' often differed.

The Anatomy Act script did not state how many paupers should be buried together. Nor did it define whether this should be in a single deep grave space or a lime pit. It was also unclear whether each body should be buried in a coffin, a coffin shell (an inferior low-grade coffin), or be carried to the grave in a recycled parish coffin (usually hinged at one end to allow the dead to be tipped out into the earth). Some basic burial facts (based on later case studies) illustrate that the statute was interpreted in diverse ways. In Cambridge six paupers were buried together, though each had a basic coffin shell;[88] in Oxford just two bodies were interred in the same burial plot – each was given their own coffin which was a very liberal treatment of the dead poor.[89] In places like Birmingham, most paupers were placed in an open lime pit.[90] They were sewn into a woollen shroud and carried in a recycled parish coffin. The dissected dead of Guy's hospital in London were meanwhile buried four to a pauper pit at one ceremony.[91] And this practice contrasts with burial practices at nearby St. Bartholomew's hospital where the governors had been very concerned about dignity for the dead since the eighteenth century.[92] Most dissection cases in this instance were buried in an elm coffin and shroud, though the poor were asked to make a small contribution too. Notions of 'decency', being 'decent', and acting 'decently' were laudable aims under the Anatomy Act. There was, however, a crucial gap in Poor Law standards and paupers' cultural expectations. The size of the grave hole and its corporeal contents made a world of difference to the bereaved anxious about their kinship standing in the local community. Local debates about a 'respectable' burial were always intense because of disagreements over whether decency meant dignity in death.

Today we are familiar with the importance of dignity in death for the bereaved.[93] It is essential, however, not to confuse the rhetoric of decency with that of dignity in the Anatomy Act. They were two very different concepts. Decency does what is best under the circumstances. Dignity respects fundamental human rights whatever the economic cost. Victorian policymakers often ignored this crucial distinction.[94] It was thus indecent to take a body for dissection until 48 hours had passed after the time of death under the Anatomy Act. But it was not undignified to cut it into so many pieces that there was little left to bury. Dissection and dismemberment cut away all identifiable dignity.

That was a material fact that few anatomists wanted to explain in detail to the poorest. This was why skin, bones, and so on, were labelled in the dissection room. Afterwards most pauper cadavers were a bundle of putrefied body parts. The deceased's remains were certainly a public health hazard. Few could be carried into an open church because of the foul smell.[95] A Christian burial under the Anatomy Act really meant a brief graveside recital of 'ashes to ashes, dust to dust' from the Book of Common Prayer. The pauper body was treated as 'matter out of place'; an object of decay, dirt, and even disease.[96] If a poor family objected on behalf of their dead relative about this indignity then the final section of the Anatomy Act had a legal sting, 'the limitations of actions' clause.

Those paupers who read or listened to the new legislation were often ignorant of Section 6 clause 5 of the Bill. This small amendment was added to the Anatomy Act on its second recommitment to Parliament in the 8 May 1832 session. Its significance was overlooked by the pauper population at the time and sometimes by subsequent historiography. The amendment stated that there was a statute of limitations for complaints. Prosecutions must take place 'within Six calendar Months next after the cause of the Action accrued'.[97] In other words, if anyone brought a case of complaint they had just six months to prove foul play, bring the action to court, and secure a guilty verdict. The problem was that most body supply scandals did not come to light until after the six-month deadline. In subsequent amendments to the legislation, anatomists were given the right to dissect bodies for up to six months, then up to a year, and finally, by the end of the Victorian period, up to two years.[98] This meant that it was very unlikely that relatives would discover in time that something had gone awry, especially when a false funeral had been staged as in *Rex versus Feist*. Few pauper complaints reached court compared to the number of bodies dissected. By the time they did, the statute of limitations clause was invoked. In any case, paupers had no money to start a lengthy prosecution process against either medical or Poor Law officials. There was no effective legal redress under the Anatomy Act. Occasionally cases were prosecuted successfully but paupers then found that the statute had no legal grasp.[99] If found guilty, the defendant should have been 'imprisoned for a term not exceeding Three Months, or fined in a sum not exceeding Fifty Pounds'.[100] Crucially these stipulations were left to 'the discretion of the Court'. Judges often ignored the statute's tariff, refusing to pass a custodial sentence. Fines were reduced on appeal, effectively endorsing misdemeanours. The practical workings of the

Anatomy Act were to prove very controversial once it became law on 1 July 1832.

The Anatomy Act after 1832 – historiography

It is now 20 years since Ruth Richardson first alerted the historical community to the meaning of death, dissection, and the destitute in Victorian times.[101] In research terms, her new findings were a milestone in medical history. Looking back, Richardson's empathy, sensitivity, and tenacity in the archives was groundbreaking. Her book was certainly a compelling read. It soon became a formative building block, the foundation stone of this and so many associated books. In many respects, though, Richardson's monograph was the victim of its own success. Everyone assumed at the time of publication that she had written the last word on the topic of the Anatomy Act. Her book was reviewed to acclaim but few in the academic community expected more substantial work to follow in the same field.[102] This was a regrettable outcome given that Richardson wrote in her book's preface that it was her intention to stimulate further work. In perhaps one of the most overlooked passages, she said, 'a great deal remains to be discovered much of which will be accomplished by detailed studies, particularly upon surviving hospital and workhouse records'.[103] The signpost was clear but subsequent research never materialised. It was easier to delve into the history of death and dying. Its Gothic narrative was a compelling read and the sources were less scattered in the archives than a shadowy dissection trade. A generation later that historical neglect has not been redressed. In revisiting the Anatomy Act, this opening chapter turns attention back to the body trade that was exposed at the Old Bailey in 1858 because the business of anatomy in so many respects remains undocumented in historiography today.[104] Although there is not much more work to be done on the passing of the Act itself, nevertheless all the areas that Richardson highlighted as still needing more work require new research. This book is based on 17,500 pauper histories drawn from asylum, infirmary, hospital, and workhouse records. They have been cross-matched to four medical schools where the dissected helped those doing the dissection to expand medical education in the metropolis and provinces throughout Victorian England. That record linkage work seeks to ask key historical questions, such as was what was intended, what actually happened? Did the legislation turn into another type of event? Over time how did those changes impact on the most vulnerable in Victorian society?

These standard but nonetheless important issues remain central to this book's revisionist perspective.

Richardson established that the Anatomy Act 'turned out to be a bureaucrat's bad dream'.[105] She was also the first to observe that the Anatomy Inspectorate was 'too small scale' to be effective.[106] It appeared that 'the supply of corpses fell short of demand' from the statute's inception. Richardson thus scrutinised surviving Anatomy Inspectorate returns. These seemed to show that the supply in London was around '36% lower...in 1841–2' compared to 1832–1833.[107] Peak figures were recorded in the '1887 and 1892 seasons, when 3,032 bodies were consigned for dissection'.[108] There are, however, no substantive details about precise body supply figures and their distribution across Poor Law union areas in central government records after the end of the 1840s. Richardson meanwhile noted that civil service reports indicated that 'In the provinces things got off to a poor start, and got worse, only improving after mid-century', though the same problem of inexact information predominates. Richardson thus estimated that 'in the course of the first century of the Anatomy Act's application about 57,000 bodes were dissected in the London anatomy schools alone'.[109] This general overview provides important context but it needs the sort of new research that Richardson called for two decades ago. Were, for instance, these official body supply figures collected by central government accurate? Did anatomists carefully record or under-report their body supply networks for reasons of political expediency? Is it true that 'less than *half a percent* came from anywhere other than institutions which housed the poor'?[110] The questions that Richardson first posed still demand substantive answers in the archives.

This book will show that historiography has tended to take at face value the picture of central government anatomy returns. They are not a reliable indicator of the body supply trade, any more than Poor Law statistics were in Victorian times. Before now, it has not been feasible to quantify the scale of body-finding networks. Empirical evidence will be provided in Chapters 4 to 7 uncovering the secret dealings of the body business in different areas of England. Regrettably it is not possible to do a comparative study for Scotland or Wales because the parish records that were collated locally do not facilitate a reconstruction of the body trade. Those limitations in the record-keeping mean that it is essential to concentrate on case studies where the material is rich and an invaluable insight can be given into the inner workings of a secret business of anatomy. In Chapter 4, at St. Bartholomew's hospital, the fourth largest anatomy school in Britain, over 6,000 bodies were traded. They were bought across London from numerous body dealers. These new findings lift the lid on a landscape of poverty that littered the dissection table in the metropolis. Outside London, other important centres of medical education were competing for corpses too. At Cambridge, analysed in Chapter 5, anatomists were determined to expand their trade to rival the big teaching hospitals in the capital. They had to outbid both London anatomists and those at Oxford, analysed in Chapter 6, to secure a chain of supply. In Oxford's case the only way to deal with the competition was to buy bodies and body parts which were sold on to students. Meanwhile, in Manchester, analysed in Chapter 7, the poor were plentiful and mortality rates high at a key centre of medical education in northern England. This meant that Manchester could buy enough bodies to educate its home students when the profession expanded in 1858 and again in 1885 once the Medical Act was extended. These four case studies will provide a new historical picture of an economy of supply in dead bodies that has been neglected. They have been chosen because they represent the typical demographic trends and geographic spread of body dealers. The time has come to interrogate the business of anatomy working in conjunction with the Poor Law to better understand how and where it burgeoned in England.

Richardson made the important historical link between the passing of the Anatomy Act in 1832 and the Poor Law Amendment Act in 1834. She argued that the former was 'an advanced clause' of the latter.[111] Her conclusion was that 'the Anatomy Act depended on the administrative machinery of the Poor Law for its implementation'. Again few historians of medicine or poverty have explored this important conclusion. Most assume that the dissection records have been destroyed at a Poor Law Union level.[112] Others point to Richardson's view that in the provinces the body trade was subdued.[113] In standard accounts local resentment is stressed. It is common to read that a concerted campaign against dissection was conducted in the regional press. Several riots in cities like Cambridge meant that guardians of the poor were unwilling to cooperate and this situation worsened over time with the coming of Poor Law democracy.[114] As we shall see in this book, there is little substantial evidence to support this viewpoint. At best, it is impressionistic and based on second-hand accounts rather than actual pauper voices or welfare records from the time.[115] It also neglects to appreciate the complexities of the Victorian information state in the way that Edward Higgs has done.[116] The hidden histories of the poorest exist in paperwork that the Anatomy Act stimulated. It decreed that all dissected cases must have a basic Christian burial. These were registered systematically after the passing of the Births, Deaths and Marriages Registration Act (1836, 7 & &, William IV c. 86). Burial and parish records now replicated dissection books. Of those that were destroyed, a duplicate copy of their work exists elsewhere in the system. Few saw the future historical potential of this record-keeping until now.

Central government records confirm that a lot of Poor Law Unions were prepared to cooperate, both in the provinces and the capital.[117] They were just very quiet about their involvement and managed their affairs discreetly. Although some refused to become regular suppliers, many more were eager to take up supply deals. In LGB files evidence has survived that shows body supply networks were extensive throughout England. It indicates that the timing of the New Poor Law and its various campaigns to reduce the poor rates were intrinsic to the operation of the Anatomy Act. This book contends that to really appreciate those practical workings, it is essential to bring together the histories of medicine and poverty. They must be mapped in parallel. Importantly, there has been a general neglect of the late-Victorian Poor Law experience.[118] This was a time when there was a concerted attempt to limit outdoor relief allowances (the small dole paid to paupers to supplement their incomes at home). That policy change meant that the destitute were offered one stark choice, either compulsory admittance into the workhouse or no more Poor Law funding. Many Poor Law Unions took in the sick and vulnerable close to death, an ideal situation in which to promote the business of anatomy. Guardians were at the time seeking ways to defray costs. A discreet supply deal was an attractive financial proposition. It is essential, given that cost-saving context, to integrate the workings of the Anatomy Act with the dynamics of the Poor Law. This is the only way to really appreciate how the business of anatomy flourished.

This chapter has already touched on the fact that bodies and body parts were traded on a regular basis. Yet historiography still refers to the 'body', 'cadaver', 'corpse', and 'dead' poor. These terms neglect the material fact of dismemberment. This book provides evidence that there has been too much emphasis on the entire body-shell being dissected and not the dismembered body parts.[119] It is likely that in years when bodies were in short supply, most medical students would have trained on body parts. The Anatomy Act was silent about this aspect of a medical education. It will be shown that this key omission in the legislation was very significant. Trainee doctors were anxious to qualify and start earning. Provided they dissected enough body parts to make up a whole body, they satisfied the basic licensing qualifications of the medical profession after the Medical Act (1858), elaborated in Chapter 3. Where that material came from, how it had been cut up, and the body parts students bought, remain misunderstood. There were three practical calculations: it was easier to purchase body parts than a whole body; buying those body parts spread the cost of a medical education; suppliers were keen to split up bodies because selling a body bit- by-bit was more profitable. This book will show that the Anatomy Act licensed a breakers-yard business. It authorised a trade in body parts in the name of medical education. Today an equivalent trade is still flourishing around the world.[120] Its lost history is therefore of some relevance in a biomedical age.[121]

This book therefore contains a new type of business history and one that underpinned medical education. It argues that this is central when re-reading the Anatomy Act too. Research links supply deals to their transportation costs; how, for instance, bodies were moved from source to dissection table. The development of the Victorian railway system was critical for the smooth operation of the Anatomy Act. All the provincial case studies show that a fast train link between supplier and medical schools was essential. At a time when preservation techniques were cruder (in a pre-refrigeration era), moving bodies quickly was a practical necessity. Rigor mortis and body decomposition were rapid. Most medical schools thus spent a lot of time negotiating their railway fees. Undertaking contracts consequently had to be profitable too. The Anatomy Act decreed that bodies (though not body parts) should be given a basic Christian burial. Given that guardians of the poor were keen to defray burial costs, many anatomists saw a financial opportunity to obtain more bodies.[122] If anatomists paid for supply fees and anatomy burials, then a smooth chain of supply was created. In reality, most undertakers contracted by officials became body dealers inside the Poor Law.

To increase compliance, the costs of obtaining and disposing of corpses were generally reimbursed by medical schools. The fact that a lot of Poor Law agencies signed up to the terms of those supply deals attests that although the profit margins were small for individual transactions, collectively the financial incentives involved were deemed worthwhile. Body supply became more widely accepted at a time when central government pushed Poor Law Unions to more profitably manage their resources at the start of what became known as 'the crusade against outdoor relief' around 1872–1873. Historians of medicine and poverty have tended to be dismissive about this changing Poor Law landscape. Its general neglect has meant that history books have yet to document the complex financial reality of pauperism statistics and official anatomy returns sent to central government. These are often misleading about Poor Law performance and achievement.[123] The Poor Law was increasingly dominated by an accountancy mentality. Welfare could be caring but it was also a profit-and-loss analysis – the capacity of the rate-base versus what could be clawed back from expenditure. These financial constraints and incentives were hidden, including those of the Anatomy Act. Records of the anatomy trade (corpses and body parts) are still documented in Poor Law accounts in ways that historians of medicine and poverty have so far not appreciated. Re-reading the Anatomy Act also means scrutinising the financial details of the Poor Law accounting system.

A revisionist perspective

Rex versus Feist raised broader historical issues that were ignored for the next 140 years until Richardson published her seminal book, and in many respects still remain so 20 years later. This book puts the Anatomy Act and its protagonists back onto the witness stand. New research will answer questions that remain neglected in historical accounts. The reconstruction of an economy of supply in four case studies (through painstaking record linkage work) makes it feasible to ask novel questions about those doing the dissecting and their dissection subjects at English medical schools.

How was the supply and demand for corpses organised in the capital compared to the regions? Could that data explain changes in medical curriculum? Given that the poor resented the Anatomy Act, its application must have been very controversial and shaped popular views of medicine. How, therefore, was it received? Was resentment pro-active or reactive? Once the initial controversy started to abate (if, in fact, it did) what was its longer-term impact at the local level? Can we link supply trends to the ways that the experience of being poor fluctuated? Did it alleviate or exacerbate death, grief, and mourning, and, if so, how? Given what is known from regional figures about the pace of urbanisation, population distribution, mortality rates, public health interventions, and so on, it is likely that bodies were easier to obtain in the north and Midlands, than in the south, east, and west. So far, only patchy statistics collected by central government on supply deals can be consulted. They provide a broad sense of body supply trends up to the end of the 1840s, and for peaks in the mid-1880s, and again in the mid-1890s, but have little substantial detail beyond that narrow scope. Who then traded what, when, where, how, and in what numbers? The reality of that performance and achievement still deserves our fullest historical attention.

Medical schools must also have had to generate and regenerate complex methods of contact and payment along supply chains between institutions and suppliers. For the first time it should be possible to assess how the economic priorities of welfare agencies and medical schools teaching anatomy converged in the interests of mutual profitability. This book therefore recounts the demography (age, sex), epidemiology (disease at death), geography (from trade at source to distribution networks by railway), and scale (who was sold, when, where, and for how much) of the body trade from Victorian times through to Edwardian England, and beyond. It ends in 1929 because the New Poor Law was repealed in that year. A detailed analysis of that Anatomy-Poor Law partnership is long overdue.

On 3 April 1858, Charles Dickens first alerted his readership to the 'Use and Abuse of the Dead' in *Household Words*.[124] His themes are not only akin to this book's new approach in the next two chapters on the thinking behind death customs (Chapter 2) versus medical education (Chapter 3), they proved to be remarkably prescient too. Dickens wrote about the long-term implications of *Rex versus Feist*. He also called attention to the need to re-read and re-think the Anatomy Act. Dickens was appalled by how 'the most sacred feelings' of the poor were made 'the subject of a secret mockery, sordid and infamous' at Newington workhouse in 1858. There was, he noted, too much 'room...left for the atrocious jugglery of undertakers, and for the dishonesty of workhouse masters.'[125] The fact that *Rex versus Feist* was lost on appeal showed that, 'Beadledom does as it pleases'.[126] Dickens's remarks went on to anticipate the recent Human Tissue Bill:

> The time has not gone by for a discussion of this subject; it never can go by until there shall have been made those further changes in our law which will not only secure the feelings of the poor from outrage, but at the same time will consult to the utmost degree possible or right in England, the interests of science upon all who live have to depend for aid in some hour of affliction.[127]

Later on in the article, he elaborated that:

The remedy for this is not the adoption of the French system, under which every person dying in a hospital by doing so bequeaths his body to the furtherance of knowledge. In this country let no man alive or dead be denied bodily freedom... if there be one dear friend, no matter of what condition, to whom the touch of the anatomist upon the dead body would be the cause of grief, then let the grief be sacred, and let the natural feeling or prejudice be reverenced.

Let it then be an ordinance, not a permission, that the unclaimed dead shall supply the needs of science and humanity, whenever the body is not that of one who, in life, prohibited its use for such a purpose. By this form of enactment, no feeling is outraged, and all mockery is swept away. The whole swarm of undertakers, who now traffic in the dead bodies of the poor, could be swept aside.[128]

For the next 150 years, Dickens's words were ignored. Central government dismissed objections. Few cared that 'undertakers combine to extort money from hospitals under the name of burial fees'.[129] Civil servants noted but did not investigate Dickens's complaint that 'the anatomist... contrives, whenever the master of a workhouse will assist in procuring each body surreptitiously' in effect 'obtaining the necessary documents to play the part of body-snatcher'.[130] The medical priority was that the body supply chain had to keep pace with anatomical demand. It never did and so a culture of denial evolved over time. Secrecy bred suspicion, distrust created double-dealing, bureaucracy cloaked duplicity, which was obscured by form-filling.

Dickens could not know at the time that the true litmus test of central government's real sentiment towards the poor proved to be the Anatomy Inspectorate. It was always seriously underfunded throughout the Victorian period. This meant that it was impossible to police the illegal body trade that had been legalised. The fact is that there was more continuity that discontinuity. Politicians never voiced this inconvenient truth. Medical professionals were publicity-shy because public sensibilities did not abate. The distrust that the Anatomy Act created and the climate in which it was passed proved to be a bad self-validating prophecy. The logic was that the poorest will resent anatomists – therefore nobody must tell the poor – *ergo* the poor suspect they are being deceived – *ipso facto* they cannot be trusted with the medical truth. It was better to let people make the wrong assumptions – create a false historical memory. That circle of deception only ended with the principle of consent in the recent Human Tissue Act. In the meantime, many, many thousands became a supply deal, a body part traded for

40 *Dying for Victorian Medicine*

profit, a dissection demonstration, an anatomical specimen ending up in an anatomy museum collection. That list might seem emotive when written down, but it was also a material fact. Dying for Victorian medicine was part of the complex mosaic of medical progress peopled by the poorest between the Anatomy Act in 1832 and the repeal of the New Poor Law in 1929.

2
Restoring the Face of the Corpse: Victorian Death and Dying

Buyer: *'The dead centre of town?'*
Agent: *'Estate agent speak for Victorian wreck by a graveyard'*
Buyer: *'At least it's within budget – where exactly in Cambridge is it?*
Agent: *'The Cross marks the spot on the map'*
Buyer: *'Looks like a crowded part of the city?'*
Agent: *'Quiet neighbours'*
Buyer: *'Depends on your point of view!'*[1]

Whether viewing a property or looking at a city-centre map, the culture and language of death are a bridge between the Victorian era and the modern world, the focus of this chapter.[2] Yet what was it like to be underneath society, to fail at life in the past, and experience bereavement in dire financial straits? To really understand the reception of the Anatomy Act, it is necessary to examine death and dying for those living in the worst deprivation in Victorian Britain.[3] Now when a person lives and dies in poverty their lonely passing can be profoundly disturbing to the general public.[4] Many local authorities pay for basic funerals from council taxes to distance the poorest from a Victorian experience. This is not simply done on public health grounds. There is an agreed consensus that taxes in a civilised society should pay to bury the dead 'decently'.[5] Yet in a credit crunch how far have we progressed? In February 2009 this question motivated a reporter for the *Guardian* newspaper to highlight that 'State-funded burials and cremations' had increased 'by up to 46% as more people die alone these days'.[6] There had been a 'sharp rise in the number of older and younger people dying without funds' for their burial in Great Britain. Up in Scotland, Edinburgh saw a '46% rise' while in Glasgow there has been a '10% increase' in the last 12 months. South of the border meanwhile it has been estimated that more than '£4m[illion]' was spent by 'councils' in the last year. The Department for Work and Pensions (hereafter DWP) has also subsidised an additional '40,000 funerals' for families on state benefits. On average a one-off payment of '£700' – half the cost – has been paid out for each basic ceremony. Overall the sums are striking. Some £28 million was paid in kind, and £4 million in full death grants, a total of £32 million in England and Wales during the financial year 2008–2009. Katie Martin, a spokesperson for the DWP, explained that:

> The problem is that the DWP have been under-funded since 1997, yet every year in April the cremation fee goes up, every January the coffin prices go up, the handle prices, the nameplate and the fittings and furnishings inside the coffin.[7]

Recent newspaper reports concluded that burials could be substandard and some still resembled 'pauper funerals'. In death, deprivation is still a fact of life in twenty-first-century Britain. Small wonder, perhaps, that the complexities of a pauper funeral and its stark symbolism, disturbed, fascinated, and horrified the Victorians too.

The Victorians lived at a time of industrial expansion and technological advancement.[8] It was a great railway age, the epoch of Empire. The passing of military heroes like the Duke of Wellington were celebrated.[9] Yet the boom and bust of the Victorian economy made the poor a visible presence in towns and cities. There were a number of deep recessions that exposed the fragility of the human condition.[10] Medical men found it difficult to cope with the pace and scale of urbanisation and its associated ill-health problems given the Victorian population explosion. Contagious diseases like cholera, typhus, typhoid, summer fevers, and measles were rife. Common complaints like diarrhoea were lethal in all the major English cities, so much so that when Lord Alfred Tennyson wrote *In Memoriam* (1849), he found a ready audience of grieving people. The poem cried out:

> That loss is common would not make My own less bitter, rather more: Too common! Never morning wore To evening, but some heart did break.[11]

For Tennyson the death of a young male friend triggered his famous religious crisis at a time when science and secularism started to erode faith and belief in an afterlife.[12] He was startled by that realisation as he stood by the grave. His sources of solace – grief and graveside ritual – gave him something to cling to amid the grim bereavement process. Recently Julie-Marie Strange has tried to explain why human beings seem to react in a concerted way by preserving long-held death customs. Quoting Jonathan Dollimore, she explains that, 'The extreme crisis of a modern philosophy of individualism is that whilst it professes to be life-affirming, it is beleaguered by virtually every death-obsession in the Western tradition'.[13] Tennyson, through poetry, found his way back to faith but for those today who cannot grasp that consolation death rituals seem to ease the initial pain after a loved one's passing. They gave, and still give, structure to a profound human experience. Family customs and comforting friends have always been helpful as well.[14] Today responses might be secular but they remain important human ones. The smallest common decencies, tiny physical expressions of empathy like holding a hand, listening to someone's pain, cooking a meal, or brewing a cup of tea are the treasured things that still connect the poorest today to the destitute who were disconsolate in Victorian times.

In Victorian studies, bridging a culture and language of death has meant that the basics of a 'pauper funeral' were established in the historical literature in the 1980s and 1990s.[15] Pioneering scholars stressed the importance of the 'common funeral' for those at the bottom of the social scale.[16] In poverty history too, the latest studies on the experience of being poor have taken new research directions.[17] Recent findings highlight that the final resting place of the dissected has been neglected.[18] Often the destitute were sold from one area, moved by railway, and then dissected and buried in another location. The bereaved had no physical human remains to inter and yet they were in emotional pain. That anatomy trade still not only merits closer scrutiny in its own right, but also demands that more research is done on the grieving process, which must have been exacerbated for the living. It is known that dissection created a climate of fear, mistrust, and resentment, but not what the differences were between a 'pauper funeral' and 'anatomy burial', and to what extent that distinction worsened abject poverty. In this context, the current chapter does two things. First, it reviews what has been documented in the historical literature on 'pauper funerals' in broad terms. Secondly, new source material is added to explore how the Poor Law and Anatomy Act worked in conjunction. Throughout this chapter a selection of newspaper reports, poor law records, and accounts of paupers in bereavement is interwoven with the secondary literature. Together they draw attention to a spectrum of poverty in death to explain why it was that the body dealers of the anatomy trade called at the doors of the destitute with such regularity in Victorian times. What follows thus provides essential context for the rest of this book.

Section one examines the coroner's records of a man who was found floating dead in the Thames. The importance of knowing his identity emphasises why a name, a face, and a physical description has always been so important for the dispossessed. To be an anonymous dissection subject was to disappear from society. Section two then reviews the recent historical literature on death and dying for paupers. Poor Law accounts are introduced to analyse the experiences and fears of those threatened by dissection. The response of two grieving parents forms parts of this discussion. They were living on the edge of absolute pauperism and had lost a child. Their emotional reaction, hidden by stoicism, builds on the latest historical work and breaks the silence that lay behind the anatomy trade. This leads into section three's appraisal of the generic term 'pauper funeral' versus that of an 'anatomy burial'. There is a brief U-turn in this chapter's chronological approach, so that experiences of both can form part of the discussion over time. It will be shown that in the Midlands, for instance, burial standards varied a lot, opening up opportunities for anatomists to exploit. Overall, Chapter 2 uses a selection of new source material to illustrate the vital importance of possessing a face, a name, a body, a shroud, and a grave in Victorian times.

The faceless corpse

In the *London Review* an article entitled 'Restoring the Face of a Corpse' was printed on 16 May 1863.[19] It praised the discipline of anatomy for pushing back the boundaries of forensic medicine. The report detailed how a leading anatomist and a coroner for the City of London had joined forces to try to establish the identity of a decomposed corpse pulled from the river Thames. The reason for restoring the 'faceless corpse' was to establish whether or not the drowned man was a murderer. There was a rumour that he had killed a prostitute, an 'unfortunate girl' called 'Emma Jackson'. The gruesome features of the case were reported in full:

> Any one who has seen the distorted, horrible, black, swollen features of a drowned and decomposing man, must know how difficult under such circumstances would be recognition – even if the distorted features were those of a familiar friend. But when such a face has been

seen living but once before, identification is almost or quite impossible...

In the Dead-House on Tower-hill – with sunken nose, with bleared swollen lips curled back over it and the chin, with protruded tongue, and flattened eyes half hidden by their huge distended lids, with rotting flesh, quite black with the gases and fluids generated by weeks of submergence, and by the floating of this lifeless body in the hot sunshine on a broad but polluted stream [of the Thames] – lay an unsightly corpse...

Through the daily press the proceedings of the inquest were by the next day made known throughout the land. The account attracted the attention of Dr. Richardson, of Manchester square, who in his early hospital duties some years ago, had occasionally used a preserving means for very bad subjects brought in for dissection.[20]

Since a central aim of this book is to try to restore the identities of those corpses sold to the anatomy trade, cases like this one are an important historical prism. They illuminate the medical profession and general public's general attitudes to the poorest in Victorian times and why that cultural backdrop enabled the body trade to flourish for so long. Turning to the detail of this case allows a dispossessed youth to step out of a history of silence.

Mr. Humphreys (city-coroner) agreed to work with Dr. Benjamin Ward Richardson (an anatomist) to try to identify the human remains. Dr. Richardson lived and worked as a medical practitioner near Marylebone.[21] He held four Poor Law medical contracts at leading dispensaries and infirmaries that cared for the poorest in London.[22] Dr. Richardson was also a paid lecturer in forensic medicine at Grosvenor Place School of Medicine (1856) adjoining St. George's hospital, where he dissected on a regular basis. He was a talented and very busy man, an elected member of the Medical Society of London by 1868. Anxious to enhance his professional standing, he became involved in the inquest in question. The *London Review* described the chemical process of reconstructing the identity of the decomposed corpse in some detail:

> The coffin shell was filled with water, 20 pails, to which had been added 20lbs of salt, so as to give the solution the specific gravity of 1.100 i.e. .50 above the gravity of the blood and the fluids of the human body, the object being to reduce by exosmosis [sic] the fluids distending the corpse by their passing over the salt. This acted well,

and in the course of an hour the face was reduced in size, but still remained as dark as ever. The water in the shell was now slightly acidulated by a pint of hydrochloric acid, with the view to get rid of the ammoniacal [sic] gas and compounds of the blood by which the dark colour was caused. The face was now raised out of the water, and two gallons of water charged to saturation with chlorine were applied by cloths to the face.[23]

Dr. Richardson then waited for the chlorine to bleach the face that had been darkened by the muddy waters of the Thames. After some time, 'the colour of the face was reduced to ashy grey'.[24] It now looked like 'the hue of ordinary blue-woven paper'. Gradually, 'the form of the natural features could now be made out'. The anatomist made a note that the corpse had 'just a boy's beard'. He then left the chlorine to bleach the entire head. On his return, he injected the head with water saturated with more 'chlorine, and chloride of zinc'. To 'give firmness to the organic tissues' of the face, he also injected 'a small proportion of sesquichloride [sic] of iron'. But the flesh was still too soft. It was impossible to properly identify the face. So the anatomist 'opened the carotid artery' and 'thus injected the face until it assumed a slight tension and firmness in the cheeks'. The face looked human again. Dr. Richardson 'then stopped, tied the vessels, and arranged the body, covering the face with a solution of chlorine and spirit'. Finally, he 'left it with the jury' to re-examine the physical evidence.

The inquest was reconvened. The corpse, its face reconstructed, was unwrapped for the coroner, jury, and chief witnesses to view. Three people who knew the murdered girl, and witnessed her murderer running away from the crime scene, were called back to give their opinion about the identity of the restored face. The first witness, 'Margaret Curley', a 'girl at the brothel', stated that the facial features did not match the murderer she had seen.[25] 'Charles Anson', a bystander from the same brothel, 'at once stated the face was one they had never seen in life'. The jury were still not convinced. The witnesses were rather ramshackle. There was some suspicion that they might have been paid by the brothel-keeper to give false testimony and suppress the scandal. The third witness was called 'Stokes the boot-black'. He was known locally as 'a very intelligent fellow' and trustworthy. He felt obliged 'to declare...he had never seen such a face before'. Stokes also stated 'positively' that the corpse 'was not that of the man he had seen before with the unfortunate Emma Jackson, and bore no resemblance' to the murderer in question.[26] The coroner concluded that the corpse 'was not that of the long-sought murderer, whom, therefore, we may justly believe to be still at large'. The *London Review* reported that:

> This is an important result...No power on earth can restore to the dead the rosy colour of the cheek, or bring back the ruddy hues of life to the pallid corpse, but it is much to recall that science may recall for a moment...some unlucky wretch whose spirit had passed into eternity...[27]

What are we to make of this case? What does it reveal about death and dying for the destitute? There were two levels of testimony – the witting and unwitting – in the coroner's court. First the witting: a prostitute was hired for sex and became a murder victim. Some weeks later a male body was found drowned in the Thames. There was strong suspicion that the murderer who had been seen leaving the crime scene had committed suicide to evade arrest. This corpse might be that person. A new forensic technique performed by an anatomist established that the body was that of someone on the brink of manhood. Witnesses stated that he was not the murderer. The coroner's case was closed. A verdict of 'accidental drowning' was recorded. The police still needed to apprehend a dangerous criminal before he murdered again on the streets of London. Nobody felt that it was necessary to establish why the young man was found in the river water. Lacking a face, identity, and personal history meant that he had no legal protection in death. He was destined for the dissection table.

So much for the witting testimony; what about the unwitting? Beyond the bare facts, the court record reveals quite a lot. Coroner's cases were public affairs. Their remit was to sift gossip and establish the true facts of sudden death. Leading anatomists read their reports with interest. In the 1860s the most interesting were published in the press on a regular basis. The reporter in this case alleged that the anatomist 'had occasionally used a preserving means for very bad subjects brought in for dissection'. This type of anatomical work was not normally public knowledge. The anatomist had close links with Poor Law medical services and forensic medicine. He was therefore in an ideal position to purchase 'unclaimed' cadavers. Dr. Richardson seems to have liked a professional challenge but we are also told that he was a pragmatist. The reporter assumed that it was known that anatomists sometimes took whatever cadavers they could get for dissection. A decomposed body found floating in the stinking river water (a cocktail of effluent) of the Thames and its tributaries was still a corporeal commodity. After

48 *Dying for Victorian Medicine*

several weeks in the water most were faceless, 'very bad subjects'. Yet, the youth's body parts could still be an educational tool. This testimony alerts us to the fact that when studying death and dying we need to think about what a lack of identity meant for the poor: whole bodies and body parts of the destitute were traded. That finding reiterates a key point of argument in the Introduction (which will be developed in Chapter 6) that historiography has tended to underestimate the extent of the financial transactions associated with the business of anatomy. Beneath the layers of witting and unwitting testimony we are also able to tease out some important research leads. These provide clues about the complex question of death norms for the destitute.

It is has already been noted above that neither the coroner nor the jury pursued the question of whether the drowned youth was thrown into the river. His death was not relevant to the homicide of the prostitute. The unexplained circumstances were documented but the court record stated the basic fact that he was 'found drowned'. Nonetheless the lad could have been killed. Pushing someone into a river was a very easy way to disguise homicide.[28] The body had also been in the water 'for some weeks'. It could have floated very far downstream. Although the case might have been a suicide, Victorian coroners sometimes passed verdicts that concealed suspected murders.[29] Their statistics were not reliable[30] and they often encountered more of the destitute than appeared in their official paperwork to central government.[31] Coroners were in an ideal position to hand over for dissection the bodies of missing persons found dead who lacked a family or were without friends. At a time when forensic medicine was in its infancy, most bodies would not have been subjected to an extensive post-mortem, given the expense involved. This made 'unclaimed' corpses of unknown persons ideal anatomy material. It is relevant that most coroners were legally, not medically, qualified. So the coroner's 'view' of the body often meant just that, looking and no more.[32] Yet there is very little research on this aspect of coroners' workloads and their links to the anatomy business for the nineteenth century (see Chapter 6).[33] Regrettably no further record survives of whether the young man was reclaimed by either his close relatives or friends. Nor is there a coroner's burial certificate to indicate where he might have been interred after the inquest. Like many poor people, he disappears from view once officialdom finished with his corpse. But this absence also makes the case significant for this chapter. The evidence, or rather lack of it, reveals that to die nameless, friendless, and faceless was to become part of an historical silence. To contextualise this contention, it is essential to review the literature on death, dying, and 'pauper funerals' below, in the second and third sections of this chapter.

The historical landscape of the dead poor

In any standard university library there are shelves of books devoted to the history of death and dying.[34] So much has been written that it can feel daunting to research the subject for the first time. Students have recently been encouraged to follow a literary route into the field of studies because it is argued fictional accounts provide an accessible introduction to key themes.[35] In Victorian times, religious beliefs, family life, issues of class, material consumption, cultural norms, and a much-debated 'death-industry' together provide important context.[36] There has been a vibrant historical literature on those who struggled to pay for funerals.[37] Most managed meagre resources according to 'a hierarchy of thrift'.[38] Working families tried to prioritise saving for a pauper funeral (paying subscriptions, for instance, to burial clubs) because it was symbolic of their social standing.[39] Paul Johnson estimates that 'by 1904 approximately 19 million people held some form of death or sickness insurance'.[40] These numbers reflect the fact that in Victorian life there was a spectrum of pauperism known to contemporaries by various slippery definitions.[41] These included 'being poor', 'the labouring poor', 'the dependent poor', 'in poverty', 'deserving pauper', and 'destitute'. Household economies among this diverse group were complex. They could change hourly, daily, and from one week to the next. It has therefore been suggested that the makeshift economies of the poorest still need to be better understood[42] which is why Amaryta Sen alerted the historical community to the need to research the 'actual achievements' of the very poorest under strained circumstances.[43] Death customs, he observes, symbolise the 'capability to function' of the vulnerable. To appreciate these new directions in historical studies it is necessary to review what is known in the general literature about death and burial to try to unravel the experiences of the poorest, those most at threat from an anatomy sale. Comparing and contrasting those on middle- and lower-income levels is an important starting point, since that research helps explain what it was really like for the destitute.

To die at home surrounded by loved ones, cared for by one's family, to not linger in pain but to accept with stoicism what fate had decreed was seen as a Victorian ideal.[44] Popular literature frequently carried consolation stories. Most bedside scenes were about keeping vigil at the 'good death'.[45] Pat Jalland pioneered this type of 'experiental history' by researching the bereavement experiences of middle-income Victorian families.[46] She concluded that they were in large part responsible for fashioning consumer aspirations. The labouring poor who lacked purchasing power could ill afford the material culture of a fast-expanding undertaking sector. Their consolation was instead the promise of an afterlife. This was expressed by a basic community celebration that mourned publicly the passing of the deceased. For those on lower-income levels, funeral bearers carried a coffin, even if low grade, to a pauper grave in which the body was placed securely. By washing the body, then placing it in a shroud, and attending a wake, the labouring poor avoided an anatomy sale. This was the cultural landscape of those who had to navigate consumer pressures in death.[47] Yet in the past there has been a tendency to elide what those on lower-income levels could afford with the needs and wants of people in absolute poverty. The Victorian scene was crowded and yet it was assumed that the poorest shared the aspirations of everyone else.[48] This was surprising given that few could afford to do so and most had alternative ways of celebrating a loved one's passing. The problem has been that their alternative viewpoints are elusive because of a history of silence among those with a strong oral culture.[49] This Victorian underworld was regrettably often hidden from a contemporary gaze.[50] Julie-Marie Strange in groundbreaking work has thus emphasised that it is important 'to shift our analytical gaze away from materialist paradigms and dichotomies between respectable and pauper funerals to consider flexible definitions of grief and mutable notions of respectability' when evaluating how the poorest buried their dead, or not.[51]

Historians concur that death was a regular visitor inside the poorest neighbourhoods of Victorian society. It created a culture of dread and resignation. Dealing with the decaying corpse depended on an assemblage of near kin, relatives, neighbours, and charitable networks within the wider religious community. This was because, though cheap, paying 'a half a penny a week' to a burial club was beyond the means of those at the bottom of the social scale confounded by life's harsh economic realities.[52] And yet respond they did in complex ways that still demand answers. Which funeral customs, for instance, were sacrosanct, and which could be compromised, remains uncertain. A synthesis of contemporary evidence suggests that it was fear of social estrangement – one's neighbours gossiping about how the bereaved had failed a loved one in death – that shaped cultural norms.[53] These, however, need to be explored still further in their historical context.[54] It is vital to know more about those who begged to bury their dead.[55]

Even after the sale of a corpse, a funeral raffle sometimes paid to inter dissected and dismembered remains; the actions of those who collected what was left merit closer historical attention (see Chapter 4). Historians thus need to re-appraise the cultural and financial subtleties of 'pauper funerals' versus those of an 'anatomical burial'. Only this type of new research can redefine to what extent the broad thrust of middle- and lower-income death cultures ever penetrated the poorest districts.[56] In what follows, a selection of Poor Law letters and newspaper accounts of 'pauper burial' have been sampled. They draw out some of the under-explored themes in the latest historical literature by using neglected sources to think about the gaps that require more work in the context of an anatomy trade that became taboo.

Turning then to the Poor Law, those who ran the system faced a basic dilemma when confronted with so many dying in their care. On the one hand, if officials were generous, then paupers might overwhelm ratepayers with claims for pauper funerals (broadly defined). On the other hand, it was difficult for householders to ignore a decaying corpse in their neighbourhood. Under the Old Poor Law (from the sixteenth to the early nineteenth century), officials were not obliged by law to bury the dead of the poor.[57] A short extract from *The Overseers Handbook* (London, 1851) summarised the practical situation once the New Poor Law was in force:

> **Burial by Overseers.** – Except in the case of dead bodies washed ashore, etc., overseers have no power to bury the bodies of persons at the cost of the poor rate. The occupier of the house in which a death has occurred in under an obligation at common law to bury the body and to defray the expenses of burial (*R. v. Stewart*, 12. A. & E. 773).[58]

The reality was much more complicated, especially in the transition from the Old to New Poor Law in the 1830s. A typical letter from Bradfield Union to the churchwardens of the parish of Burghfield exemplifies the dilemmas in Berkshire:

January 10th 1839

Gents,

One of your paupers of the name of John Lee is now lying in the poorhouse of the Union situated in our Parish [Bradfield], and we are very distressed for Burial in our Churchyard even for our own parishioners, I take the liberty of requesting you as a favour to remove him to his own parish for interment, I think you cannot imagine this an immoderate request as the expense of one of the Rate payers sending their horse to convey him home in the hearse will be very trifling and of rare occurrence to any Parish individually, compared to the obligatory expense we must shortly be put to, in finding a cemetery for twenty-eight Parishes who continue to pour in their aged and infirm into the Poor house, I may add on purpose to end their days.[59]

Bradfield parish had become the centre of Bradfield Union when the surrounding parishes had been combined under New Poor Law in 1834. But there was no extra burial space when the poor from some 28 parishes came flooding into the workhouse to get welfare. Strategically some poorer families decided to commit their elderly relatives to the workhouse when they were close to death. This saved families on the breadline the cost of a pauper burial. If they had died at home the onus was on the house occupants to bury their dead under common law. Once in the workhouse pauper burial became the financial responsibility of local ratepayers. It is evidence like this that reinforces Roy Porter's view that for the poorest 'dwelling on death was...not a morbid fixation, but honest realism'.[60] A stand-off between Poor Law officials was serious from the point of view of the labouring poor in bereavement in the Midlands during the 1840s. In 1842, for instance, Brackley Union in Northamptonshire got its clerk to write a letter on the same subject to the Poor Law Commission at Somerset House in London:

9th June 1842
Burial of Paupers

Gent,

I am directed by the board of Guardians to inform you that an Aged Pauper belonging to Mixbury lately died in the Union Workhouse [at Brackley] and being refused burial by the clergyman of that parish [Mixbury] he was interred at Brackley, which has caused additional expense and brought the case more immediately under the Notice of the Board, which resolved that I should write and press the subject upon your Attention in order that a clause may be inserted into the now pending Bill in Parliament to guard against similar cases for the future which besides being an Onus upon the Parish where the Workhouse happens to be situated [i.e., in Brackley town] is distressing to the Feelings of the surviving Relatives and Friends of the deceased Paupers.

Yours &c
Robert Weston, clerk[61]

Letters of this type were commonplace, so that by 1844 relieving officers on behalf of guardians of the poor were then given the authority to issue burial orders in each Poor Law Union area.[62] The *Overseers Handbook* explained:

> But s. 31 of the Poor Law Amendment Act, 1844, empowers the guardians to bury the body of any poor person who may be within the Union or parish in which they act, and the section authorises them to delegate their powers to their officers.
> In order to meet cases in which direction must be given promptly, boards of guardians often adopt a resolution authorising their relieving officers to give directions for the burial of poor persons at the cost of the guardians, and to report any such case to the guardians at the next meeting. Where the application is made to the overseers to defray the expenses of burial, the applicant should at once be referred to the relieving officer, as, except, in the instances afterwards mentioned, the overseers cannot legally charge the expenses of burial in their accounts, even if the case be one of sudden and urgent necessity.[63]

Although the New Poor Law made provision for guardians to pay the cost of burial, crucially this was still not a statutory requirement. The legal wording of the new guidelines was evasive and is worth repeating – 'boards of guardians **often adopt** a resolution'.[64] Burial powers were optional. An order by central government was not an official amendment to the original legislation. The *Poor Law Handbook* in 1903 still emphasised this procedural point: 'Guardians are **not bound** to bury the body a person dying in their Union'.[65] They had the discretion to approve or ignore the guidelines, a characteristic way in which the New Poor Law operated. They could and often did bury the dead of the poor on public health grounds. They did not, however, have to do so in every case, especially in hard times when residents defaulted on rates. Money was spent when dangerous diseases that were highly contagious attacked the community, for example during cholera outbreaks.[66] But for more common complaints, like fever, pauper burial was not a public health priority. If there were fewer funds to pay for pauper funerals then they could be, and often were, cancelled.

Naturally the poorest were anxious to bury their dead. They often asserted their right to burial to a relieving officer or guardian of the poor locally, and then to central government if requests were refused.[67] But the reply was invariably that pauper burial was a customary obligation, not a legal stipulation. Chapter 1 touched on this gap between rhetoric and reality. The Anatomy Act was reliant on the strict operation of the New Poor Law. If no legal right to pauper burial existed, the anatomy trade could operate in practice. This was the basic bureaucratic framework that penalised the poorest for the crime of poverty. It could be harsh. Underneath there were often many human dilemmas in death. A letter written between the Poor Law clerks of Hardingstone and Brackley unions in Northamptonshire about two orphaned children in August 1840 is instructive:

> Dear Sir,
>
> I am directed to inform you of the death of Jane Foxley lately residing in Towcester but belonging to Milton [Milton Malsor near Towcester] and that this board [Hardingstone near Northampton] has granted Outrelief of Two Loaves (1s. 4d) weekly to her Daughter Ann aged 10 who will be taken good care of by her Aunt and Uncle, if your Board [Brackley] will sanction such an allowance –
>
> The deceased's son Frederick aged 8 years will be taken into the workhouse till you give directions for his removal to your Union, which I trust shortly to receive – The Uncle and Aunt are very Industrial Persons and will I have no doubt do their Duty towards their niece and would also take Charge of the Boy, if they had sufficient Accommodation which they have not.[68]

Death had split up this young family. Calculative reciprocity meant that relatives could afford to take in Ann Foxley.[69] She had already proven her material worth by nursing her dying mother on a meagre outdoor relief allowance of two loaves. She would soon go into service to earn extra money. The orphaned boy was instead admitted into the workhouse at Brackley. There he might get an apprenticeship or fieldwork funded by a farmer guardian. Jane Foxley did not have enough money to subscribe to a burial club. After her death, her family applied for a pauper burial so as to avoid an anatomy sale. There were some guardians at least who were persuaded by the argument that the poorest should be permitted to bury their dead with dignity. But what did that burial entail; what rituals did those in destitution believe were sacrosanct?

A short article that appeared in the *Times* on 20 August 1868 provides important clues about the financial position of the poorest at death's door.[70] It explained that a medical officer of health named Dr. Bateson and a sanitary inspector called Mr. Smith had applied to a local magistrate's court for a removal order to bury a corpse on behalf of St. George's parish in Southwark. Dr. Bateson had been called to 'a small house in Oakley place, Upper Grange-road, Old Kent Road' in the East End of London. He found there a 'corpse of a girl lying in a room inhabited by a family of poor people'. Dr. Bateson testified to magistrates that: 'the girl died on Saturday fortnight and ever since had been in the same room where the family lived and slept'. The medical officer and sanitary inspector together surveyed the premises. They discovered that 'there were only two rooms in the house, and the other room was tenanted by the family whose lives were all endangered by the corpse being kept in the house'. The removal order was being made on public health grounds because 'the parents' were 'refusing to bury the corpse'.

The newspaper report confirmed that the magistrates accepted the public health argument.[71] It then revealed that the medical officer and sanitary inspector only 'after considerable difficulty' managed to remove the corpse from the premises. Further enquiries revealed 'that the reason the parents kept the corpse so long was to get up a raffle to pay the expenses of the funeral and *wake* [sic]'. The poor family had been trying to makeshift their meagre means and needed extra time to pull together enough money to bury their daughter with dignity. The funeral raffle at a local public house was a common way to raise funds.[72] This short extract thus gives us a brief but important insight into the lives of the poorest in death. More though needs to be known if we are to address Julie-Marie Strange's call for 'a broad, flexible and colourful landscape of death' for those in absolute poverty.[73] She has convincingly argued that we need to look beyond material constraints that have 'largely denied a sense of humanity to the poor and working class bereaved'. What little they had, they held on to. Their funeral rites might have been meagre compared to middle-income funeral mores but they were in fact highly valued – sometimes more so, because the smallest gesture gave rise to intense emotional expression that had no other outlet. In the short newspaper extract quoted above, all the family could afford to do was hold on to the decaying corpse. To outsiders and Poor Law officials this was offensive on public health grounds.

But to the poor family in question it was a financial victory to still be in possession of the body. And more so, given that the corpse was found in Southwark district next to Newington workhouse where so many bodies had been sold for dissection to Guy's hospital in the mid-Victorian period. *Rex versus Feist,* 1858, discussed in Chapter 1, created an infamous legal precedent. This family's fear of removal, dissection, and dismemberment was not irrational. Possession of the body kept some control within this family's close circle; such was the power of the pauper funeral and its value for those who dreaded the alternative. As one anonymous aged pauper in Northamptonshire wrote with eloquent desperation on a scrap of paper to guardians in Brackley Poor Law Union after having been committed to a local asylum where she feared she would die:

> If you please *Sir*
> i'm sending you these
> few lines to say that I
> am fit to come out, and
> if you be so kind *Sir*
> to put the case before
> the Board of Guardians
> Yours truly
> Martha Turby
>
> Send for me as soon
> as you can.[74]

Nobody responded. Lacking kin, when she died her corpse disappeared out of the area.

Briefly reflecting on Chapter 1, evidence was presented about how the poor came to Newington workhouse to dress the corpse, to make sure the deceased's name was chalked on the coffin, to watch the lid being nailed down, to walk with their loved one a final time to their pauper grave. A focus on material consumption (the appearance and trappings of death) often undervalues these small and yet very significant gestures of grief. This is why Julie-Marie Strange argues that the poor's silence and stoicism was often composure under the worst of circumstances.[75] Many dissembled about the timing of death. This delayed having to transact an anatomy sale because of dire financial circumstances. Yet this viewpoint should not distract us from the sense that the poor could be motivated in myriad ways too. Sometimes they made a pragmatic calculation to save burial money and sold a corpse. Frequently there was no other financial choice. Occasionally the calculating profited from the death of a loved one (a theme developed in Chapters 4–7). Setting aside, however, the range of financial motivations, the fact of being placed in the position of having to even consider the possibility of becoming an anatomy supplier still attests to the grind of the deepest poverty in Victorian society. There seems, therefore, to have been an implicit understanding that a 'pauper funeral' was valuable and an 'anatomical burial' terrible. The next section will explore that fracturing of understanding to unravel its importance in the poorest communities. Meanwhile, a final story in this section illuminates what it must have been like to grieve on the edge of absolute pauperism, threatened by an anatomy sale when confronted by the death of a child. The story of Mary and Arthur Morton is symbolic of many cases that will be encountered in the rest of this book.

On 10 November 1849, the *Northern Star and National Trades' Journal* published in Leeds started a campaign to highlight the plight of bereaved paupers.[76] The editor was an exponent of the work of Thomas Martin Wheeler (Secretary to the National Charter Association and National Land Company). Although the tone of the articles had a strong political bias in favour of the poorest, the reporter also evoked the emotional experience of being poor at death's door in the North of England. He recounted the sad story of Arthur and Mary Morton. Their case history is illustrative of how easy it must have been to misjudge the emotional reaction of the poor and why historiography still needs to delve deeper into the set of circumstances surrounding the complexities of death and dying for the destitute in Victorian times.

In 1846, Mary Morton became pregnant. Nine months later she gave birth to a boy.[77] Unfortunately it was a difficult birth and she fell ill. Nonetheless, she nursed her child. The infant, however, became sick and subsequently died. The cause of death was unknown but the doctor thought it was a severe late-summer fever. Mary remained in her sick bed. She did not have sepsis (childbed fever), which would have been fatal. She did suffer though from repeated pains in her abdomen after the birth, leaving Arthur (her husband) to arrange a pauper funeral for their dead infant. The newspaper account then gives us important insights into the grieving parents' hidden misery and heart-rending pain:

> Return we to the house of desolation and mourning; during the time of Mary's illness, Arthur was too much absorbed by grief to attend

to any domestic cares; he had fallen into a state of torpid apathy, more fearful to contemplate than his previous moroseness. By the doctor's agency a nurse had been provided and all the arrangements completed for the child's funeral, and it was not until the corpse was being borne from the house that he [Arthur] showed any signs of being conscious of the loss that had befallen him.[78]

Arthur Morton was a naturally sullen man, of very few words, brought low by grief. He was in shock. The casual observer, however, would not have known the emotional turmoil beneath the surface of his abrupt and brusque manner. Only the fact that he had no energy for even basic household chores belied what was really happening at home. He was a man on the edge of a nervous breakdown. Arthur was distraught by the time the small coffin reached the churchyard in Brompton, on the outskirts of Leeds. Only then did he betray his true emotions in public:

> *dead* [sic], and sunk senseless on the coffin – with difficulty the [funeral] bearers conveyed him to his room. The pauper funeral then proceeded, and the body of the prized and petted infant – the child of many hopes – was laid in its mother earth without a single mourner to weep over its earthly fate; no father's tear to water its lowly grave – no mother's sob to waft her prayer to Heaven and bed a welcome for her babe.[79]

The reporting tone is very sentimental but it also gives us some important facts. Like the café encounter in the preface to this book, the reporter stated that Arthur Morton called out for *'My beautous boy'*. A pauper in the past and a lady in an Oxford café today seem to have shared a terrible human experience; it is the same language, the same pained expression, the same image of beauty being voiced. The Morton story also reveals that only the Poor Law funeral bearers walked to the grave. Nobody, not the father (too distraught) or mother (bedridden), or relatives (absent, living elsewhere), or friends (few wanted to intrude) were present. It appeared to be a bleak pauper funeral, an arbitrary event in the life of the community. On closer inspection, the reporter noted that it was in fact a family tragedy, deeply felt by both grieving parents, who could not face the grave scene. The newspaper reporter asked contemporaries not to judge what they saw but to think of the pain hidden at home: 'Oh! The mockery of human ceremonies, the hired ostentatious action of grief! Can they recall the dead to life? Can they assuage a single pang of those whose hearts bleed in secret?'[80] Again the bias is evident and yet as a proportion of this poor family's total income, a pauper funeral was expensive and very necessary in human terms.

The reporter told his readership that Mary Morton was 'the disconsolate mother, grieved in silence as only a mother can grieve, that she was denied the privilege of seeing the last duties performed to her lost child and her sick bed was indeed a bed of weariness'.[81] She was not there but the pauper funeral still gave the parents a sense of some closure. Meanwhile, 'Arthur – he who should have been her shield and a consolation in this, their day of mutual tribulation – he was a frantic madman, raving continually of past joys, embittering the sorrow of the present hour, by insane reminiscences of bygone hopes'.[82] Over time, 'slowly, did his mind recover its former tone – the unceasing attentions of his sick wife alone prevents his falling a victim to insanity and to what a world of misery did he awaken'. Once the child was buried and Mary got out of her sick bed to nurse Arthur, 'the benevolence of the doctor towards them ceased'. Now they discovered that 'Mary's illness caused her to lose employment, and deprived of her scanty earnings, charity was their only resource'. As for Arthur, he became 'a link in that great chain of outcast humanity which is continually clanking in our ears'. This, claimed the reporter, was the fate of those unfortunates who lived on society's margins and they were numerous. Certainly the periphery where the poor resided was a crowded space in Victorian life. The newspaper account, though written in an emotive style, alerts its readership to one basic fact. A pauper funeral was a considerable achievement for Arthur and Mary Morton under the circumstances.

Is this account credible or a sensationalised view of the poorest to sell newspapers? The exterior story of the death of Arthur and Mary Morton's infant is easily misjudged. Many contemporaries would have been disinterested in such a commonplace event in the lives of the poorest. Few would have looked beyond the basic fact of poverty. Bystanders at the funeral would have seen how unsentimental the poor could be about their children.[83] Objectively then the facts here seemed damning. Yet the reporter states that their offspring was 'the child of many hopes'.[84] The phrasing appears to be deliberate. One potential reading is that Arthur and Mary had a lot of difficulty conceiving and carrying a child to term. It is easy to assume that all of the poor could breed when they wanted to. At the root of the Mortons' private pain there seems to lurk a sadder unspoken story. Mary certainly fell very sick after the difficult birth. Pregnancy had not been an easy experience. She did not have the physical or emotional strength to attend the funeral. Her financial support crumbled when Arthur collapsed because he was the breadwinner. She therefore had to get out of her sick bed to nurse him. Arthur's recovery was not helped by the fact that medical relief ceased at the graveside. Arthur and Mary Morton had to quite literally pick up the remnants of their family life and put each other back together again. It is unlikely that they would have faked grief, since they could ill afford to lose any regular income. No record survives of the rest of their lives. They were typical and soon forgotten. Arthur and Mary Morton did have options, but they were terrible ones. They could have sold their child for dissection, but they did not. If the Poor Law had been less generous they would have had to transact an anatomy sale. A relieving officer would have acted as a go-between. The normal procedure was that the clerk to the guardians liaised with the Dead House porter at the workhouse and a Leeds anatomy school contact.[85] An undertaker would have discreetly collected the body. A fee of up to a month's wage for such a young child would have been some compensation to see them through the worst grief. It is painful to imagine what that might have done to their mental well-being.

Lost stories like these – the ones that seem to be about resilience in the end but were fraught with pain along the way – litter the anatomy trade. Yet they remain under-researched. The subject was 'taboo', as Richardson points out, but enough evidence survives for historians to try to speak about pauper realities.[86] Correspondence has been left locally and centrally to begin to re-engage with pauper strategies, performed in death. In the final section the complex nature of 'pauper funerals' versus 'anatomy burials' are explored in-depth. Taking this chronological U-turn brings together some of the missing aspects of death and dying for the destitute within the confines and changing landscape of the New Poor Law.

The pauper funeral

Figaro in London was a satirical journal sold in the capital when the Anatomy Act and Poor Law were being passed. On 29 April 1837, a sardonic article about the anatomy trade made the headlines. Entitled 'More Mutilation', it read:

MORE MUTILATION

Scarcely has the excitement subsided ... when others spring up around in all directions of an equally promising character. Heads, arms, legs and toes are being now dug up in every suburban cesspool, and coroners are calling inquests and pocketing their guineas [fees], and then returning verdicts of 'Found Dissected'!

The other day somebody stumbled over a human nose; and now we see this morning by the [*Morning*] *Chronicle* that an arm has been picked up in the moat just before the Penitentiary. There is, of course, an immense deal of agitation about it, and people say that it is decidedly a case of *'a female having come to harm'*, but, we, on the contrary, have a strong suspicion of *'arm having come away from female'*. However the local authorities [Poor Law and Police] are *'going at it'*; and are arresting every body without an arm, in the hopes of finding to whom the limb belongs. They make every one of the poor old pensioners, who happen to have lost such an article, go with them to the bone-house [at the workhouse], and have an arm fitted on, but as yet there is no sign of a successful fit. Those without an arm are rather absurdly supposed by the local authorities to have had a *'finger in it, or a hand in it'*; but how can this be, when their hands and fingers are usually cut off with the arm, we of course are not sagacious [astute] enough to determine.[87]

It was unclear whether the lady in question had lost her arm following a surgical amputation. Local rumours spread that hospital porters often sold limbs to medical students unable to buy a whole corpse for their anatomy training (see Chapter 6). There was meanwhile a strong suspicion that the true 'fit' would not be found among the living, but the dead. This tale was the story of a poor Cinderella with a Gothic twist. London churchyards were full to capacity. It was easy to cut off a limb sticking out of the shallow ground and sell it for a quick fee. It was moreover not illegal now that the Anatomy Act had been passed authorising corpses of the poor to be supplied to anatomists in the capital. Written with dry humour, the ironic tone in this short extract is entertaining, informative, and grisly. It hints heavily about a taboo body business, one run by anatomists and Poor Law officials with police cooperation and the input of some coroners that was starting to come to the attention of the radical press by 1837.

Two years later, Dr. George Alfred Walker published his famous *Gatherings from Graveyards* (1839) which described the appalling burial conditions in a selection of parish graveyards across London.[88] The *Lancet* soon reported that Dr. Walker had 'succeeded in awakening an unusual degree of public attention to the subject of intramural interments'.[89] He had highlighted the space crisis causing public health problems for London's residents. Walker's findings were published in major national newspapers like the *Times* (always a key critic of the Anatomy Act and Poor Law) and most provincial newspapers, such as the *Midland Free Press* (a radical paper). In Southwark,[90] Dr. Walker explained how easy it was to get hold of a limb:

> A body partly decomposed was dug up and placed on the surface, at the side slightly covered in earth; a mourner stepped upon it, and the loosened skin peeled off, he slipped forward and had nearly fallen into the grave.[91]

Dr. Walker drew attention to just how deep pauper graves could be and how dangerous because they remained open until filled up. One example must stand in for the many. Walker published a report that first appeared in the *Weekly Despatch* on 9 September 1830. It told the story of two fish dealers from Billingsgate market killed in a pauper grave. 'Thomas Oakes' fell into a pauper pit 'twenty feet deep'. 'Edward Luddett' jumped in with a ladder to try and rescue him but they both perished. At the subsequent coroner's hearing held in the workhouse of St. Botolph's parish in Aldgate evidence was heard that:

> Such graves as those [pauper graves] were kept open until there were seventeen or eighteen bodies in them...It was not the custom to put any earth between the coffin in those graves, except in cases where the person died from a contagious disease, and in that case some slaked lime and a thin layer of earth were put down to separate them.[92]

Dr. Walker argued that it was self-evident that there were public health risks from noxious fumes and a risk of higher infection rates in poorer districts where the biggest common graves were left open. Human remains were mixed with animal and vegetable refuse that was deposited in the pits on a daily basis. The decomposed material gave off a noxious 'miasma'. In his account, Walker was also concerned to highlight the plight of the poorest – again he did so in Gothic tones that turned his work into a best-seller. In one typical case in 1839 he visited Bethnal Green Union in the East End where the poor were numerous. At No. 33 St. Clement's Lane, Walker attended a very sick poor man – 'his health was broken, his spirits depressed, and he was fast merging into that low form of fever of which this locality has furnished so many'.[93] Outside his window was a '*"Green Ground"* – a grave yard a few feet from the house'.[94] The sick man told him: '*"Ah that grave is just made for a poor fellow who died in this house, in the room above me"'*. He had died from ' *"typhus fever... they have kept him twelve days* [without burying the body] *and now they are going to put him under my nose by way of warning to me"'*.[95] This was the fate of the poorest denied a pauper funeral.[96]

Julian Litton has helpfully researched in some detail surviving undertaker records from the time.[97] Expensive catalogues were often produced to advertise wares. They set out industry standards and are therefore instructive about the plight of paupers vis-à-vis the undertaking trade. In 1838, J. Turner Esq. published a 'nine-page catalogue' publicising that he was a '*Coffin Maker, Plate Chaser, Furnishing Undertaker and funeral Featherman*'. His merchandise lists are instructive. They reveal what paupers could not afford and just how extensive funeral paraphernalia was in the decade in which the Anatomy Act and Poor Law were passed. The catalogue described: '111 coffins of thirty different types, twenty-sizes of off-the-peg shroud (each available in four different qualities of material), fifteen styles of ruffling, winding sheets, mattresses, coffin furniture and palls'. If someone could afford it, he also supplied 'hearses, coaches and horses' and crucially to 'any part of the country'. A basic coffin cost '17 shillings'.[98] To put this in context, the fee was equivalent to a week's wage for a skilled sewing machinist or engineer living in London.[99] It purchased 'a good inch elm Coffin, smoothed, oiled and finished [i.e. waterproof so body fluids did not leak before burial], one row round of black or white nails [to keep the body secure], a plate of inscription, four handles, lined and pillow [to identify the face and for the comfort of the corpse]'.[100] The most expensive item in the catalogue was a coffin costing '£9' which bought a '1 ½ inch Oak case'. But by the time that the funeral had been elaborately furnished, attendants paid for, ostentatious transportation arranged, the undertaking bill had risen to a staggering '£50'. This was more than a year's salary for the average artisan. Litton estimates that in this case the undertaker's profit was about '10 guineas', evidence which explains why Charles Dickens thought that the undertaking trade exploited everyone, especially the poorest who could afford it least.

Poorer families could not hope to buy into this consumer range but many did try to purchase some small aspects of its material culture. For instance, 'a six foot shroud of *common quality* costing 2s. 8d' was one of the more affordable items in Turner's catalogue.[101] It was more likely that a pauper would approach the type of undertaker who supplied St. Bartholomew's hospital across the road from the Old Bailey, near St. Paul's cathedral. A surviving trade card in the hospital archives simply states: 'Mr B. Hewett, Little Gray's Inn Lane – Coffin Elm, Plate, lined and with nails – Cap, Pillow & C. within the distance of St. Giles – £2-0-0 – without Coffin £1-0-0'.[102] These basic commodities expressed some distance, however narrow, from absolute destitution. They symbolise why most paupers valued a parish burial where the deceased with all their body parts intact would be interred, even if wrapped in a winding sheet rather than a coffin. There was no guarantee that this would happen following complete dissection and dismemberment. This material fact explains why so many paupers wrote letters (or got someone to draft them on their behalf) to obtain burial on the parish in the nineteenth century. They recognised the difference between 'pauper funerals' and an 'anatomy burial'. There thus remains an important gap in the historical literature on 'anatomy burials' and their complexities under the New Poor Law.

Mortality studies have tended to misconstrue the nature of pauper rhetoric in Victorian times.[103] This is because of a general neglect of pauper letters in the history of death and dying.[104] The way for instance the poorest used the term 'pauper funeral' at the time has seldom been scrutinised.[105] A closer inspection of the letters written by, and about, paupers in Poor Law archives is revealing.[106] Sources suggest that it was not 'pauper burial' (being buried on the parish) that many feared; most were frightened by the prospect of 'anatomy burial' (a material mess). Representative Poor Law records in Northamptonshire indicate common expectations amongst the labouring poor in the area:

Funeral dress complete with cape 24 yards [of material]	£1 4s. 0d.
Coffin Broad Cloth 10/-6 a yard	£1 6s. 3d.
Flannel for lining 1/- a yard	£0 12s. 0d.
Black lining 4 yards	£0 5s. 4d.
Slate handles 9 [of them]	£0 2s. 3d.[107]

This undertaking bill, dated 4 March 1848, was for a typical 'pauper funeral', presented by Mary Church, 'Linen Draper, Silk Mercer, Hosier and Haberdasher, and Hatter and Funerals neatly Furnished'. She ran a pauper funeral business from the Market Place in Kettering town centre. Small artisans, shoemakers, and a wide variety of agricultural labourers claimed a 'cheap deal' coffin from the workhouse and Mary Church provided the basic extras. The standards in Kettering were not, however, those across the county.

In Brackley 'pauper burial' was a much 'shabbier affair' (on the A5, old Roman Road crossing between Northampton and Oxford). A letter from the Relieving Officer at Kettering to his equivalent at Brackley

Union highlights the tensions that a typical pauper family faced. It is quoted here in full because it gives us important insights into the narrative form, the rhetorical strategies, basic orthography, and financial constraints that are often generated in Poor Law files in respect of 'a pauper funeral':

> To the Relieving Officer of Brackley
>
> The Object of this letter being to acquaint you of the situation of John Wiggins the Bearer who Resides in this Union of Who is at this time totally Destitute of Employment and has been for the last seven weeks through a Factory at which he usually works being under Repair – And by an Accident which happened in August last – to his Oldest Child a girl of about 12 years of age – which by the Breaking in of a board Fell among the wheels of a Paper Mill and was dreadfully mangled – and at that time the mother was hardly recovered from her lying-in – and by reason on these Circumstances his Rent in Arrear and his Landlord threatens to take his Goods for non-payment of Rent and his Baker refuses to Serve any more Bread upon Credit now this Case we Would beg to leave to Recommend to your Serious Notice or in a short time you will be under the necessity of Receiving him and his family consisting of five small children which would be attended with a Great Expense You were under the Expense of burying the child above, tho with the greatest reluctance and Caused a Scandal in his Neighbourhood at the way the Child was left a Week in the House and then to be Intrd'd [interred] so Shabbily I wish that you would now send Word to John Higgins And I must Report that Another of his Children is likely to die according to the Word of Dr. Thomas, and we will be Placed at Great Inconvenience and Expense Unless you take the child back and offer it more than the usual Funeral.
>
> Duty to you &c.
> [Then written over the page]
>
> I Believe that above to be a True Statement of the Case, of John Wiggins
> B. Marton, Relieving Officer, Kettering.[108]

This lengthy letter claims to be between two men of equal educational standing who are responsible for investigating pauper claims for poor relief in their area. Its style suggests that they had been Overseers of the poor under the Old Poor Law. The letter writer has a basic but by no means sophisticated education. His orthography echoes the way that pauper claimants often wrote to press for poor relief. The rambling turn of phrase, the repeated use of capital letters to emphasis points, a lack of punctuation stand out. Even though Mr. B. Marton is the Relieving Officer for Kettering Union, his letter style belies how close he is to the pauper community in terms of his basic expression. He certainly writes in a tone that reflects the reality of local conditions. The detail is not officious but it is threatening. There is a strong suggestion that worse welfare claims will follow if the Relieving Officer in Brackley Union does not act now. The family may be forced to return home, which would be an expensive drain on the rates. This type of posturing is very akin to the sorts of rhetorical strategies that paupers adopted in their letters elsewhere during the Old and New Poor Law. Marton is above the threshold of relative poverty but he seems to understand how quickly one can fall below it into absolute pauperism. In a town like Kettering, artisans, over their working life cycle, often came into the remit of the Poor Law. His attitude is unsurprising but instructive.

There is then a lot of detailed information in the letter. The Relieving Officer has adopted a common pauper strategy. Authenticity was established by presenting the facts in a forthright and fulsome manner. The eldest girl had an industrial accident from which she died. The father may have been subsequently disabled (this is implied in a follow-up letter). Although shoemaking was a major employer in Kettering town, the eldest girl worked in a paper mill. She was 'dreadfully mangled' by machine wheels. There was considerable disagreement between the guardians of the poor at Kettering and Brackley about what should constitute a 'pauper funeral'. The twelve-year-old girl was laid out at home for 'a Week'. Her corpse putrefied on the premises while Poor Law officials argued about the cost of burial, what should be paid, and who would pay it. Kettering treated the dead liberally; evidently Brackley did not. In the end guardians in Brackley Union paid for a pauper funeral but we are told this was 'with the greatest reluctance' and it 'Caused a Scandal in this Neighbourhood [around Kettering]'. Local people were affronted that a young girl had been 'Intr'd [interred] so Shabbily'. At the time, 'the mother was hardly recovered from her lying-in' because she had just given birth to her fifth child, so she could not work to help pay for a pauper funeral. The letter stressed that sickness was a common occurrence in the family: 'another of his children is likely to die' and this fact had been verified by 'the Word of Dr. Thomas' the local Poor Law doctor. The Relieving Officer in Kettering wanted his equivalent in

Brackley to take over responsibility for the sick child: a small outlay on medical relief would avoid another pauper funeral. The latter, though basic, must have been worth having. The family meanwhile lacked credit and could not feed themselves. The baker refused to supply any more bread. Their rent was in serious arrears and they would be evicted. John Wiggins was out of work and did not expect to be re-employed soon. To stress the urgency of the case, the letter was carried in person to Brackley town. Wiggins travelled nearly 30 miles, almost a day's walk for a man who may have been disabled.

It is correspondence like this that opens a window on a world-without-welfare and the value of basic 'pauper funeral' rites, however shabby. Poor Law records emphasise not only the important regional and intra-regional differences in death customs but the crucial distinction between a 'pauper funeral' and that of an 'anatomy burial'. The two were very different rites of passage. As we have already seen, however, sometimes the two have been elided in the historiography.[109] This is understandable given that the Anatomy Act first appropriated the term 'pauper funeral' to describe how paupers would be buried after dissection. It did so to reassure the general public that the burial of human remains after an 'anatomical examination' was 'decent'. The phrasing did not elaborate in detail that 'a pauper funeral' (being buried on the parish) was not the same thing as an 'anatomy burial' (put in a pauper pit). Chapter 1 set in context that there was often little left to bury because the human remains were fragmented. Preservation fluid injected into the carotid artery delayed disintegration of the human remains but afterwards they were a material mess. Past a certain point in the dissection process, when skin, flesh, and tissue had been handled many times, there was just a pile of disintegrating bones, brain, trunk, and decomposing flesh. Care was taken to label the human material but it was in effect putrefied. It was never explained that an 'anatomy burial' dealt with perishing human waste, whereas a 'pauper funeral' preserved the basic body by placing it in the ground before decay set in. This was the corporeal reality. Yet parliamentary language and everyday speech referred to the generic term 'pauper funeral' as we saw above.

The confusion was one of common parlance as well. In everyday conversations the poor knew what they were talking about when they spoke about material changes in their circumstances because they shared each other's reality.[110] There was no need to explain the nuances of 'a pauper funeral' when they were there in person to take part. Being buried on the parish meant having possession of the body. The grieving could stand in close proximity to the corpse. Neighbours, friends, and relatives formed a kinship group that walked together to the grave as they followed the horse and trap, or rode in a hearse. This procession performed a sense of belonging for the living and the dead. These visible actions spoke volumes. Alternatively, if loved ones had been compelled to sell a corpse for anatomy then they had given up the body. Often there was no cadaver to bury. Many had been transported by railway out of the area. The bereaved were seldom given any information about the destination of corpses or their grave site. It was therefore impossible for these paupers to keep up appearances in the same way. Absence betrayed their desperate situation. So how did they express respect for the dead when confronted by friends, neighbours, and gossip about the missing body? They had only one option, namely a colloquial nod of the head. 'He's had a pauper funeral' was a familiar refrain in mid-Northamptonshire in the mid-1880s.[111] That complicit act – an unspoken assumption, never advertised, nonetheless understood – expressed a pauper's reality, and perhaps an increasing one given the abundant evidence in Poor Law case files held by central government and local newspaper reports generated in the West Midlands, a selection of which now follow on 'anatomical burials'.

In the late-Victorian period when the Poor Law became harsher (at the inception of the crusade against outdoor relief around 1873) boards of guardians reviewed their medical outdoor relief lists.[112] Previously it had been the custom in many areas to pay for 'pauper funerals' on medical orders issued by a Relieving Officer. The West Bromwich Union was one of the first to agree that paupers in need of medical care had to come into the workhouse. If they could not afford a 'pauper funeral' then their loved one's dead body would be treated as 'unclaimed'.[113] Their welfare debt to society would be repaid by selling the corpse for a small fee to 'Queen's College anatomy school in Birmingham'. Mr. Hampton, an outspoken guardian who sympathised with the poor, raised objections about this change of policy but he was defeated by the majority on the board of guardians. He therefore decided to expose the anatomy trade in the newspapers. On 13 March 1877, the *Birmingham Daily Post* reported that Hampton alleged that bodies taken for the workhouse for dissection were manhandled in the West Bromwich Union. This allegation was seconded by 'Mr Ward', another guardian. At a stormy meeting both men demanded to know specifics – how exactly were pauper cadavers sent to the anatomy school and what burial rites were they given? Significantly, the guardians did not refer to these procedures as 'anatomy burials'. Instead their discourse made deliberate and repeated use of the more familiar colloquial term, a 'pauper funeral'. The Chairman of the board of guardians, in defending his position, said that:

> The body to which [they] referred was properly dressed, put in a shroud, and removed in a shell to the Dead-House. It was lifted into the coffin in the shroud, and was taken away to Birmingham in the coffin. He thought it was not necessary to have the body disturbed [i.e. exhumed to check that these facts were correct]. He did not see the conveyance but was under the impression that it was a proper hearse.[114]

The Chairman emphasised that on the final journey of each pauper corpse from West Bromwich to the anatomy school at Birmingham it looked like a 'pauper funeral' was taking place (notice the use of the generic term). Usually this would have been done at night to avoid bad publicity but now that the anatomy trade was public knowledge, the guardians were careful to stress that standard rites were paid for each time. Later, the Chairman confirmed that the material remains were buried at a 'pauper funeral' (not the more accurate term 'anatomy burial'). He withheld details of where each corpse was buried, the state of the body, and avoided the question of dignity once dissection was finished. Instead he stressed that burial was either in 'consecrated ground' if a Christian burial or in a 'burial ground connected with the religious denomination to which the person belonged when alive' (Roman Catholic, Baptist, Methodist, Quaker, and so on). The term 'pauper funeral' was thus a convenient fail-safe for Poor Law officials in West Bromwich. It was utilised because it was also known to the poor in the area. Using it avoided discussing the material details of the anatomy trade. Another case nearby in the late-1880s shows how further research and discourse analysis can tease out these complexities of 'pauper funeral' versus 'anatomical burial' in the West Midlands.

In the winter of 1889, a pauper named James Clarke died in the workhouse in Walsall Union in the Midlands.[115] The clerk made a note that he was 'a tramp' with 'no relations'. His corpse was subsequently sold for anatomy to 'Queen's College Anatomical School'. The body was dissected and dismembered by Professor Bertram C. A. Windle (Chair of Anatomy). Once the cadaver had been dissected down to its extremities it was placed in a 'deal box' (a very basic coffin shell). It was then buried in 'common ground' at 'Witton cemetery...in a multiple grave on 31st January 1890'. A fortnight later, James Clarke's widow knocked on the door of the workhouse in Walsall Union. She was told by the master 'that her husband was dead and the body had been removed'. Thinking this meant that he had received a 'pauper funeral' (had been buried on the parish) she 'went away satisfied'. The master had not lied. Strictly speaking there had been a publicly funded burial. He did not, however, elaborate that the generic term 'pauper funeral' disguised 'an anatomy burial'. He had reverted to common parlance, using a rhetorical strategy embedded in the Anatomy Act and familiar among the poor. He did so for a mixture of reasons that included efficiency, sensitivity, and anxiety about bad publicity.

A week later, James Clarke's daughter called at the workhouse. Mrs. Margaret Longmine (née Clarke) had been estranged from her father, did not know that he was destitute, and had been searching for him for some time.[116] She too was told that he had died and been given 'a pauper funeral'. Upset by the news, Margaret Longmine wanted to view the grave. This request alarmed the workhouse master. The daughter insisted that she must be shown his burial site because she wanted 'to give her father a proper headstone'. The master could not do this without alerting her to the fact that the body was now located in Birmingham, not Walsall. Guardians had not advertised the fact that they had agreed to be suppliers to the anatomy trade. Hoping to avoid public exposure, the master wrote to Professor Windle at the anatomy school in Birmingham, outlining the bereaved daughter's case and attaching a copy of her original letter. An alarmed Windle asked the Anatomy Inspectorate to give him legal advice:

> When they [the daughter and her husband] write to me again, as they probably will do, I shall inform them of the fact, which they do not at present seem to have grasped that their father's body has been submitted to *complete* dissection and that the remaining *fragments* have been duly buried.[117]

In confidence Windle wrote that only the trunk and material extremities remained. A simple 'pauper funeral' (the rendering of 'ashes to ashes, dust to dust') had been ordered. There was no church service or loved one present to mourn James Clarke's passing. His unmarked grave was in Witton cemetery, opened by Birmingham City Council in 1863. The city authorities had allocated a burial area to Queen's College Anatomy School for their dissection cases. Windle admitted that he was 'concerned' about the indignity of 'pauper burial' (the same generic term once more) in Birmingham. There was just enough human material to fill a coffin shell. The Anatomy Inspectorate replied that 'he had acted correctly' within the strict letter of the law. Windle asked them to verify that 'the wife not the daughter has the first legal claim on the body' and that he could 'refer further queries to London'.[118] A civil servant replied that he was correct on both counts. They would continue to give him legal advice but he should suppress any scandal. He must not elaborate about the difference between a 'pauper funeral' and an 'anatomy burial' in print. This was in the national interest.

Delving deeper into the files, an extraordinary finding emerges. In the Local Government Board book in question there is also a copy of an official memorandum, numbered B7315A, dated 28 November 1889.[119] This has been pinned by a civil servant to the inside of the case file of James Clarke. It has been placed there to remind everyone connected to the Anatomy Inspectorate 'that the work of the Anatomy Department is covered by the Official Secrets Act 1889 [52 and 53 Vict. c. 52]'. To our modern eyes this action seems heavy-handed. The Victorian information state was evidently concerned to suppress knowledge of the anatomy trade. Bad publicity would be detrimental and so legislation now decreed that there must be no further official discussion. James Clarke's daughter was confronted by a bureaucratic wall of silence. The case reveals that there was a world of difference between a 'pauper funeral' referred to throughout the correspondence and the 'anatomy burial' that had taken place. Windle was convinced that James Clarke's daughter (given her education and marital status) would appreciate the distinction between the two types of rites. He intended to relate the detail but then never replied further on instructions from central government. In the file he admitted that it was probably better that the bereaved did not understand what had taken place. The case was closed in 1890. Today it reveals much about the value of a 'pauper funeral' versus an 'anatomy burial'. For many, the latter was too awful to write down or speak out loud.

Perhaps paupers did retreat when they heard the Anatomy Act poster on the wall being performed in Chapter 1. Poor Law records suggest that over time rhetorical strategies developed that must have somehow lessened the pain of what had to be faced in destitution. The poorest in Northamptonshire protested to obtain a 'pauper funeral' first (whether in kind, part-paid, or in full). When they failed, a nod of the head was an act of kindness and complicity. It was certainly less brutal than the terrible truth of an 'anatomy burial'. Anatomists, with the full cooperation of many poor law officials encountered in this book, were willing to exploit that complicity by never explaining what was hidden from public knowledge. In Northamptonshire and elsewhere in the Midlands, Burial Board records show that the average grave space in the 'common' section of graveyards cost about 'a shilling' per foot to dig.[120] That seam of earth kept the poorest in a common grave among their community. Having enough money meant that a faceless corpse did not float downstream; body parts were not sold after operative surgery; limbs did not peek from a narrow seam of soil to be sold by the inch from overcrowded graveyards; paupers were not compelled by poverty to enter the workhouse close to death to ease the burden of a funeral for their families. In local villages, provincial towns, and major cities the details differed but the basic dilemma to avoid an anatomy sale and substandard burial was the same everywhere. This pauper landscape of death and dying is known but not intimately, certainly not as much as it should be by engaging with Poor Law sources that underpinned the Anatomy Act.

Conclusion

This chapter set out to explore the complexities of death and dying from the perspective of the poorest to set in context what it meant to supply the anatomy trade. Faceless and anonymous people who sank into absolute poverty explain how easy it was to lose one's identity and become vulnerable to dissection in death. There has been a narrow and yet influential historiography on pauper funerals which now interrogates pauper's silence and stoicism, as well as their material constraints to better understand the experience of being poor. Poverty historians have indicated that there is a wealth of new material in Poor Law accounts that can help to illuminate lost lives, the grieving process, and the emotional range that paupers must have experienced in bereavement. The Mortons' story is just one among many in the nineteenth century. Pauper funerals were consequently complex and varied. In some areas the Poor Law was generous, in others not. This made a great deal of difference to those threatened by dissection. The coming of Poor Law democracy did start to change attitudes but, as we shall see in the rest of this book, anatomists also found ways to maintain their supply lines until the New Poor Law was disbanded in 1929. Meanwhile, each corpse was a financial crisis for the poorest families to try to overcome. Thousands often failed, resulting in an 'anatomy burial'.

It is to this theme that the next chapter now turns. The time has come to follow the body of a pauper into the dissection room. In Chapter 3 new research rediscovers what it was like to be dissected and to do the dissecting. The corpse was that of

> a man whose name is unknown...age unknown...died at West Street, Smithfield...died 20th November 1834...body removed by the beadles for dissection...November 23rd 1834 at 6pm...sold to St. Bartholomew's hospital.[121]

How did he get there; what happened; did someone kill him or was it a natural death; why did he die; where was he found; was he loved or just abandoned; who sold him and for how much; was a coroner involved; did the beadles of the hospital make a profit from the supply deal; how was he dissected; did it take long; what was left when the medical students had finished; where was he buried; is there still a marked public grave; how did he fall into a history of silence? This long list of questions deserves more in-depth historical scrutiny in the next chapter.

3
A Dissection Room Drama: English Medical Education

Two men dressed in black on a bleak winter's night in November 1834 carried the corpse of an 'unknown' man into St. Bartholomew's hospital's dissection room.[1] An anatomist was standing by to receive the cadaver. He made a new entry in his anatomy register. The corpse was numbered 181. This reflected how many cadavers had been bought since the Anatomy Act in 1832. The number was then written on several labels. Identifying tags were tied around the head, torso, and limbs. All cadavers were monitored in this way. It meant that few students could take away body parts to sell on, a problem in years when human material was in short supply. Once the corpse had been dissected and dismembered 'down to its extremities' the numbered parts were sewn or tied together by the demonstrator in anatomy and his assistant. At St. Bartholomew's hospital over 6,000 pauper bodies were processed like this. These rare accounts reveal a hidden English anatomy trade that underpinned medical education in Victorian times.

Body supply figures, set out in Chapters 4 to 7, substantiate an idea that more of the poor were dissected than has previously been suggested. To set those figures in context, it is important to focus attention onto medical education.[2] Historical accounts often look down on corpses laid out on the dissection slab. The aim in this chapter is to look around the dissection room, to glimpse the interior world that the poorest entered and in which medical students trained in human anatomy.[3]

Through the historical prism of paperwork associated with the anatomy trade, it is possible to say more about the dissection room drama. For, as Michael Sappol's excellent research has shown, human anatomy has always been a 'performance' of power relations.[4] In the nineteenth century the medical sciences needed to establish their professional authority over the corpse to promote new educational standards. Eager newspaper editors, with their eyes fixed on improving sales figures, helped their cause by being determined to get behind the dissection room door. Stories were commissioned detailing the dramatic potential of human anatomy. Only a handful of British historians have written in detail about the 'avid audience' that was created by the press campaign to popularise the melodrama of anatomy during Victorian times.[5] Yet the high sales figures of popular medical textbooks, like *Gray's Anatomy* (1854), reflected this theatrical trend.[6] The body was to be cut, exposed, and staged, for medical and public enquiry, provided its identity had been erased by poverty. This new reading material helped to forge a sense of corporate identity among the medical fraternity. Anatomy thus 'burst its disciplinary boundaries' into 'microbiology, zoology, anthropology, health reform, neurology, pathology, gynaecology, endocrinology, etc'.[7] That flowering of medical knowledge soon captured the public imagination.

Anatomy museums were one public venue in which the popular appeal of human anatomy and its teaching delivery was broadened.[8] Collections of human specimens and skeletons were displayed alongside dissected and stuffed animals. Patrons usually funded purchases to sustain medical curriculum. The viewer engaged with, but remained distant from, the lancet marks on the corpse. Wax models instead were studied. Popular anatomy books picked up on this public engagement trend. The genre soon tried to widen its appeal. Better profit margins in publishing could be made from exposing the hidden sides of a medical education to entertain a growing readership. It is not possible in a book of this length to study every facet of the complex popular culture that sensationalised the dissection of the human body. Ruth Richardson has, in any case, traced resentment by the labouring poor at the cheaper end of the popular journal and newspaper market.[9] The aim in this chapter is to sample instead more mainstream publications whose middle-income readership became medicine's chief consumers. They not only bought into the popular anatomy genre, but their offspring paid to become medical students.[10] If they were attracted by the dissection room drama then this consensus could help to create 'the growing cultural authority of medical science' over the corpse.[11] The cultural attitudes of personnel who managed the smooth functioning of the business of anatomy are neglected too. Many welfare agencies that supplied the anatomy trade were administered by middle-class ratepayers and their paid officials.[12] Their reading materials would have shaped Poor Law opinions about dissection. These must have impacted on the success or failure of the operation of the anatomy trade regionally. Attitudes would have influenced negotiations, made and remade, between anatomy schools and their suppliers.[13] Yet, the transmission of ideas about dissection between welfare agencies (asylums, infirmaries, workhouses) and the medical fraternity (students, doctors, anatomists) remains understudied.

Trainee doctors in the system, by looking down onto the cadaver, could hold up their professional heads in a society keen to read about the dissection room drama. The ways in which this 'popular anatomy' became 'embodied in social identity' are under-explored in Britain compared to America. Sappol has shown convincingly that the act of reading about an anatomy demonstration in newspaper reports and fictional stories dramatised and yet distanced the reader from the horrific detail of the actual experience.[14] Medical science took advantage of common prejudices about the poor to promote its corporate agenda in the United States. There consequently developed a vibrant body trade in which the poorest and racially stereotyped were exploited in death.[15] This chapter, in sections one and two, will develop Sappol's agenda by engaging with a general history of medical education and its licensing system in England by the mid-Victorian period. It illustrates that medical scenery with typical portrayals of medical students. These are taken from popular literature at the time. Section three then returns to the corpse carried into St. Bartholomew's hospital in 1834. A paper trail reconnects researchers today to a generation of trainee doctors. The aim is to look around the actual dissection room. Section four will finally examine the difficulties anatomists faced. A selection of common press stories gives a flavour of recurrent themes. Together they reveal the mundane, rather than melodramatic, interior of the dissection room. Although a genre of popular anatomy set out to entertain its reading public, its unwitting testimony hinted at a more disconcerting human drama. Stories tantalised by sending a sharp tingle of dread up the spine. That gothic portrayal is compared in section four with the recruitment literature distributed by medical schools and advertised in the press to promote the medical profession. This exercise will show that there was a significant difference between the glamour and performance of cutting the cadaver in popular anatomy texts, compared to the de-humanising experience. The real dissection room drama is found underneath the business of anatomy. It was one peopled by corpses, eyes fixed upwards, and trainee doctors, gazing downwards, onto a dissection room table standing centre stage in medical circles in Victorian times. It is vital therefore to appreciate that scenery and to understand its setting before encountering the bodies that were dissected in the rest of this book. We start with an overview of the development of, and attitudes to, medical education in the nineteenth century.

Still they gazed and still the wonder grew
That one small head could carry all he knew[16]

Medical students were often the subject of derision in nineteenth-century popular culture. It is easy to see why, given the changes taking place in medicine. In many respects they were an easy target for contemporary criticism. No more than a cursory glance is needed to get a sense of the paraphernalia published about human anatomy teaching. In the popular press, numerous editorials discussed whether medical training was fit for its purpose. Leading journals read by a cross-section of the middle class were also forthright in their scrutiny. Typical headlines included: 'Is Medicine Really a Science'; 'Medical Students – the returns of the numbers and their meaning'; 'The Inner Life of a London Hospital'; and 'What are living beings?'[17] Before looking at one or two of these sources in more detail, we first need to set them in context by reviewing the broader trends in medical education that were underway.

When the Anatomy Act came into force in 1832, medicine had gone through a number of transitions.[18] In broad terms this meant that whereas in the eighteenth century most poor patients were treated at home, in their parish, or sent to provincial voluntary hospitals, by the nineteenth century a more institutionalised and hospital-based system of medical care was coming into vogue. Medical students who paid for private anatomy lessons soon found that educational provision was now based at medical schools usually located next to major teaching hospitals. Gradually there emerged a consensus that there was a need to regulate new licensing qualifications. Recently Keir Waddington has written an important appraisal of general educational trends.[19] He makes a noteworthy observation that new professional standards evolved in complex, incremental ways and that 'student demand, licensing requirements, individual lecturers, medical knowledge and worries about institutional inadequacies all combined to shape teaching'.[20] Although some form of hospital training was a central feature of medical education, complex licensing qualifications in reality complicated the career decisions of many new doctors.[21] It was not until the Medical Act of 1858 was enhanced by the passing of the Medical Amendment Act in 1885–1886 that licensing was clarified, a process explored later in this chapter.[22] In the meantime, better health care became a litmus test of ones social standing and economic spending power. That consumer trend meant that although medical services were patchy and regionalised, practitioners also began to respond to economic pressures to provide more professional services in a unified way. Once demographic trends sparked widespread support in the press for better health care services, it was only a matter of time before restructuring of the medical fraternity followed suit. Population pressure was a catalyst in the history of medical consumption and educational improvements.

After 1750 there was a marked population explosion in England.[23] Tony Wrigley estimates that:

> In 1680 the population of Britain was about 6.5 million, or 7.6 per cent of the west European total of about 86 million. Yet in 1840 the British share had risen to 10.5 per cent (18.5 million out of a total of 177 million). By 1860 the comparable totals were 23.1 and 197 million and the British percentage had reached 11.7, an increase of almost 60 per cent compared with the situation 180 years earlier. Since 1860 there has been a further rise in the British share of the west European total, but it has been much slower and more modest. In 1990 the population of Britain was 56 million, 13.1 per cent of the west European total of 429 million.[24]

This demography facilitated an expansion of medical education to meet growing health care needs; more especially, the Old Poor Law – the welfare system through which sickness claims made by the labouring sort and poor were paid – became driven by an ideological imperative that basic health care was fundamental to the well-being of everyone in the English community. In practical terms, this meant that the desire to expand medical services on the part of physicians, surgeons, and apothecaries found a ready market from the top to the bottom of society. The recent work of Steven King on cures, medical payments, and treatments, paid by overseers on behalf of ratepayers for the sick poor across seven English counties reveals a key finding for the period 1750 to 1834.[25] King's remarkable new discovery is that while there were regional differences in aggregated expenditure on sickness relief, most communities spent a considerable proportion of their welfare funds, 'not less than 25%, on medical care broadly defined'. That trend mirrored a growing interest in medicine, health care, and medical education among the middling sort who paid for basic welfare provision. Medical fashions thus began to flourish at a time of mobility and urbanisation.[26] There was consequently a growing demand for more qualified doctors. Taking the cure at Bath or visiting Oxford to buy the latest medical remedies were just two consumer trends that drove the growing range of services available throughout late-Georgian England.

Before long, quacks, fakers, and charlatans – the medical entrepreneurs of the long eighteenth century – moved in to try to make a fast profit in medical markets.[27] This outcome was both a problem and an opportunity for the medical fraternity. Irregulars obviously needed to be identified and marginalised or they would gain a strong financial foothold in the medical marketplace at a time when doctors relied on fee-paying patients. The unscrupulous also needed to be stopped from selling dangerous remedies. Many irregulars, though, developed good distribution networks for their harmless, common herbal remedies. The difficulty for the medical fraternity was that these popular cures were helping to expand the market in medical consumables. It was financially advantageous and yet profiteers exploited health-care knowledge that had been the exclusive privilege of the educated elite. For strong commercial reasons medical education thus became an important economic weapon in the tussle over the future financial control of the medical marketplace and its general accessibility. Although historians have questioned rising levels of consumption by exploring the thirst for better scientific knowledge, nonetheless it was financial considerations that were the chief catalyst for the reform of the medical licensing system throughout England.[28]

The reform of English medical education was in turn divisive because it involved restructuring the different ranks of the profession. In the sixteenth century, three basic ranks of licensed medical practitioner predominated, namely physician, barber-surgeon, and apothecary.[29] By the eighteenth century members of the Royal College of Physicians were still the top rank. It was the most expensive route into medicine but was highly valued and therefore worth the financial outlay. Meanwhile the lower grades of medicine were changing in ways that would impact on the economy of supply in dead bodies by the nineteenth century. In 1745, the split in the United Company of Barber-Surgeons has been seen as a formative moment in medical reform.[30] Division was a catalyst for medical educational changes over the course of the next three generations.[31] Essentially, surgeons wanted better professional recognition of their skills. They thus broke free of the barbers to avoid being tainted by the accusation that they were a jack of all trades and master of none.[32] By 1800, the Company of Surgeons had gained a Royal Charter and thereafter they were known as the Royal College of Surgeons. Soon they were issuing their own diplomas. This new license meant that a surgeon needed greater access to more corpses. Yet murders and murderers – the latter being the chief source of cadavers – always lagged behind student demand. Medical education, it was argued, could not advance without legislation to secure more dead bodies. The third rank of medicine was meanwhile held by those affiliated to the Worshipful Company of Apothecaries. Historians agree that the apothecaries had been more forward-thinking than surgeons.[33] In 1617, they pressed for and obtained a Royal Charter from King James I. By the eighteenth century they dominated the manufacture, licensing, and selling of pharmaceutical products in medical care. Naturally the apothecaries were very alarmed by the rise of the medical entrepreneurs peddling cheap herbal remedies.[34] It was for this reason that they lobbied Parliament to pass the Apothecaries Act in 1815. That statute anticipated the stricter licensing standards of the Medical Act (1858).

Once Queen Victoria was crowned the traditional professional standing of physician, barber-surgeon, and apothecary, had divided into the tripartite ranks of physician, surgeon, and apothecary. It is important to stress, though, that the formal distinction between surgeon and apothecary was very blurred, so much so that most advertised themselves as 'surgeon-apothecaries' in early Victorian society. They have been described as 'emergent general practitioners' who chose to expand their apprenticeships and expertise because their ambition was to reshape the field of medicine in their financial favour.[35] This concerted change meant that the centre of clinical instruction and lectures was now located in London. That trend brought about a key shift in what Susan Lawrence describes as 'the business of educating – not training – the senses' of future doctors. The evolution of licensing standards was at the heart of this philosophy.[36] A consideration of those changes and what they meant for the anatomy business by the mid-Victorian period is important for the later case studies in this book. But before entering the dissection room proper, it is necessary to briefly understand the training requirements that necessitated human anatomy teaching.

Training in medicine

To become a physician, a medical student needed patronage.[37] They would have to pay for a university education in the liberal arts. Pupils walked the wards of an early modern hospital with a leading fellow of the Royal College of Physicians. It was also common to study abroad at one of the great centres of human anatomy teaching like Bologna,

Leiden, or Paris. Returning to London, a medical student would be expected to then attend a variety of demonstrations and lectures. This meant purchasing extra lessons in anatomy, midwifery, and surgery, usually at private anatomy schools. At the same time they had to serve their hours in service by studying the theory and practice of medical care. Most physicians were hands-off in their style of medicine. Schooled in the arts of physic, these were the gentlemen of the medical world. Their prestigious qualification was respected by all ranks of medicine and the general public. That fact notwithstanding, it was surgeons rather than physicians who led reforms in medical education by the early nineteenth century. Few medical students had the social connections or private income to train as a physician. Most concentrated on the licensing qualifications offered by the Royal College of Surgeons because they were less expensive and more accessible for the middling sort. There was, though, concern about European competition. On the continent, corpses were in more plentiful supply because giving one's body to anatomy was a civic duty. English surgeons thus needed to promote their licensing standards and resolve body supply problems to maintain their share of an expanding field of European medicine.

In Victorian England medical students could be licensed in one of three key ways.[38] To recapitulate, they could study for a diploma with the Royal College of Physicians (MRCP), but this involved a long apprenticeship – generally over seven years – and was expensive. Alternatively they could opt to become a 'five-year pupil' apprenticed to a surgeon, studying for the licensed diploma issued by the Royal College of Surgeons (MRCS), a less expensive but still costly qualification. A third trend was to follow the route into 'a new type of medicine', basic but practical, which was becoming more accessible. In voluntary hospitals 'observation, physical examination, statistics, anatomy and pathology' were being opened up to students, a 'move towards institutionalised teaching'.[39] Until the hospital system was up and running the apothecaries provided the most common qualification route in early Victorian times.[40]

It was significant that after the Apothecaries Act of 1815, it became more common to study medicine in two stages. All medical students first obtained an entry-level qualification. This involved serving a five-year apprenticeship and completing six months on the wards of a voluntary hospital. Those who followed this professional route were awarded the License of the Society of Apothecaries (LSA). Aspiring doctors attended regular lectures in 'anatomy, botany, chemistry, *material medica*, and the theory and practice of physic'.[41] Roy Porter estimates that 'by the 1830s more than 400 students per year were taking the LSA'. Those who obtained the LSA could then opt to undertake further advanced study. A second more skilled qualification entailed studying for either an MRCP or MRCS. This achievement was dependent on an individual medical student's financial resources and natural skills. The Apothecaries Act thus codified and refined new licensing standards. Reform broke down the old tripartite divisions of medicine, rendered obsolete as consumer fashions and scientific knowledge connected to the human body became more sophisticated. For some time, however, it was unclear in many voluntary hospitals whether the governors, surgeons, or lecturers controlled the training of medical students. The Apothecaries Act thus prompted the Royal College of Surgeons to anticipate further licensing changes. In the 1830s they insisted that anatomy courses had to be conducted in hospitals and overseen by a designated MRCS examiner to merit official recognition. What was happening is that, as Roger Cooter so aptly puts it, 'the cake got cut' – medical specialisation became a feature of education in London.[42]

London hospitals in turn responded to licensing restructuring by copying the Paris 'anatomical-clinical model' of medical education.[43] In France, the poor were used to train doctors because dissection and accurate diagnosis was held to be in the national interest. Many English surgeons had trained abroad by the 1830s. This created improved medical networks and better knowledge transfer opportunities. The dissemination of ideas led to a new type of teaching delivery on English hospital wards. Gradually those who ran the English system gained honorary or salaried appointments in voluntary hospitals and costly apprenticeships were replaced by formal anatomy demonstrations and lectures. Clinical experience with patients on the wards became the norm. In this way, continental standards were slowly institutionalised. Attendance at one school of medicine connected to a particular hospital solidified in medical circles and it was there that students now studied for their licensing qualifications provided there were enough bodies to dissect. Across the capital the figure of the 'hospital student' soon caught the public imagination.

In the 1830s and 1840s the state of medical education and the growing number of doctors was the subject of public debate in the popular press. In some areas there were disputes about a lack of access to a qualified medical man. Most doctors were educated in major cities and so the countryside often lacked basic health care services. Critics also questioned the scientific authority of the fastest growing profession in

Britain. Concern was expressed that patients might become objects of experimentation and prurient interest. Others noted that standards of clinical investigation and instruction had improved in leading teaching hospitals but the Poor Law continued to attract the least qualified men. This was unfair to the poor but it did potentially increase the pool of dead bodies for trainee doctors! Commentators thus continued to be concerned about the general rowdy conduct of medical students. One typical example stands in for many published at the time. In June 1840, the *Penny Satirist* mocked the 'Medical Student in search of a Supper'. It began:

> Of all classes of human beings, indigenous to this country, we know no class so peculiar as that of medical students…. If the reader has ever been in the vicinity of one of the large hospitals, or in the immediate neighbourhood of one, he must know that these places are literally infested with these important members of society. If the hospital be a large one, there is scarcely a house within a mile of it that has not one student at least in it but the generality of them are entirely tenanted therein.[44]

The writer elaborated that a 'medical student can be properly observed at three distinct intervals':

> First, *at home*, the window of the front parlour in which he lodges is thrown up as far as its dimensions will admit, it being about seven o'clock in the evening and a hot summer's day, and the young gentleman may be observed seated. His mouth is decorated by a lavish havannah [cigar]; his arm is lazily leaning over the window-ledge – one leg is upon a chair, *a la Françoise* and if his knee is neat, the other is half-dangling out of the window, on the table stand a majestic pewter-pot bearing the heraldic insignia of the hardier publican, on the top of which stands *Quain's Anatomy*….[45]

Medical students were criticised for their high spirits. Drunken fights were often reported in the press.[46] Concern was expressed that trainee doctors were paying for cigars and beer, rather than their education fees. The depiction of a pewter-pot propping open an anatomy book is a striking image that hints at disorderly and drunken behaviour. Sampling similar accounts it is clear that the general public were ambivalent about students walking hospital wards because of their reputation in those terms.[47] The article then turned to the medical student's general pose:

> Secondly, *abroad*. His aspect is now altered. He is now revelling in all the luxuries of a bear-skin taglioni-coat, his glossy curls are contained within the circumference of a D'orsay gossamer; and there is a degree of quizzicality about his pale countenance, which, when the circumstance of his having a great biccory stick is taken into account, would fully satisfy the idea that he was intent upon a *lark*.[48]

The medical student was the dandy about town. He had an effeminate character, in search of fun, entertainment, and japes. His style was continental. He wore an Italian greatcoat with a Parisian collar and carried a silver-topped cane. Certainly time spent training in Paris had shaped his fashion-conscious look. The article then probed further:

> Thirdly, in the *Dissecting-room*. This last locality and its temporary tenants, the medical students, are matters of great interest. If the reader can exercise his ideality [sic] so far as to imagine a long high room, in which are about a dozen longitudinal deal tables, he will recognise a dissecting room. In the centre of the room is a large blazing fire, around which stand or sit, about a dozen young men, some in fashionable attire, others dressed in the babillments [clothing] of dissectors – the glazed cap, the flannel jacket, the glazed sleeves and apron. Some preparatory to their anatomical labours are enjoying a puff of Indian weed [cannabis]; whilst others may be observed attentively looking over their shoulders of those persecuting the cause of science.[49]

The reporter told his readers that at first he did not want to look down at the horrors on the dissection table. He then thought that he had a duty to his readership to do so:

> Amongst the latter might be observed in the dissecting room, a young man in the garb of one in the pursuit we have mentioned. He was attentively engaged in depriving a leg of its muscles; and interspersing his labours by an occasional perusal of some anatomical book which lay open before him. The monotony of study was relieved by an occasional piece of snuff, taken from a *potato* [sic] box, with which, together with the *subjects*, the table was furnished.[50]

There is important witting and unwitting testimony in this popular depiction.

The medical student's daily working life in the dissection room was a monotonous experience. Long hours were spent dissecting and dismembering body parts. Practical demonstrations had to be checked and re-checked against anatomical manuals to learn the intricacies of the human body. Later in the chapter we will see that this impression had quite a lot of basis in reality. Narcotic substances like cannabis and snuff meanwhile alleviated the noxious fumes of cleansing and preservation fluids that stopped the human tissue decomposing. They also kept bored students awake during the long hours spent in a crowded dissection room. Addictive drugs, self-dosing, and privileged access to pharmaceuticals were seen as common throughout the entry level of the profession. Whether or not the general public appreciated why medical students used chemical stimulants is not stated in the extract, but the satirical tone suggests that drug use was associated with youthful pranks.

The reader is then given an important glimpse behind the dissection room door. There were a dozen tables and the same number of medical students in the room but there were not a dozen subjects. Some students had to look over the shoulders of others because there were not enough cadavers for everyone. Since dissection work was intricate and time-consuming, it must have been exhausting to concentrate during an eight-hour session. The extract tells the reader about the design of the dissection room stage. It was a narrow room, ventilated by a skylight. Students could not see any natural daylight through an eye-level window. Fresh air was in short supply in winter. A fire heated the chilly room. The skylight kept prying eyes from penetrating the secret world of the dissection room drama. This basic design emphasised that the scene was mundane, rather than sensational. In the scientific performance of the corpse, the student actors put on a professional costume. The uniform of the dissector explains why so many medical students preferred to dress like dandies once their working day was finished. Students wore a 'glazed' apron, cap, and sleeves, and a 'flannel jacket' on top because blood, mucous, and tissue could be washed down clean. The clothing again suggests that dissection was an impersonal experience. Putting on fashionable attire marked a mental and physical transition to the outside world. Corpses, though central to this drama, remain anonymous, even though the poorest were leading actors in the cast of the dissection room.

In a follow-up issue in November 1840, the *Penny Satirist* reported on basic dissection room conditions across the capital. It published an anonymous account written by an art student about dissection rooms he had known in the 1830s. He gained entry to improve his life-drawing skills around the time that the new anatomy legislation was enacted. This first hand account is noteworthy both for its details and because it deepens our historical appreciation of basic dissection room design. It reads:

> The dissecting room was underground and there was a museum of skeletons and hearts, livers, legs and lights upstairs; a dreary place it was at dusk, I assure you. And downstairs there usually were half a dozen dead bodies, stolen from churchyards or bought on the sly ... besides a number of amputated limbs, such as heads, arms & legs, &c., in various stages of *scientific preparation* [sic], under the operation of the scalping knife. Here I drew – But as the students were at work in the early part of the day, and I wanted to draw the parts, which the knife had laid open, I was compelled to wait till they had ceased.[51]

The art student recounted that the main advantage of working in a basement was that medical students could attend day and night without being seen. The art student also explained that there was another important reason for being underground.

> The dissecting room was not down stairs but down *ladder* [sic]. It was simply a ladder through a species of trap door that we made our descent and in one room were the operators, and in another room the sort of back kitchen with a water pipe and a sink, where the bodies were washed. In this sink there was generally a body lying, and the water running upon it, as at the *Morgue* in Paris, where the exposed corpses have a continuous shower falling upon them to keep them cool [i.e. preserved].[52]

In the early nineteenth century, preservation techniques were cruder. It was rare though to read an explicit account in the press recounting in detail the practical ways that French anatomy had influenced English methods. Most dissection rooms copied the way that the 'unclaimed' dead were handled in the Public Morgue of Paris.[53] Bodies fished from the Seine (homeless, suicides, and so on) were laid out for display on black marble slabs. On viewing days, white flesh against the dark marble made the person's identity visible. The Morgue was open seven days a week. The citizens of Paris were encouraged to view the dead through a large window to name the cadavers. Above each body was a tap and water pumped from the Seine flowed over the corpse to preserve the human remains to facilitate proper identification. Clothes were washed and hung on pegs in case they might be recognised by visitors. After about six weeks, all bodies were dissected to ensure that death certificates and public health records contained accurate diagnosis. An anatomy register was kept detailing as much as possible about the dispossessed person – length of body, size of head, clothing description. It was the duty of the Morgue registrar to answer letters from families writing in about missing persons from all over France. The English art student evidently knew about this system and admired it. Aspects had been adopted at the unnamed anatomy schools he visited on a regular basis around London in the 1830s. The need to keep the bodies fresh meant that it was more practical to wash the corpses continually in basement sinks close to the drains. They could then be buried unseen, outside in the backyard. Yet a record of this work was to be kept secret, not made public: a big difference between France and England. The art student also told readers that he stayed up most nights to draw the corpses. He was sometimes afraid, admitting that:

> I was always glad when I got to the top of that ladder, but still I had part of the museum to pass through on the way out. The displays were sometimes as frightening as the cadavers – there were skeletons there and mummies, stuffed men and women – I seldom looked at them – sometimes to show my daring and courage, I did stand and take a peep, and even touched them – but not very scientifically.[54]

He was always relieved to 'find myself alive in the street, in the company of living folk'.[55]

Common designs staged a corporeal drama in the dissection room. Its popular depiction gave medical students a mixed reputation. Testing its veracity involves getting much closer to the theatrical performance of the anatomical body on the table. It is time to return to the cadaver of a 'man unknown...age unknown' brought into the dissection room of St. Bartholomew's hospital in London on a winter's night in 1834.[56] It was this man who began Chapter 3, and we now follow the corpse into a hidden, anatomical world. The time has come to put back his identity by working backwards in the archives from death to life.

Death by poison – the perspective of the poorest

On 13 November 1834, 'a man of respectable appearance' knocked on the door of a common lodging house in West Street, Smithfield.[57] He told the landlady that 'times were not with me, as they used to be'. The stranger asked to hire a room for the week. The landlady replied that he must pay her a week's rent in advance. She was a shrewd businesswoman and suspected the stranger had financial problems. The man agreed to her terms provided his privacy was respected. He stated that he must not be disturbed after the deal was done. The landlady accepted his requirements, charging him 'extra' to be 'accommodated in a separate room'. The bargain struck, she pocketed the cash and asked no more questions. The stranger's sad story was his own affair. London was filled with people who had fallen on hard times. Pauperism was an endemic social problem in the capital when the Poor Law Amendment Act was passed in 1834. If the man could not pay his rent the next week, then he would have to walk to the workhouse to get basic welfare or face debtor's prison.

Seven days later, on the next rent day, the lodging house occupants woke to the sound of 'the bell being rung violently' above their heads. Running upstairs, the landlady's common servant found the new arrival 'lying stretched out on the floor in a strangely convulsed state'. Frightened by the sight of the stranger frothing at the mouth, she rushed out the door for a doctor. A medical man came immediately. Looking around the room he found 'by the basin a phial, which contained a preparation of opium, and which was labelled *poison* [sic]' indicating an overdose. The doctor arranged for the man to be admitted as an emergency case to St. Bartholomew's hospital for the sick poor where medical staff had a lot of experience in rescuing attempted suicides. The man's stomach needed to be pumped to expel the poison. Although the hospital was very close to the lodging house, across the road from the Smithfield market area, the man was dead on arrival.

Coronial and medical records can be used to reconstruct what happened to the corpse before it was buried.[58] It was first collected by the beadles and taken to the Dead House within the precincts of the hospital. Officials then called in Sergeant William Payne, the coroner for the City of London. He often used the hospital Dead House as a coroner's court. The coroner searched the pockets of the unidentified man. He found 'a handkerchief, marked with the name of Burgess and four pawnbrokers' tickets, 'three in the name of Thompson, and one in that of Andrews'.[59] One of the coroner's officers was then instructed to find out about the ownership of the goods. He returned to say that there had been 'three pledges' for 'silver milk-pots pawned at different pawn shops' on 13 November 1834. The coroner concluded that the pawn money had been used to pay the week's rent in advance at the lodging house. Nobody, though, knew the man's real name.

The 'view' of the body by a coroner's jury was postponed. Further enquiries were made in the locality over the next 72 hours. There was no question that it was a suicide. It was standard procedure, however, to check in poison cases if anyone had visited the deceased in the last hours before their demise. Enquiries were also made to try to locate where any poison might have been purchased in the vicinity. Provided there was no evidence of a suspicious death (misadventure, manslaughter, or murder) the coroner could use his legal powers to pass a verdict of suicide without the need for a costly post-mortem, By 6 p.m. on 23 November (ten days after the stranger rented the lodging house room, three days after death) the case was closed by the coroner. It now passed back into the jurisdiction of the hospital staff.

The man had died nameless and friendless. Nobody came forward to claim his cadaver for burial. Officially it was now 'abandoned'. This made it a valuable commodity for medical purposes. The coroner had not needed to cut up the body to establish the true cause of death. The corpse had a few lancet incisions. It was thus very fresh. The man's precise age was unknown but he was not old. His cadaver could be an excellent teaching tool for medical students. At St. Bartholomew's 'a course on comparative anatomy was introduced' into the medical curriculum in 1834. Students were eager to dissect bodies not affected by old age or decay.[60] According to hospital regulations, the beadles who walked the wards were charged with removing any dead bodies. They had the legal right to supply the cadaver to anatomists for dissection. This was justified on the basis that the cost of medical care could not be recovered from the man's family or kinship network. Guardians of the poor did have the legal option of intervening by asking that any health care expenses be taken out of their regular subscription payment for poor patients to the hospital. The man, however, had not claimed poor relief and died alone. Nobody came forward to pay for his medical expenses or pauper funeral. Since St. Bartholomew's was a voluntary hospital, it had strict regulations about recovering from the destitute their welfare indebtedness on the dissection table.

The stranger's corpse was next taken to the dissection room. Late November was an ideal time to dissect. The chilly weather kept the body fresh longer. Most students dissected in wintry conditions. The corpse of the 'unknown' man was cut up over the next three days. Waddington explains that sometimes '[i]n winter, students at St. Bartholomew's were literally frozen out of the dissection room' because it was so chilly inside that they turned purple, their cold breath whitening the icy air.[61] A contemporary account written by James Paget when a medical student survives in the hospital archives recounting dissection room conditions during the 1834 season:

> I entered at St. Bartholomew's on the 3rd or 4th of October 1834. ... The dead-house (it was never called by any better name) was a miserable kind of shed, stone floored, damp and dirty, where all around stood a table on which examinations were made. And they were usually in the roughest and least instructive way; and unless one of the physicians were present, nothing was carefully looked at, nothing was taught. Pathology, in any fair sense of the word, was hardly considered.[62]

Dissections were stage-managed over several days by a demonstrator in anatomy. Paget described what he learned in the hospital environment in the winter of 1834:

> The mere knowledge that I learned from them in my first year was, I think, much less than I learned by reading and by work in the dissection room, the dead-house, and the outpatient room. I did as much dissection as I well could, on most days, in the hours then usual, 10 or 11 to 1.30 – reading "Stanley's Anatomy" and the "Dublin Dissector", which was then an advanced book; and, at home, the translation of "Cloquet's Anatomy" which very few then ventured on.[63]

The emergency admission of the 'unknown' suicide in late November 1834 was a vital teaching tool for students like Paget. In that case, the demonstrator wrote in the dissection register that the unidentified man's corpse went under 'a post-mortem examination' until 'having only the extremities entire'. At the finish, the cause of death was confirmed as 'died of effects of poison'. Again Paget explained that: 'The Demonstrator was supposed to go through the whole of the anatomy that could be taught in dissection but he could omit what he did not like or did not know'.[64] On this occasion there was a complete dissection and dismemberment of the corpse. Procedures were now followed closely according to the regulations of the Anatomy Act. A death certificate was counter-signed by 'C. L. Parker', an apothecary from the hospital. His job was to do a second forensic test. He did a pharmacological analysis of the stomach contents. This was to ensure that the hospital diagnosis was correct. An entry stated that Parker confirmed 'death by opium poisoning'. The anatomy demonstrator next reported to the Anatomy Inspectorate that the body had undergone 'a complete dissection'. Certificates of notification were also sent to the coroner, a local clergyman, and undertaker contracted for pauper funerals.

After these formalities were completed, a final entry was made in the anatomy register. It closed the case as far as the anatomists were concerned:

November 20th 1834

A man whose name is unknown...age unknown...died at 5 West Street Smithfield...died 20th November...body removed by the beadles for dissection November 23rd 6 p.m....[65]

Seven days later another entry was added to the register. It stated:

The name of the individual was subsequently ascertained to be John Unsworth and the body was delivered up to his friends. The body was buried in the parish of St. Martin, Ironmonger Lane, November 30th.[66]

This outcome was unusual. Most cadavers were never collected.

Further record linkage work shows that the stranger chose to die alone, having fallen into absolute poverty. He pawned his last few possessions to get enough money to die with dignity in private. There was some suspicion that out of desperation he might have stolen some of the pawned goods. His friends could not confirm or deny this fact because they had not seen him for some time. They did recount that he kept moving around, changing his name to avoid debtor's prison. One week he was 'Mr Burgess', the next 'Mr Thompson' or 'Mr Andrews'.[67] Later the anatomists noted that he was typical of the sort of stranger who hired rooms in common lodging houses – desperately sinking into depression, down on their luck, determined to die anonymous. It was the one place where loneliness was guaranteed.

The man's friends evidently did not give up on him. They reported that he was a missing person and searched the locality; then a neighbour saw his coroner's case reported in the *Times* on 26 November 1834. They came to the hospital to claim his body only to find that it had already been dissected and dismembered. His torso, limbs, and head had been severed. The demonstrator in anatomy helped the friends of John Unsworth to tie together his human remains. These were then wrapped in a winding sheet made from cotton. The bones had been sewn up inside his yellowing skin, now in the advanced stages of decomposition. During dissection, organs had been removed and tissue samples taken. The remaining torso was now a ragbag of human material. It was, however, treated with as much dignity as the anatomical process permitted at the end. John Unsworth was then carried by his friends for burial through the capital's streets to Cheapside. The irony of the resting-place name was not lost on the duty-anatomist in his ledger remarks.

Normally, dissected bodies were buried by three undertakers employed by anatomists at St. Bartholomew's hospital.[68] In 1834, H. J. Nicholson of 14 Willow Street, Rochester Row in Westminster interred corpses from the West End. Meanwhile W. Dentwart from St. Sepulchre's had responsibility for burying cadavers around St. Giles in the East End. This meant that Mr. Teale, who worked from premises at 61 Shoreditch, 'burial days being Mondays–Wednesdays & Fridays', interred all dissection cases from the City of London area. All anatomy subjects were buried at cost, just '£1-0d-0s' per funeral. Burial bills were paid out of petty cash reserves generated from student fees collected by the medical school at the hospital. On this occasion the friends of John Unsworth pooled their makeshift resources to provide a pauper funeral.

These basic but hidden aspects of a medical education are seldom the focus of detailed historical discussion. Little research has tried to look up from the perspective of the corpse on the dissection table, rather than down onto the performance of the medical student cutting the cadaver. The purpose of this book is to try to shift that medical gaze to better appreciate why the Victorians were very concerned about this type of death event. Many felt it tarnished the professional image of medicine, a theme elaborated in the case studies that follow in the rest of this book. In popular literature dissection was at first depicted as a grotesque experience with Gothic overtones that repulsed but thrilled its readership. Crucially the genre removed readers from the dehumanising event for everyone in the dissection room. Once the ranks of the medical profession expanded, however, the real difficulties of the anatomist could no longer be disguised in the theatrical performance of the dissection room. A public campaign began to voice what was really happening behind closed doors. Newspapers claimed that dissection involved exploiting the poorest and was disturbing the young minds of impressionable medical students. It is to that commentary that this chapter now turns, since adverse publicity confronted anatomists once the anatomy trade expanded to meet high student demand.

The difficulties of the anatomist

In October 1858, *Chamber's Journal* published an article entitled the 'Difficulties of the Anatomist'. It explained that:

> A Great deal of discussion is now taking place in London and elsewhere as to the best methods of educating young men for the medical profession. Of course doctors differ on this, as on most other subjects, but on one point they are all agreed – namely that all scientific medicine and surgery has anatomy for its basis, and that without a good knowledge of the structure of the body, a man can no more be a safe medical practitioner than a house can be a safe dwelling without a foundation. Now this anatomical knowledge can be acquired only in one way...by spending days, months, years, in the dissection room till the student of anatomy not only masters the details, but at least even thinks *anatomically* [sic] and can with very little effort apply his practical experience...Notwithstanding all this, the prejudice against dissection has been and is so strong, as either to make men content to a mere smattering of anatomy, or drive them into the most terrible and degrading means of obtaining material for investigation.[69]

The article reported that the passing of the Medical Act had not resolved historic body supply problems; quite the reverse. It had ignited debate in the popular press about the teaching of human anatomy and its central role in medical education. The journal article asked its middle-income readership to consider carefully, 'if society throws this grave responsibility on one class of its members, it is surely the duty of those who govern society to provide adequate means for supplying this great want in medical education'.[70] This ongoing attitude – despite the fact that the illegal anatomy trade had been legalised in 1832 – reflected the fact that for centuries the art of dissection had been 'tolerated' even though in the popular imagination it had a darker reputation as a 'beautiful but seductive science'.[71] It was considered dangerous in the hands of the unskilled or ill-educated. Anatomists, according to *The Hospital Pupil's Guide* (1818), should only be permitted to carry out their dissection work with 'the right exercise of reason' and restraint.[72] Across Europe, notably in France and Germany, the state regulation of medical education and human anatomy had taken hold in the early nineteenth century. In England, by contrast, Enlightenment values, though fashionable among the educated elite, had many critics because of Northern European sensibilities about the sanctity of the corpse. Throughout the English provinces death customs were diverse. This complicated the reception of medical education. Although it was important to obtain either an MCRP or MCRS or LSA diploma by studying human anatomy, evidence survives that some students purchased their qualifications by paying higher fees to surgeons prepared to ignore a lack of attendance at dissection classes. Regulations could be interpreted in different areas, in diverse ways. Sometimes the individuals involved agreed to look the other way if a student found dissection work too harrowing.[73]

It was a fact of early Victorian life that London teaching hospitals competed with private anatomy schools and medical museums.[74] Rivalries over fee-paying medical students were sometimes intense. These were often exacerbated by a lack of supply of dead bodies to dissect. In the national press there was considerable public discussion about the psychological impact of the anatomical sciences on the impressionable minds of young medical students. There were three key strands to these arguments. Medicine was first seen as a difficult profession since it involved confronting dearth, death, and disease on a daily basis. Dissection, it was argued, prepared medical students for the type of harrowing work that they would have to face on hospital wards. Those who thought dissection prepared students for the dreadful side of medical practice also recognised that it was a dangerous profession. Corpses were often contagious with fatal diseases like cholera.[75] Cutting them up needed to be regulated, especially given the emotional impact of dissection. It could sully the vulnerable minds of trainee doctors. Again, James Paget, when a medical student at St. Bartholomew's hospital, recalled catching typhus from poor patients. It taught him the dangers of being a doctor for the sick poor and cutting up their corpses. He wrote:

> In the first three months of 1839 the monotony [of medical training] was disturbed by a severe attack of typhus caught in a poor house in Lambeth where I was examining, with Havers [a fellow student], the body of a woman whose child lay ill with the fever in the next room. I was terribly ill, but with the wise guidance of Dr. Latham and Dr. Burrows [his teachers], and with kindly nursing, recovered unharmed.[76]

A century earlier he would not have been so fortunate, as the sad circumstances surrounding the death of Charles Darwin, senior (1758–1778) suggest.[77] The story of his demise was often retold in medical circles to warn medical students about the dangers of dissection work. The journal *Medical and Philosophical Commentaries* explained what happened when Darwin sustained a fatal cut while dissecting a corpse in 1778:

> About the end of April, Mr. Darwin had employed the greatest part of a day in accurately dissecting the brain of a child which had died of hydrocephalus, and which he had attended during its life. That very evening he was seized with severe headache. This, however, did not prevent him from being present in the Medical Society, where he mentioned to Dr. Duncan the dissection he had made, and promised the next day to furnish him with an account of all the circumstances in writing. But the next day, to his headache there supervened other febrile symptoms. And, in a short time, from the hemorrhages, petechial [sic] eruption, and foetid [sic] loose stools which occurred, his disease manifested a very putrescent tendency.[78]

It was this type of dangerous work in the dissection room that concerned those entering the profession. Conditions remained dangerous until disinfectant and preservation techniques improved later in the nineteenth century.

A second school of criticism meanwhile held that dissection was prurient, 'it meddled in things it should not', disturbing the mentality of the medical student and the scientific status quo.[79] Its worst aspects could be mitigated by learning about the human body in the anatomy museum from specimens, rather than training in a material science on the dissection table. Distancing trainee doctors from the actual human body was seen as a healthy thing. Anatomists were encouraged to stress to medical students that the corpse was a machine, not a person. Only a clinician's mentality could mitigate the material facts of dissection.

A third perspective, by contrast, stressed that anatomy was the instrument of creation.[80] Cutting up the human body revealed the hand of divinity at work. The human body inside was composed of the skeleton, organs, and life-support systems, beneath the surface of the skin. Opinion differed about whether anatomy proved the story of creation. Some searched the human body to prove evolutionary development compared to that of the animal kingdom. These two perspectives were known as 'creationist' versus 'transcendental' anatomy respectively.[81]

Regardless of which intellectual stance one took at the time, this third school of thought did promote a viewpoint that everyone else accepted. There was general agreement that cutting up a corpse in the dissection room was an experience that trainee doctors never forgot and it could be a harrowing encounter.[82] That common reaction prompted many trainee doctors to make career changes before they had fully qualified. To understand why trainee doctors were so repelled by the dissection room, it is necessary to examine in closer detail public debates about human anatomy. Our guide is Charles Dickens once more because he never accepted that the reception of popular anatomy reflected what really happened behind the dissection room door. In November 1850, Dickens opened *Household Words* by giving readers a glimpse of a world of medical education that was to become a watchword for scientific progress and new professional standards in Victorian times:

> The first day of October is a great day for the doctors...To the medical folks of these three kingdoms but more pre-eminently – does the day especially belong. To them it is the opening of a new year – the commencement of a new activity...
>
> There is at the London medical schools an assemblage of doctors in all stages of growth – from the raw country student in green coat and highlows, to the staid hospital professor in black scholastic gown, through all the intermediate niceties of fast students and slow students, reading students with specs and note-books, and smoking students with cigar-cases and imperials [cigarettes].[83]

Dickens was determined to point out that a better scientific understanding of the human body derived from the poorest on the margins of Victorian life. The article stated:

> But the first of October is no longer preceded by the forays of the *resurrectionist* [sic]; no longer clouded by the lack of means for pursuing the branch of study on which the superstructure of medical knowledge must be raised. A population of two millions has ever some members dropping from its ranks solitary and unknown – the waifs and strays of society – without friends to know or to mourn their fate. Almost always paupers, often criminals, though their lives may have been useless, or worse, they seem to make, when the fitful struggle is over some atonement after death. The wreck of their former selves is offered at the shrine of science for a while, and when thereafter they are gathered to the kindred dust of the graveyard, they may

sleep none the less calmly for having contributed no mean help to the advancement of that branch of human knowledge which has its ovation on the first of October – the great day for the doctors.[84]

Dickens's ironic tone reflects not only his social conscience but common nineteenth-century attitudes to poverty. He remarked that contemporaries mistakenly believed that those who fell into destitution did so because the 'undeserving' chose to turn to a life of crime and so in death the dissected were expected to atone for their moral failings in life. Dickens mocked this twisted tale of redemption. He refused to believe that it applied to everyone. Certainly he appreciated that the fall from relative to absolute poverty was often arbitrary. Most were the unwilling victims of financial misfortune. Medicine, he pointed out, was rather hypocritical about the 'waifs and strays' it needed to improve education standards.

One reason for Dickens's disquiet about hidden aspects of a medical education is that the general public knew very little about the inside of a dissection room. It remained a mystery and therefore a tale of half-truths and scare-mongering. The Anatomy Act discouraged any disclosures about how a dissection room was equipped, the way it was run, and what actually happened to the human corpse when it was cut up. Anatomists were naturally very wary about going into detail since the statute had stirred up strong public reactions. In the 1830s anti-dissection demonstrations and riots in cities like Sheffield and Manchester became renowned. The medical profession knew that their anatomy work had offended some people's deeply embedded cultural sensibilities. By 1871, when the Anatomy Act was extended, public interest in the body trade intensified in the press. New regulations permitted anatomists to cut up corpses for up to six months. The *Daily News* was soon at the vanguard of a public campaign to confront the cultural cost of anatomical legislation. Given that it was the 'third most popular newspaper in the mid-nineteenth century' and its first editor was Charles Dickens, its editorial line had an important impact on popular perceptions. Leading article writers were determined to expose the inside of the average dissection room. One typical episode revealed hidden aspects of a standard medical education by the late-Victorian era.

On 16 January 1877, the medical correspondent of the *Daily News* reported on 'Anatomy in the City' of London.[85] The paper carried an account of the prosecution of Mr. Thomas Cooke, surgeon, employed by Westminster Hospital, who resided at No. 31 New Bridge Street, Blackfriars. Cooke was charged at the Guildhall Police Court with 'unlawfully' carrying 'on the business or process of anatomy so as to cause effluvium which was a nuisance and injurious to the health of the inhabitants' next door to his house in Blackfriars. It was alleged that Mrs. Alice Benbow, the 'housekeeper' at No. 32 New Bridge Street, his neighbour, had:

> lately experienced a very offensive smell which had commenced about two months ago and that smell came from the roof of No. 31. They had never been clear of the smell until Saturday last when a dead body was removed about 10 o'clock at night. She saw the coffin leave the place...Mr Cooke was a surgeon at Westminster hospital. He was a demonstrator in anatomy. He was the author of many scientific works on the subject, and was a gentleman well-known in the profession...He carried on a school of anatomy...and it was his intention to carry on his business in a house of that kind until it was decided it was a nuisance...She had suffered from the smell for some time, and her child who was 12 years old, and was very nervous, had suffered also. She would not stay in the room by herself. Her servant had also suffered from the smells.[86]

Albert Harris, a court officer, was appointed to make a site visit to investigate the allegations and report back to the police court. He recounted that:

> He rang the bell which was answered by the housekeeper. He asked her which was the room of the school of anatomy and she replied that she was not permitted to tell. He however went to the third floor and knocked on the door, when he was told to come in. There he saw six or seven gentleman. There were two tables in the room, on one of which was the trunk of a human being and on the other were some pieces of flesh. On the right side of the door there was a large tank half full of liquor and some flesh in it. There was but very little smell in the room. He remained in the room a quarter of an hour or twenty minutes.[87]

Under cross-examination, Harris explained that the anatomy room had 'windows' which were 'kept open' while he was there 'but the smell was not offensive'. He commented that there was a 'strange smell' and it did linger but it had two likely causes. One of the men standing at the dissection table 'was smoking' and there was 'a smell of carbolic acid' in the air during his visit. Both were common in a dissection room.

Smoke cleared away offensive smells when the body was initially cut open. Disinfecting fluid was washed over the torso and body parts to make sure that students did not get infected by the corpse. The dissection room had moreover been lined with 'boards' along the party wall between No. 31 and No. 32 to try to contain any offensive smells following complaints from Mrs. Benbow. Harris reported that there was a 'skylight' in the roof of the room and it too was lined with boards. He discovered though that this was to prevent the neighbours looking into the dissection room. The boards did not let noxious fumes escape outside because 'they were not high enough' – they did not stretch out onto the roof – to act as a ventilation shaft. Harris noted that when he visited the premises next door the 'strange smell' was 'much more offensive' on the other side of the party wall but he could not explain why this was so.

Further enquiries were then made along the street of terrace houses. At No. 34, the 'housekeeper', named John Pegg, had 'lived there for two years'.[88] He told the court that 'he lived on the third floor and slept on the fourth floor' and so shared the party wall with No. 32. Pegg testified that 'when the windows were opened a nasty, dead, faint smell came into the room, which made his wife feel sick'. He recounted that 'it made him feel queer but not as bad as his wife'. Lodging in the other rooms of No. 34 were 'two clerks', both respectable men. Christopher Wright Howlet was clerk to the Church of England Sunday School Institute and James McDougall was employed by the British Empire Mutual Life Assurance Company. They lived and worked from the premises in the 'back room' but 'frequently had to go down to the lower parts of the house because they could not stand the smell upstairs' at night. The Judge asked the witnesses to comment on how long the smell lasted and when it started. Pegg, the housekeeper, said that 'the smell came in every now and then, and had done so for eight or nine weeks'. The others commented on a 'stench' that had recently 'got worse'. The implication was that every time a human body was brought in for dissection the offensive smell returned.

Mr. Lane, the barrister defending Dr. Cooke, the anatomist, then addressed the court. He told the jury that:

> In the premises there was one corpse, the body of a baboon, and a rabbit. The defendant had received the extension of time from the sanitary inspector... But when the report of the case reached the newspapers the sanitary inspector had required the defendant to bury the [human] body immediately... Consequently the [human] body was

buried on Saturday night; otherwise it would not have been removed at present for fear that its removal would prejudice the case.[89]

He admitted that complaints had been received from the neighbours 'but he was prepared to prove that it was no nuisance'. Lane refuted Mrs. Benbow's allegation that, having gained admittance, she 'had fallen over a corpse and fainted' in the hall of No. 31.

Sir Thomas Gabriel, the presiding judge, decided to adjourn matters until a week later to give enough time to call more witnesses to testify. On Saturday, 20 January 1877, the case was reconvened. Witnesses recounted that 'two hours after midnight' certain 'coffins' were brought 'in a cab' to Dr. Cooke's house. Evidence was heard that the neighbours thought something odd was going on but they did not know what was happening until they saw an account of the anatomy school being discussed in the *Daily News*. Meanwhile John Luff, a neighbour from No. 36, was called to the witness stand. He told the court that he worked at Hanwell Asylum where 'his room faced the Dead House in which post-mortems were made'. Luff stated that the 'smell' coming from No. 32 'was of the same order, only worse, than the one he had experienced at Hanwell'. The court enquiries now focused more closely on the medical fraternity.

It was the medical expertise of James Johnston Brown, MRCS (a former colleague of Dr. Cooke) that proved the most telling. Brown and Cooke had both worked together at a private anatomy school in Stamford Street before the Medical Act was passed.[90] There Brown had taught 'anatomy and obstetric medicine' alongside Cooke. In Brown's professional opinion, 'it was impossible to carry on anatomy without giving out an offensive exhalation'. Cutting up a 'corpse' or a 'baboon' into parts involved a lengthy and smelly procedure. The defence strongly refuted this claim. During the rest of the hearing, the defence summoned a number of expert medical witnesses to try to counter Dr. Brown's damaging testimony. 'Mr Charles Hawkins, one of the inspectors for London under the Anatomy Act...Dr Sedgwick Saunders, medical officer of health...and Mr. William Doughty, one of Mr. Cooke's students' all testified that there was 'little smell'. The defence also called Mr. Alexander Reed, of the Westminster hospital, who worked in the medical school where Dr. Cooke was employed to teach anatomy during the day. He explained to the court that 'he injected the body for Dr. Cooke's dissection'. There was no need to store and use noxious chemicals at home. Beforehand disinfection and preservation were done routinely at the hospital. The bodies were then despatched by hansom cab to Dr. Cooke to dissect in Blackfriars.

At the conclusion of the trial, Sir Thomas Gabriel rejected the case for the defence. Instead he directed the jury to pass a verdict that 'a nuisance had been perpetrated' and dissection had to stop on the domestic premises.[91] Dr. Thomas Cooke told the court that he would not appeal the decision. Later he did write a subsequent letter to the *Daily News* justifying his position. It stated that: 'The circumstances connected with the smell i.e. its occurrence, its cessation, its recurrence, and even its varying intensity, accurately coincided, not with the facts, but with the idea present in Mr and Mrs Benbow's minds'.[92] He could see that it was impossible to overcome his neighbours' prejudices about the scientific study of anatomy and he had thus taken the decision to remove his school 'even at a loss of £500, or more'. He reiterated that the medical officer for the City of London conceded afterwards that 'he put his nose within an inch of the body and remained half an hour in the room, yet neither perceived any smell, except that of pure carbolic acid and nor saw anything to find fault with'.

One or two letters appeared in the next editions refuting or supporting his remarks. But soon the case faded from public view. It had, though, provided some fascinating firsthand insights into the working life of a dissecting room, a typical medical student's training conditions, and public fears about the Anatomy Act and its extension by 1871. There is no doubt that by the time Dr. Cooke's case got to court, his private anatomy school was doomed. The Anatomy Inspectorate seldom supported anatomists exposed to public derision. Civil servants were always very concerned that one scandal might create a wave of scare-mongering, and it was therefore judged best to lose a case than expose more facts for public consumption. It was moreover very difficult for anatomists to carry out their work in private. They were under constant surveillance by their neighbours once the Anatomy Act came into force.

It was noteworthy that despite the detailed evidence during the case there was little discussion about why the medical students were prepared to pay extra fees to Dr. Cooke to dissect at night. The trial evidence should have alerted the court to a hidden anatomy trade. In many respects what the witnesses did not say in a packed public space was as important as the testimony taken down in evidence. Most medical students lacked bodies to dissect and some were prepared to pay privately for corpses. They could then train more quickly and start earning to pay back the expenses of their education. This was the unspoken underside of a medical education. For this reason anatomists in the early part of the nineteenth century had dissected in basements and below ground. They bought bodies and body parts wherever and whenever they could.

That body trade went on expanding to match medical student numbers, a theme throughout this book.

Comparing and contrasting contemporary accounts of dissecting rooms like this is fraught with difficulties about how representative they are of generic designs. Yet they also permit a glimpse of the actual dissection room setting, so that research can start to build up a picture of what a medical student's experience would have been like. What is surprising about evidence gathered from the popular press is that the actual equipment and basic outlay had changed very little over the course of the nineteenth century. In design terms it did not seem to matter whether the anatomy session was located in a basement or near the roof of a house. The interior was basic, the equipment crude, and for reasons of hygiene the furniture was hard-wearing. It had to be infused with carbolic acid or washed down with chlorine, and therefore needed to be sturdy. Inside, there was little aesthetic appeal. The impression the layout created in the minds of medical students was a formative one they never forgot. Victorian records generated within medical schools explain in more detail why this was so everywhere.

The Dead House at St. Bartholomew's hospital, as we have seen, was described in the 1830s as 'a miserable kind of shed'.[93] At Oxford meanwhile a temporary room made of corrugated iron was used until 1885.[94] Working conditions were crude and cold. Corpses were displayed on large wooden tables and dissected in rotation. In one corner stood an immersion basin filled with preserving fluid. Since it was not big enough to accommodate a whole body, all corpses were dismembered into parts. This equipment and layout was not improved until the 1890s when a leading architect, H. Wilkson Moore, was brought in to re-design the anatomy department and oversee the construction of purpose-built facilities. Nearby at the Radcliffe Infirmary a local coroner started a campaign to close the dissection room and Dead House next door because he claimed that it had 'foul drains' and was 'unwholesome' to work in.[95] Several hospital porters caught blood-poisoning from handling cadavers. They died from weeping wounds on corpses diseased with cholera. Others often redistributed decomposing body parts from fatal operative surgeries. Autopsies, post-mortems, and dissections were unregulated and unhygienic. So concerned was the coroner that he paid for a mortuary room to be converted in Oxford city-centre from his salary in the 1890s.[96] At Cambridge anatomists transported bodies at night from the Dead House at the rear of the Old Addenbrookes hospital on Trumpington Street to the anatomy department located inside the precincts of the expanded Downing site.[97] Again it had the same basic design, essentially a bare room with a stone floor. Corpses were also rotated on wooden tables laid out in a room roofed with corrugated iron. It was always cold inside too. By the 1890s conditions did improve but the majority of medical students still had very basic facilities. This type of primary evidence confirms that when accounts were discussed in the popular press they were often accurate representations of basic design features and common working practices. Similar sources provide further evidence.

Reynold's Magazine decided to commission Mr. B. Freeman, a doctor, to write about the dissection rooms he had known as a medical student between the 1830s and 1860s. He wrote that, despite the Medical Act, standards had changed very little:

> It was the last night of dissection for the course and most students in the graduating class were anxious to cease [sic] the opportunity the Professor's absence afforded by going out to attend little matters which always seem to have accumulated towards the close of the session...So there were but Clair, Eldridge, Waltham and myself left to commence the dissecting operation.
>
> The dissecting room...It was lit, a long narrow hall, with high windows, having ventilators attached to the upper part, through which the wind moaned dismally...The floor was covered with dark oil-cloth. At one side of the room stood a large cabinet, with glass doors, through which could have been seen labelled skeletons of almost every bird or beast that had name.
>
> Further on were human relics. Here might be seen the tiny skull of the week-old infant, side-by-side with the deformed chest and crooked spine of the untimely dead. Here might also have been seen every phase of animal and vegetable life; while along the walls were coloured models of the principal organs of the body, as acted upon while under the different courses of remedial agents. Last, not least, among many other objects of note that were there to illustrate the pathology of disease, hung human skeletons...[98]

This evocative description reveals that the study of human anatomy had diversified. It encompassed comparative, morbid, and pathological studies as the century progressed. Yet it also helps to set in context that the layout of the hospital dissection room bore a marked resemblance to similar premises elsewhere. Certainly those dissection rooms that did open their doors to public scrutiny seemed to have altered very little. Given that fact it is important to briefly compare these common and rather crude experiences to the recruitment material advertised in the press that attracted medical students into the profession. For as we shall see there was a significant gap between the image-making associated with the dissection room drama and the reality of being a medical student reliant on the body trade for one's professional qualifications.

Picturing the performance of dissection

One of the fascinating aspects of studying the history of anatomy in the nineteenth century is that stock pictures staged in the dissection room proliferated in the Victorian press. Some well-known depictions have been used as modern-day book covers.[99] This final section revisits those anatomical images that were in circulation, and which served to popularise human anatomy. Attractive images were created for an expanding audience of fee-paying medical students. Education was expensive and qualifications arduous, so positive images of what Victorians called 'well favoured females' being dissected promoted a message that anatomists hoped would have widespread appeal to those wishing to train as doctors. This advertising trend, of course, soon gave rise to 'the male gaze *par excellence*'.[100] Importantly, anatomical pictures also served to elevate the status of medical practitioners. The reason this promotional work was necessary was that medical practitioners doing the dissection, and the sources of their medical knowledge, were still disputed by the general public who resented the Anatomy Act. This meant that while the medical student saw an attractive female on the dissection table, he was as yet unaware that the anatomist often had a different set of perspectives (and it was *he*, until women started to train in medicine in the later nineteenth century). There was within the profession a certain level of under-confidence about dissection's place in the new medical sciences. Ironically, images of allurement expressed ambivalence about contested forms of male medical authority over the corpse.

The latent message of the anatomical imagery was complex and multi-layered in print. The fact that sanitised pictures were produced of buxom females being dissected attests not just to anatomists trying to put a positive spin on their work and use of clean methods, but to their publicity-shy habits, as well as the general level of disquiet about the unsavoury body trade, seen as the underside of medical education. What this meant for medical students was that those who naively believed in what Alison Bashford has termed 'a classic formulation of female powerlessness and male invasion; men's desire for the female body conflated with desire for knowledge' of dissection, must have got a shock on their first day at an anatomical demonstration.[101] It is curious then that although images in ballad material, popular cartoons, standard anatomy treatises, and medical textbooks have been examined by medical historians, those who were so closely associated with the body trade have not been subjected to the same level of scrutiny. And this neglect is important. For if, as Bashford contends, the dissected body was associated with impurity, pollution, and sexual danger (a viewpoint widely accepted), then what she calls the 'symbolic economy of the gaze' by medical students and anatomists must have been affected by the literal economy of supply in the dead poor.[102] The sex (male, female) and types of bodies bought (young, middle-aged, elderly) for dissection would have exploded the myth of the romantic image by confronting the dissector with a stark corporeal reality. There is no doubt that the explicit pictures that were disseminated bore little resemblance to actual medical students' experiences in the Victorian dissection room.

Ludmilla Jordanova has examined sexual images of female corpses and what they reveal about issues of gender and medicine from the eighteenth to the twentieth century.[103] She found that those dissected were 'unmistakenly sexualised' by a male gaze and this gave rise to a symbolism of desire.[104] Dissectors were promised a hands-on opportunity to feel, touch, and penetrate the interior of female corpses with the lancet. The boundary of life and death was repugnant, but it was also necessary to breach the medical frontier of the skin barrier in the name of the public good. Male students overcame any scruples because they were given an opportunity to look at the corpse for as long as they wanted without fear of criticism or retribution.[105] In Victorian society, picturing dissection legitimised medical voyeurism. Each lifeless body was literally powerless to voice a complaint about being objectified by medicine and science in the private world of anatomy. Yet this cultural interpretation is not without its ironies.

Victorian anatomy books predominately depicted the male body and its musculature because it was easier to dissect. That trend also reflected the fact that fewer women were purchased for dissection than men. This meant that the sexualised imagery that was in circulation was highly misleading. It did not convey the ratio of women to men in the body trade: 1:2 in the capital compared to 1:3 in the provinces (refer Chapters 4–7). There was also a lot of human waste produced by the process of dissection. Most bodies were broken up for sale; others were mutilated before they were purchased; whole corpses were in short supply. The pictures of anatomy and how dissection was staged over a buxom female were fiction. Anatomists were guilty of misinforming their reading public. Yet for a lot of medical students romantic images were their first entry point into the dissection room. It is estimated that: 'The 'first Medical Register (1859) contained the names of about 15,000 doctors – a century later there were six times as many'.[106] Approximately 90,000 medical students thus needed to dissect a corpse over a two-year teaching cycle. There were self-evidently a lot of curious and disappointed stares from those promised a picture that would delight the eye.

Comparing and contrasting stock pictures with some of the first photographic depictions of dissection rooms around the country is an instructive exercise. Not only did these artificial and contrived depictions help to disguise the anatomy trade, they gave further encouragement to the medical profession to close ranks and never speak about the true nature of the body business. Fortunately those images seen by medical students on a regular basis (predominately line and chalk drawings) and those that recorded their actual experiences (early photographs) have survived in the Wellcome Trust Library in London. What follows is based on a sample exercise and it reiterates an important central theme of this book, that regardless of whether the corpse was male or female, dissection turned them into 'matter out of place'.[107] Personal identity and sexuality were eroded by complete corporeal disintegration. The medical student attracted by the allurement of an image was instead the instrument of the anatomical destruction of their own gullibility.

Illustration 3.1 depicts 'the dissection of a beautiful young woman directed by J. Ch. G. Lucas (1814–1885) in order to determine the ideal female proportions' published in 1864.[108] It is a chalk drawing and was viewed by many medical students at London hospitals since it was widely in circulation. The first thing that one notices about the image is that the chalk medium softens the contours of the nubile female on the dissection table. Her long and luxurious hair resembles traditional images of the fallen woman, a Mary Magdalene figure in Western painting. She is also very similar to femme fatales in Pre-Raphaelite paintings, notably William Holman Hunt's *The Awakening Conscious* (1851–1853). Like that image the woman on the dissection slab is shapely. Here though she is not alive. She represents what Bashford (among others) has described as 'the Victorian cult of the beautiful dead'.[109] This explains why her head has been deliberately pulled back into an awkward pose. That pose pushes upwards and thus forward the neck and torso for the viewer. The art historian Bram Dijkstra elaborates that such images were

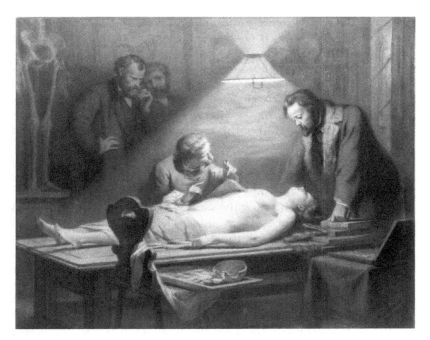

Illustration 3.1 ©Wellcome Trust Image Collection, Slide Number, L0013321, 'The dissection of a young beautiful woman directed by J. CH. G. Lucas (1814–1885) in order to determine the ideal female proportions', chalk drawing by J. H. Hasselhorst, (1864).

stereotypical and often depicted femme fatales of questionable morality positioned in physically impossible positions. This genre of painting became known among painters of the time as 'the woman with the broken back', where the figure is sometimes painted 'asleep' but more usually 'dead for male delectation'.[110] The image tends to be sexualised into a familiar morality tale to attract and delight medical students. This is why above the woman's head a light shines. It makes her the vanishing point in the picture. She is given a central role in the dissection room drama, turning her life story into a degrading experience where her sense of self is a disappearing act for medicine. That message may be too subtle for some students. Youth sees attractive fleshy tones and the contours of her shapely corpse. This perspective device draws the viewer to the picture's centre. This anatomy performance is being staged for its medical audience.

Behind the female figure stands a medical man fixed on the dissection of the corpse. There is no blood on the table, despite the fact that a demonstrator in anatomy holds back the flesh beneath the woman's right breast with a pair of pincers. Before the 1890s it was rare for anatomists to study breast tissue in its own right. Scientific opinion held incorrectly that the fleshy composition of the breast was the same as the rest of the body. This is why the rib cage has been exposed in this way. Nonetheless the effect of this focus is to accentuate the left breast for the viewer, a perspective trick that academic painters often used.[111] It is a classical and bloodless image. In reality, the anatomist would have been covered in blood as he cut the fresh corpse. There was no suction tube, only cloth swabs to keep the flesh clean. To distract the viewer's eye from questioning the sanitised scene, a second man looks down from the head of the dissection table onto the breast area. It is noteworthy that if one traces the perspective lines from his eyes onto the cadaver it is clear that he is staring at the right breast; the flap of exposed skin prevents him seeing what the demonstrator of anatomy can glimpse inside the rib cage. Again the breasts have become objectified and subject to scientific enquiry. This perspective effect reinforces that the bloodless corpse has been normalised for medical student consumption and to elevate the authority of anatomists over the corpse (discussed above).

Moving then around the room, another man stands behind the demonstrator of anatomy in shadow but still visible to the viewer. His dress code reflects his medical status, similar to the man looking over the head of the table. Both wear heavy Victorian coats trimmed with fur and velvet respectively. They are hands-off and therefore seem to represent physicians. To emphasis this point the first man leans on some anatomical books that symbolise his medical authority. The second male figure thoughtfully smokes a pipe to dispel the smell and aid his concentration. Beside the second man, further in the shadows, stands another male figure of lower medical rank. He is engaged in a conversation, quietly spoken. This is not a sacred space but it is a place where the theatrical performance of anatomy pervades the atmosphere. To emphasis this point, around the room the acting demonstrator has staged his equipment. There is an open box filled with clean instruments, bookshelves containing skulls and specimen jars, and a skeleton standing on a plinth. Beside the corpse stands a wooden chair on which some bones, half a skull, and part of a shroud to cover the body are depicted. The viewing experience is like having a box at the theatre in which one stares down into a carefully designed stage set. There is certainly no sense of looking up from the perspective of the corpse. Yet the dissection room drama would be a non-event without the sad case history of people fallen on hard times, like John Unsworth's earlier in this chapter and that of a servant girl such as Sarah Ashton elaborated in the next.

Overall then, the image is one of concentration, learning, a medical gaze, and close proximity to a nubile woman. It is a bloodless scene, relatively uncluttered, and unhurried. Although the image is a chalk drawing of black charcoal on white paper, the skin tones seem to be very fresh. There is no physical sense of tissue discolouration from preservation fluids being injected into the carotid artery. Importantly there are no medical students present in the room, although two of the men do seem to be of a lower rank. The medical men present have special access to the choicest female bodies, even though this book will show they were always in very short supply. In other words, this image bears very little relation to an actual dissection room scene. Even popular anatomy stories did not romanticise a medical education in human anatomy to this extent. Small wonder some medical students were so shocked by their first dissection experience had they seen and believed this type of image. Charles Darwin once observed that the 'dullness' and 'disagreement concomitants of anatomical work, drove him away from the dissection room' when studying at Edinburgh University in the mid-1820s.[112] Given such attitudes, it is important to compare this imagery with actual photographs from the mid-Victorian period onwards once the Medical Act came into force.

Moving then to Illustration 3.2, the viewer encounters the material reality of the dissection room that medical students experienced in Victorian times.[113] The image shows the interior of the Department of Anatomy at Cambridge University (see also Chapter 5). Nine dissection tables are arranged in rows across the hangar like room. They are lit from above by three triangular-shaped skylights that run the full length of the roof. There are five windows around the rear of the room, but the light cast down onto the corpses essentially illuminates their material remains. Nine cadavers are in various stages of being dissected. This is noteworthy since most popular anatomy stories stressed that a minimum of 12 bodies was common when in fact getting enough bodies was always, for all periods, a constant headache. The image of a male figure in the foreground is still discernible because the corpse, though wrapped in chemical-soaked bandages to preserve the tissue, is still intact. As the eye moves around the room from front to back the identity of each corpse becomes more abstract because they have been

Illustration 3.2 ©Wellcome Trust Image Collection, Slide Number, L0002687, Victorian Photograph of 'The Interior of the Department of Anatomy at Cambridge University, 1888'.

dissected and dismembered. Under each table are two stools, one each for the demonstrator of anatomy and his assistant who did the material preparations. During each session medical students stood around a table looking over the cadaver. On average there were six students per corpse in the 1870s, eight by the 1880s, and twelve by the 1890s.[114] It was an increasingly crowded and busy working space in which the equipment was crude. Daily the technical staff washed down the floor, wooden tables, and so on. Here though we have a bloodless, spacious scene, lacking medical students who had to press forward to see procedures. Surviving papers of the anatomy department at the time this photograph was taken reveal that the somewhat sanitised image had also literally been recently cleaned up.

Professor Alexander Macalister was one of the most gifted anatomists of the Victorian era. In 1883 he was appointed to the Chair of Anatomy at Cambridge University. On taking up his appointment, he made some private notes about the dissection room. They reveal that the average medical student's experience of human anatomy teaching had not changed since the passing of the Medical Act. He wrote:

Teaching of Anatomy
1. Faults of the present system.
2. Waste of material, more than half is sheer waste.
3. No check on slack men, or attendance and work.
4. Desultory and unmethodical teaching.
5. No regular instruction in method.
6. Has the effect of driving men into the hands of [private] coaches.[115]

Macalister, as we shall see in Chapter 5, became one of the leading reformers of medical education. His notes indicate why trainee doctors often paid extra fees to attend private dissection rooms, like that of Dr. Cooke discussed earlier in this chapter. Illustration 3.2 is bloodless because Macalister had instructed his staff to clean up the premises on his arrival. All bodies were now to be labelled and kept together on one table, thus the ordered layout that one sees in the photograph. There was in fact never enough material to go round. Most medical students were opportunists because they had to acquire extra human material to qualify within the two-year teaching cycle under the Medical Act regulations. So even this empty image still fails to convey the actual scramble for bodies.

Similar images from equivalent medical schools support this observation. Illustration 3.3, for instance, is an image of the dissection room at Edinburgh University taken in 1889.[116] Though a slightly smaller room, its layout is remarkably akin. Likewise in Illustration 3.4 we encounter the inside of the dissection room at Newcastle Medical School in 1897.[117] It too duplicates the dissection room design at Cambridge. Our final image, Illustration 3.5, is of 'five students and/or teachers dissecting a cadaver' at University College London around 1900.[118] All show that the experience was becoming more and more routine, and had little romance. In not a single surviving photograph is it possible to identify a nubile female body, again reiterating the body supply demography explored later in this book's case studies. Each emphasises that men were always purchased first because fewer women were sold for dissection. Male medical students who believed otherwise, if not disappointed, were certainly wrong-footed by the profession's stock images in Victorian times. The real dissection room drama was masculine – male students training on men – as in so many photographs – and it was very mundane work.

Illustration 3.3 ©Wellcome Trust Image Collection, Slide Number, L0013441, Victorian Photograph of 'The interior of Edinburgh University dissection room, 1889'.

Illustration 3.4 ©Wellcome Trust Image Collection, Slide Number, L0014980, Victorian Photograph of 'The Interior of the Dissecting Room, Medical School, Newcastle-Upon-Tyne, 1897'.

Illustration 3.5 ©Wellcome Trust Image Collection, Slide Number, L0039195, Victorian Photograph of 'The Interior of a dissecting room: five students and/or teachers dissect a corpse at University College, London' [undated] but late-Victorian.

Conclusion

An editorial in the *Fortnightly Review* in June 1886 reflected changing attitudes to dissection:

> If the State were to undertake the medical guardianship of its subjects, and doctors were to be Government officials, paid not by individuals but out of the public purse, on a scale strictly commensurate with their activity and success, the sick would probably be just as well cared for as at present, and their attendants would have a position of greater freedom, and at the same time greater dignity. Promotion in the service would be strictly according to merit as estimated by the medical body itself and special consideration would be given to original investigation... It might be enacted that a careful and complete autopsy of all dead bodies without distinction should be made by thoroughly qualified officers expressly appointed for the purpose

and full records of such examinations should be kept and should be issued to members of the medical profession at frequent intervals.[119]

There was no need though to use the bodies of the general population when the poorest were in such plentiful supply. This reassured readers who were anxious that 'dissections of the dead body shows nothing more than the structure of the machine'. Medicine needed to advance but that progress was to rely above all on the poorest in the dissection room drama hidden from public view.

All medical schools, regardless of their size, location, scientific reputation, medical curriculum, or laboratory facilities, needed a regular supply of dead bodies. The Medical Act (Extension) in 1885 required all doctors to qualify in both medicine and surgery to satisfy registration qualifications overseen by the General Medical Council. This meant that the anatomy trade was intrinsic to future educational provision. The demand for corpses became intense. Reconstructing the economy of supply is therefore vital to really appreciate the timing of the expansion of medical education throughout England. That context also raises important questions about the role of dead bodies in certain kinds of professionalisation. Specialists in competing fields of medicine who carried out research involving the close study of human material had to become associated with the body trade. An economy of supply complicated working relationships. Should, for instance, an anatomist, coroner, hospital lecturer, medical student, or laboratory researcher in pathology have the first call on available supply lines? Although historians of medicine have known about the general importance of the economy of supply, it is seldom documented in detail for individual medical schools. This book's central aim is to redress that neglect.

Across Britain, approximately 1,200 medical students studied at 11 major teaching hospitals, and on average 350 trainee doctors attended 9 provincial hospitals on a regular basis when the Medical Act was extended. All needed bodies to dissect. Waddington has moreover calculated that: 'In 1886, there were 439 new entrants to medicine registered in England'. Of those, '77 students' attended St. Bartholomew's hospital making it 'the largest medical school in the capital and the fourth largest in Britain'.[120] In comparative terms, 'Edinburgh had 280 students, Glasgow with 137, and Cambridge with 90, were the three largest medical schools in the country' by 1886.[121]

In a book of this length it is not feasible to research Scotland because of the different types of parish records that make a reconstruction of the Scottish economy of supply unfeasible at this research stage. This book thus focuses on case studies of St. Bartholomew's and Cambridge, the third largest and biggest provincial medical school in England respectively. These are complemented by research on Oxford in the Midlands and Manchester in the North of England, a smaller and much larger school respectively, which had distinctive economies of supply, compared to London. That historical prism informs what was happening everywhere once the Medical Act was extended. All the case studies that follow show the reliance of the anatomy trade on the New Poor Law. A combination of legislation penalised the poorest and in concert created a body business that underpinned the progress of medical education. Caught in the middle was the medical student anxious to qualify and start earning. New doctor and desperate pauper are part of the rich mosaic of medical progress in the nineteenth century.

Looking back and looking forward

Part I has re-read the Anatomy Act. Some of the destitute have been dignified with a name and a case history. This has been balanced by accounts of average medical students anxious to get hold of corpses. To add colour and complexion, the importance of death customs have set opposing views in context. In the process it has been possible to follow the body of a dispossessed person into the dissection room, retracing the circumstances of their death and what it meant to die alone in Victorian times. This has afforded an important glimpse of the interior of the dissection room drama in which anatomy was performed. Although dissection was a melodrama in the popular press, it was also a harrowing, routine, and tedious educational experience. After the Medical Act each medical school developed in a way that suited its institutional context, student numbers, and financial circumstances. The expansion of the medical profession was consequently complex. It was underpinned by an economy of supply in cadavers. Before now it has not been possible to retrace the body trade that the Anatomy Act legalised. Retracing the business of anatomy involves deploying a new methodology and sophisticated record-linkage techniques. It is recoverable from the official records of the Victorian information state. The scene is now set to turn our attention in Part II of this book to the finer detail of that anatomy trade and those who effectively died for Victorian medicine.

Part II
An English Anatomy Trade

4
Dealing in the Dispossessed Poor: St. Bartholomew's Hospital

The capital's streets were busier than usual at dusk on the first Wednesday of July 1836. People congregated outside overcrowded and stagnant lodging houses. They stood by the open door or on the street corner. Most were unable to sleep. It was a breathless, sultry summer's night. Then, all of a sudden, a 'terrific hail and thunderstorm' broke out.[1] London newspapers reported that at about 'one o'clock in the morning' the sky lit up 'from east to west' in 'one stream of liquid fire, the electric fluid playing in the most fantastic forms, and the rain falling in torrents'. By six o'clock, 'hailstones in many places being the size of walnuts' pounded the street-scene for 'thirty minutes'. The damage to property was extensive as the eye of the storm passed along the River Thames. Eyewitnesses said they saw 'every conservatory' and 'greenhouse' smash in the vicinity. In the fashionable districts, plate glass windows fell 'like dominos' into the main thoroughfares. Mother Nature's unpredictable force wreaked havoc, killing many early morning commuters. Few of the bereaved families had anticipated the need to save for this rainy day.

At first light, 'Larman...a lad of sixteen years of age' was struck by lightning on the streets.[2] Passersby brought him to the door of St. Bartholomew's hospital. Doctors on the early shift found him 'in an insensible state' and 'without hope of recovery'; he became one of many fatalities. Around 'half past seven', the storm gathered even more pace. An unnamed clerk hurried on foot to a City bank. He was trying to escape the storm's ferocity but was struck by another huge streak of lightning. Instantly he fell to the ground, 'bleeding from a wound in the head and much scorched'. A passing omnibus driver carried his smoking body to St. Bartholomew's emergency room. The surgeon on duty cut through the man's charred clothes but he was dead on arrival. A 'few pamphlets' were found in the clerk's pockets containing general business information but no personal identification. Over the working day, the extreme weather conditions filled up the hospital's Dead House with 'unclaimed' corpses. Many cadavers were very fresh and young, ripe for dissection. The vagaries of the British weather were a boon to St. Bartholomew's trade in the dead.

St. Bartholomew's anatomy registers provide an historical prism of common complaints that killed the poorest in the capital.[3] Aged blood was often frozen by ice-cold winters. Classic hypothermia cases populated the dissection table. Dense smog choked many suffering from typical lung diseases, like asthma, pneumonia, and tuberculosis. Regular storms increased traffic accidents. A 'dreadful hurricane' in December 1836 caused havoc just six months after the exceptional lightning storm described above and consequently more bodies became available for anatomy.[4] In the congestion, frightened horses spooked by braking carriages skidded to avoid falling debris, colliding with commuters. These everyday deaths contributed to the hospital's reputation for hands-on human anatomy teaching. Those morbidity profiles will be elaborated more fully later. Meanwhile over the next 50 years the number of medical students steadily increased. Chapter 3 explained that there were '439 new entrants...registered in England' by 1886.[5] Of these, 'seventy-seven were at St. Bartholomew's, making it the largest school in the capital and the fourth largest in Britain'. That achievement reflected an economic trend. Lots of the local poor died in destitution near the hospital. In desperation, bereaved families transacted anatomy supply deals. Hospital staff recognised that dearth was an indignity in death. In their defence, it was said that by dissecting the dispossessed better medical standards would benefit those living in poverty too. Achieving that goal meant regular body-trafficking by hospital staff and numerous suppliers. A set of wicker baskets that looked like large laundry hampers was left on permanent standby just inside the hospital gates for passing body dealers to fill up and return by nightfall.[6] Bodies were also bought in person or at the back of the 'Fortune of War pub on Giltspur Street, Smithfield – opposite' the hospital premises.[7] The only way to appreciate the scale of that hidden business is to delve inside St. Bartholomew's dissection trade. First, however, a brief discussion of the hospital's history sets both the story that opened this chapter and the economy of supply that was created in a London context.

The hospital's brief history

From the twelfth to the twenty-first century, St. Bartholomew's hospital has been a beacon of health care at the centre of the square mile of the City of London. Founded in 1123, it gained a deserved reputation for medical advancement and learning in the medieval world. Standard histories stress that it was seen 'as one of the five leading London institutions renowned for the care it extended to the sick poor'.[8] The local church of St. Bartholomew the Great was one of a number of surrounding religious houses and guilds that supported the work of the hospital down the centuries. Statesmen like Sir Walter Mildmay, Chancellor of the Exchequer to Elizabeth I, made generous charitable donations and legacies to boost the hospital's endowment. Investments paid for leading medical men to walk its wards. Staff, like William Harvey, contributed to the hospital's aura of innovation and tradition in the seventeenth century.[9] By 1726, anatomical and pathological study of the human body was very much in vogue among the educated elite that attended the hospital. The Enlightenment drive to improve medical standards on the wards resulted in 'two rooms... set aside for post-mortems in the dead house'.[10] From these basic beginnings emerged a more modern style of medical education throughout the latter half of the eighteenth century.

St. Bartholomew's governors were determined to engage with 'a new form of active benevolence to help the ill-defined *'deserving'* poor' by 1800.[11] That charitable endeavour was driven by evangelicalism arising out of business philanthropy. Concern was expressed about the need to redress the worst aspects of urbanisation. Three dread words – death, dearth, and disease – shaped the meagre lives of the common lot. Opening up the doors to the sick poor served not only the growing needs of commerce but an expanding medical fraternity too. An emergency room treated everyone free-of-charge; nonetheless general admission to the wards required the written support of a sponsor who subscribed to the hospital's voluntary funds. In this way St. Bartholomew's governing body balanced its budget and kept its doors open to those most in need. By the 1820s, St. Bartholomew's could justifiably boast that it was 'the largest school' of medicine 'in London'.[12] It opened a state-of-the-art lecture theatre in 1791. Thirty years later, a purpose-built anatomical theatre was in full operation.[13] Onto that medical stage walked teaching staff determined to dissect more corpses. After the Anatomy Act, the reform of education standards at the hospital was slow and piecemeal.

St. Bartholomew's nevertheless played a pivotal role in the rolling out of new professional standards in early Victorian medicine.

The location of St. Bartholomew's was very fortuitous. It was in an ideal central London position to attract generous charity support and develop an economy of supply in the dead. Both factors enabled staff to increase the recruitment of more medical students to sustain human anatomy study. In the surrounding streets the homeless slept in alleyways or occupied rooms in the poorer lodging houses in the rougher neighbourhoods.[14] It was a fact of life that the destitute crowded cheek by jowl. Later in this chapter the street geography of that body business will be examined in more detail. General trends show that the daily life of the dissection room was dependent on those dying on its doorstep and hospital wards. Smithfield meat market, for instance, was opposite the hospital entrance.[15] Traders and clients often supplied the Dead-House. On market days, crowds gathered to trade their animals. The famous wholesale meat market was a busy medical thoroughfare. Heart attacks, premature births, brain clots, stroke victims, street-stabbings, prostitutes with sexually transmitted diseases, severe fever cases, childhood measles, even those trodden by large animals were all intrinsic to the economy of supply.

Every September anatomists also benefited from the annual St. Bartholomew's Fair coming to the capital.[16] It was usually staged outside the hospital premises and has been described as a 'four day Saturnalia'.[17] Estimates vary but contemporary accounts suggest that 'twenty-two shows', featuring acrobats, dwarfs, gypsies, performing animals, and exotica (including the 'monstrous') were staged for delectation.[18] Performers were accompanied by countless street-sellers peddling food, household goods, and trinkets. The honest trader, light-fingered pickpockets, and unscrupulous vendors crowded together with pleasure-seeking customers from across the social spectrum. The feeble who came in search of entertainment in their dull lives often collapsed from exhaustion, ill health, and over-excitement. The fair proved to be ideal for acquiring anatomy subjects. As we shall see in subsequent chapters, travelling fairs were always important sources of 'unclaimed' corpses throughout the nineteenth century.

St. Bartholomew's was similarly well-placed to take advantage of corpses generated by legal cases. Daily the criminal fraternity climbed into the dock at the Old Bailey near the hospital. After the Anatomy Act it was no longer legal for anatomy schools to automatically acquire the bodies of executed murderers for dissection. Nevertheless, faceless corpses found floating in the Thames and homicide and manslaughter victims were obtained with regularity by anatomists (see Chapter 2).

Coroners' cases were always an important source of anatomy supply. The coroner's court was affiliated closely with the dissection room at St. Bartholomew's. Mr. Payne, the City of London coroner, staged his court hearings in the Dead House of the hospital, a common practice that continued into the later Victorian period. Modern forensic medicine did not interfere with these traditional arrangements until Public Health reform decreed that purpose-built mortuaries should replace hospital Dead Houses.[19] Close ties between anatomists and coroners have been underrated by historians of medicine.[20] That oversight is significant because it neglects to take into account the medico-legal competition for corpses. Official authority over cadavers was integral for professional norms. Anatomists and coroners needed exclusive access to dead bodies to improve their proficiency as the century progressed.

It is important to appreciate that each coroner's hearing threatened heavy costs which could be recovered by selling on bodies after a standard inquest. The 'view' of the dead body by a coroner's jury was taken quite literally in Victorian times.[21] It was an external, physical examination. Obvious bruising or fatal wounds were identified and matched to witness testimony. Post-mortem reports were expensive and therefore infrequent. Juries instead congregated in public houses. Corpses were displayed on the bar. Verdicts were settled as quickly as possible and without incurring undue expense. Abandoned bodies found in the street were seldom cut to investigate the cause of death. Beggars, drowning cases, and drunks were common cases requiring little detailed forensic evidence. Those cadavers were very valuable to anatomists. Nonetheless the body trade was patchy. It depended on supply levels and these were shaped by specific background poverty conditions. In some places, coroners cooperated closely with anatomists for financial reasons, in others not.[22] The social history of coroners provides important clues about the symbiotic relationships that developed between anatomists and coroners.[23]

St. Bartholomew's hospital in the meantime charged fees to bury the dead that died on its wards. This was waived if the body was handed over for dissection. The standard charges on admission were too much for many trying to escape their daily misery:

Patient Charges

The Beadle for giving a notice of death to the patient's friends	1s 0d
The Porter for a certificate to the Parish where the patient is buried	1s 0d

To the Bearers for carrying the corpse to the Hospital gate	2s	0d
The Matron for use of the black cloth	1s	0d
The Steward for certifying the patient's death	1s	0d
Sub-Total	**6s**	**0d**
Fees to be taken on the admission of patients	2s	6d
The Sisters of the Cutting Ward may have	1s	0d
The helper there may take	1s	0d
The other Sister may take	1s	0d
The Beadle for carrying the patient to the ward	0s	6d
The helper	0s	6d
Total	**12s**	**0d**

Burial Fees

Coffin and shroud	6s	6d
Ground	3s	6d
Minister	2s	6d
Bearers	4s	6d
Black Cloth	1s	0d
Notice of death	1s	0d
Registering	1s	0d
Total	**19s**	**6d**

By Order of the Governing Body: Nineteen shillings and six pence is the amount of burial fees to be deposited by each patient on admission.[24]

Twelve shillings was needed to gain hospital admission. The beadles however often insisted that patients pay the full fee of 19 shillings and sixpence. A better-off ratepayer could provide written surety for someone in the last stages of a fatal illness. In most cases the hospital preferred to see the cash deposited up front to balance its books. For so many people the dissection table was taboo, but in poverty it was also inescapable.

St. Bartholomew's wards were crowded by those caught up in the hectic fever of City finance too. Jerry White quotes a typical case that featured in the *Annual Register* of 1862. A man walked out into oncoming traffic in the street rather than face family ruin: 'Suddenly he left the crowd on the pavement, walked up to a coal wagon...and placed his head deliberately in front of the fore-wheel. The next instant the wheel passed over his head, and crushed it completely flat'.[25] The stress of misfortune was one of the commonest causes of attempted suicide and mental illness. At the worst financial moments in fraught-filled lives, charitable care was often extended to ordinary people who could ill afford the personal losses and associated health problems brought about by a run of bad luck. Having placed their blind faith in the money-men, many found themselves 'walking a narrow ledge between...peace of mind and lunacy'.[26] It was rare for just one individual to be dragged into the vortex of pauperism. Micro-economies – kinship networks of grandparents, parents, children, aunts, uncles, close friends, and neighbours – all suffered together.[27] The ill-fated tried in vain to makeshift the remnants of their insecure lives. In deprivation, there was simply nothing left to pay for a pauper funeral. Suicide rates in the City of London and its surrounding districts were unsurprisingly higher in years when investment returns plummeted.[28] The crash of the Northern and Central Bank in 1836 and the Barings Bank smash in 1890 were two of the most high-profile insolvency cases that ruined many small investors who could not face a life of poverty.[29]

This chapter captures those who fell through the financial cracks into a Victorian underworld. New evidence rediscovers those bodies that were abandoned or the unfortunate whose relatives had to apply to have hospital fees waived in return for handing over their loved one's dead body for dissection. We begin with a typical case of suicide, this time involving a young woman sold for dissection, for one of the unusual features of St. Bartholomew's body business was the high number of young females acquired for dissection, compared to elsewhere. Natural disasters and financial ruin were not the only reasons for falling into destitution and despair. A stormy courtship brought many people into the ambit of the Anatomy Act. Sarah Ashton's short life symbolises the sad circumstances of many failed courtships among the capital's servant population. In this way, this chapter engages with the economy of supply from a human perspective by examining its facts and figures for the first time at a significant teaching hospital.

Suicide from seduction

On 8 March 1835, the dead body of Sarah Ashton, aged 24, was brought into St. Bartholomew's hospital. It was examined by Mr. Payne, the City of London coroner. He found that Sarah had 'destroyed herself by swallowing oxalic acid under very distressing circumstances'.[30] Further enquiries established that the deceased had lived at 109 Gray's Inn Lane, where she worked as a 'servant to the lodgers who occupied the first floor' of a large house. By all accounts, Sarah was a hard worker but she had fallen in love with a cad called Richard Sutton, a clerk to a solicitor in Gray's Inn. The day before her successful suicide attempt, Sarah had opened the door at lunchtime. On her employer's step stood a young woman in an anxious state of mind. She warned Sarah that they had both fallen in love with the same man. The stranger said that 'she had been courted by him [Richard Sutton] for three years and that Sunday next the bans for their marriage were to be published' in the parish. She thought that Sarah should know the truth and take warning not to pursue the romance.

Knowledge of her false and fickle lover was a devastating blow to Sarah Ashton. She confessed to her friend 'Martha Clark... charwoman... after obtaining a promise of secrecy' that she was already 'pregnant, and had been since December by Richard Sutton.'[31] Now four months gone, she hoped foolishly that Sutton would not abandon her. Sarah was convinced that 'her sweetheart... Richard Sutton' must keep his promise. She told Martha that 'it was all right, for Richard Sutton would marry her'. By evening, she was crestfallen. Sutton refused to support Sarah and simply walked away. In desperation, she tried to procure an abortion and when that did not work committed suicide by swallowing more poison. She could not face the shame of her physical predicament and inevitable unemployment once her pregnancy was exposed.

At the subsequent coroner's hearing, Richard Sutton told the court that he was 20 years old and worked as a legal clerk to Mr. Joseph Dunn, the barrister.[32] He had 'been keeping company with the deceased for about 16 months'. He was also 'keeping company with a young woman living near Russell Square' (the female caller on the doorstep). Sutton claimed that 'the deceased knew of this'. He told the court that he explained to Sarah that 'he would always do the best he could for her, but that he could not keep her, which is the truth'. He confessed that before having sexual relations he 'did not say then he would not marry her'. He justified having second thoughts about a betrothal when he learned that the deceased already 'had a child 14 months old'. The court officer was ordered by the coroner to check his testimony: 'it turned out that the deceased had confessed as much to the charwoman'. Sarah had been thrown over by another young man once before and although she kept the child (a concealed birth) it was later sent away for adoption. This context explained her deep distress at her second failed courtship. Sarah had gone into service to make a fresh start. She clearly did not want to repeat her past misdeeds. When she met Richard Sutton just two months after the birth of her first child, she thought that she had found love with a better man, her 'sweetheart'. He courted her, seemed sincere, and offered her the promise of secure affection.

Summing up, the coroner described the personal conduct of Richard Sutton as 'most base, in paying attention to two females at the same time'.[33] The coroner also expressed the view that Sutton 'ought to be scouted from society'. He 'was sorry the law did not permit of a criminal proceeding against him – it was to be hoped that other young women would never listen to him again after this'. Sutton was admonished but walked free from court. Sarah's corpse meanwhile remained 'unclaimed'. Her cadaver was that of a 24-year-old woman, four months pregnant, lying in the Dead House at St. Bartholomew's. It was thus a very valuable teaching tool for medical students. Her reputation was ruined, she had no close family, and as a servant who had committed suicide, there was nobody willing to take legal responsibility for her burial. Once released by the coroner, her body was passed on, like many others, to anatomists for dissection. Sarah's short life is symbolic of many females who became part of the economy of supply at St. Bartholomew's. Had she lived, she would have been unemployed and on the streets, like the majority of women who afterwards were dissected.

There was just one stark choice for a woman like Sarah thrown out into the street. Like many servants she could sign up to the oldest profession in history, prostitution.[34] It was either sell your body on the streets to survive, or die in despair ending up on the dissection table. Workhouses did provide emergency care for women in labour but most 'respectable' females were keen to avoid the social stigma of poor relief. Charities likewise tried to rescue pregnant women from a life of vice but accommodation could never keep pace with demand. Sex before marriage was considered a social evil. Moral attitudes and religious beliefs were often harsh on pregnant, single servants. Sarah's nearest workhouse would in any case have been Holborn Union in Gray's Inn Lane, near her former employer's house. The master, Mr. Hewett was a major supplier of cadavers in the pay of St. Bartholomew's anatomy business.[35] Holborn had a strict policy of caring for women in labour and then asking them to leave as soon as possible after the birth. This avoided long-term claims on welfare funds. Deaths in pregnancy or from the misfortune of being on the streets were sold on. St. Bartholomew's anatomy registers contain countless women who were former prostitutes.[36]

Historians are divided about the social ramifications of a prostitute's life in Victorian London.[37] The chief difficulty is that no one knows how many prostitutes there were in the capital. Estimates varied from '8,000 (police) to 80,000 (evangelists) and sometimes more'.[38] Evidence does survive of some women using prostitution as an economic tool to improve their social mobility. Some women employed their earnings to move from a career on the street to serving customers in a shop. Others secured a client who paid to lift them out of poverty on a permanent basis. Many prostitutes became comfortable mistresses. There were no success stories for the women listed in the St. Bartholomew's dissection registers. They were driven to street prostitution by endemic poverty. These females were in the worst circumstances. All were at the sharp end of a spectrum of prostitution. In death, the anatomy table was their fate. Often they had lost touch with friends and family. Few in any case could afford to be sentimental about the death of a relative who worked the streets. Body 'number 84', a '22 year old female' called 'Charlotte Burton', for instance, was a prostitute that lodged on 'Curtain Road in Shoreditch' in the East End.[39] She tried to give birth to a concealed pregnancy without medical assistance. Soon after birth her child died and she suffered 'puerperal convulsions'. Dirty living conditions and lack of an adequate midwife caused sepsis, or its more common name, childbed fever. Charlotte died on the night of the '18th February, 1834' in abject poverty. Her fellow lodgers sold her body for dissection on '19th February at 8pm'. 'Mr Teale the undertaker' made the body deal and took the cadaver to the Dead House at St. Bartholomew's. For the next '21 days' her body was dissected and dismembered down to its extremities. It was a 'prized' teaching tool since it was young, displaying the signs of parturition. Bits of Charlotte that were left were buried in the 'Hospital Burial Ground, Seward Street' along 'Goswell Street, old St. Luke's parish'[40] Similarly on the '8th November 1836', a street prostitute, 'age unknown', named 'Mary Sullivan' was an emergency admission to the hospital.[41] Two days later she died of 'syphilis', a diagnosis confirmed after dissection. The beadles removed her body on the '10th November 1836 at 7pm' to the Dead House. Her sexual condition was so common that her corpse was dissected very quickly. Students were wary of catching a contagious disease. A slip of the lancet nicking the skin was a well-known risk.[42] There were no latex gloves to protect the dissector. Mary Sullivan's body was 'cut to its extremities' for '2 days' before interment in the same common graveyard. No opportunity to learn was wasted.

The winter months were hard times for prostitutes on London's streets. Sex workers relied on alcohol to dull their senses and inure them to the rough clientele. This was the typical story of a 'young woman' called 'Eliza Williams', body 'number 42' in the 1837 anatomy season.[43] She was an emergency admission on the '19th January' that year. After death, for '9 days' her body was dissected and dismembered. Like so many in the records at that time she died from the last stages of 'syphilis'. Sometimes it was very difficult to give a precise cause of death when a lifetime of street prostitution took its toll. Death certificates tended to state the immediate cause of death but not the run of medical complaints that led to that fatal point. Only detailed archive work illuminates lost medical histories like that of a '47 year old rough woman' named 'Anne Gearing'.[44] She came from a notoriously tough neighbourhood in 'St. Giles' parish and died alone in the road on the '10th October 1847', body 'number 3' that season. Evidence suggests that she plied her trade outside London theatres along Drury Lane and hence was sold for anatomy to St. Bartholomew's. Anne Gearing's corpse was traded on '4 days' later on '14th October' by 'Mr John Dix the undertaker'. Dix was a paid intermediary and regular supplier to the hospital. The Anatomy Inspectorate described him in a report dated 4 November 1843 as 'well recommended for his zeal, integrity and judgement'.[45] Officialdom saw him as a major and trusted body trader across London.[46] Dix moved between street deals, hospitals, and Dead Houses in need of cadavers in the supply chain. The evidence suggests that Anne Gearing was Irish. She had been an older street prostitute for some years, though her final cause of death was described as 'delirium tremens' in the anatomy register. A life of vice turned her into an alcoholic, dying in the gutter.

Not everyone had a long street life. A '24 year old female' known as 'Marie West' of 'unknown address' drifted between lodging houses and alleyways where she met her clients for sex.[47] She had the misfortune to contract a bad dose of 'venereal disease', which rotted her womb, eventually killing her on '23rd September 1850'. The entry for her body, 'number 62' of that teaching year, records that within '2 days' her cadaver was sold and over the next '3 days' dissected entirely. It is worth though keeping in mind that prostitution was not an exclusive female activity. On '2nd November 1851' a young lad called 'Thomas Barton' aged '17' of Bunhill Road, St. Lukes' parish was brought by an unnamed person to St. Bartholomew's as an emergency admission.[48] The doctors could only numb his excruciating pain with laudanum. He died of 'carcinoma – penis' that night. Removed from the wards by the beadles to the Dead House, within '2 days' his body was a ragbag of human material. The record is unclear about precisely how he contracted the cancer – it could have started as prostate trouble and migrated to his penis – but the more likely possibility was participation in some form of street prostitution. The anatomists always accurately recorded enlarged prostate glands and since none was mentioned in this case its omission is noteworthy.

Children did not escape this fate either. In 1856, two poor women each sold on their 'still born child', in the early winter cold.[49] These entries attest to the value of the rich case studies in the surviving anatomy registers. Not until the Deaths and Births Registration Act (1926) did stillbirths have to be registered for a burial with a proper death certificate signed by a qualified doctor.[50] These cases cannot be recovered from surviving parish records.[51] The surviving entries in the St. Bartholomew's archives thus give us an important glimpse of a Victorian maternal underworld. The records suggest that it was prostitutes and unfortunate women, those who had fallen on the hardest of times, who evaded legal enquiry. Dissection registers show the real reasons behind numerous spontaneous miscarriages and stillbirths (abortions, accidents, domestic violence, prolapsed wombs, and sometimes infanticide). Some women made a pragmatic calculation when they transacted stillbirth supply deals. The speed of the transaction and its discreet nature certainly hid the shameful dealings. The stillbirth that some never forgot was surely the cruellest fact of poverty. Two selected examples stand in for many in the dissection records.

At the start of 1856, two consecutive stillbirth entries reveal that discreet deals were made by neighbourhood intermediaries in the pay of anatomists.[52] Body 'number 53' was the 'stillborn child...offspring of Elizabeth Wakefields'. She lodged at 'No. 9 Lamb Court, Clerkenwell' and sold her stillbirth '9 days' after labour to an intermediary named 'E. Llewellyn'. In the intervening period, Elizabeth Wakefields could not makeshift her meagre economy to pay for a pauper funeral. A dissection sale was a pragmatic calculation. Likewise, the 'offspring of Susan Miles', a common lodger at '131 St. John's Street', body 'number 54' in the same anatomy register, was recorded as having died on the '17th January 1856'. This time it was a much quicker sale transacted by another intermediary known as 'G. G. Gardiner'. Within '2 days' of death it was a done deal, dissected over the next 48 hours, and buried by 21 January 1856. All these stillbirths were a boon to medical students. Often a capital 'L' pencil entry denoted the demonstration of unusual medical features to a lecture-hall full of students. The cases were ones that were seldom seen either on the wards or during an average doctor's home visits in the vicinity. Corpses with encephalitis (water on the brain), a hole in the heart, conjoined twins, limb malformations, gross deformities caused by malnourishment all figured in the records. A memo styled *Removal from Wards* confirmed that dead bodies were not to be taken away throughout the night 'as they might tend to disturb the quiet of the ward', except those *'special cases* being removed irrespective of the hour, as hitherto'.[53] Certain disabilities were to be prioritised 'where speedy removal was advisable, being removed as soon as possible after death irrespective of the hour'.

They were entered in an *Abnormalities and Deformities* register (a practice that Chapter 5 will expand upon at Cambridge). Later some cases were displayed in the hospital's Anatomy Museum. Thus a Report to the General Court dated 24 February 1870 stated that the role of the Demonstrator in Morbid Anatomy was to:

> draw up or cause to be drawn up, a faithful account of the pathological appearances (with a proper reference to the clinical record of the case in the ward book) in a *Book* to be kept for that purpose, which shall be laid before the Treasurer and Almoners on the first Thursday in January and July....
>
> Ye shall furnish to the Curator of the Museum [of anatomy] a short record of all *specimens* which you may send to the Museum from the Post-Mortem room.[54]

These selected examples from the registers are not statistically significant. They do however give a flavour of those lives that were staple subjects of the dissection table. In the archives they emerge out of a history of silence, their contribution to medicine seen in more detail. Turning then to that economy of supply, it provides a new view of the practical operation of the business of anatomy at St. Bartholomew's.

Economy of body supply: general trends

Figure 4.1 depicts the number of cadavers purchased for dissection. It shows that altogether 5,063 bodies were recorded over two trading periods (2,656, 1832–1872 and 2,407, 1885–1929).[55] There is then a missing register for the period 1873 to 1884 inclusive, which this chapter later investigates. In the meantime, the findings from the surviving registers show that on average 60 bodies were purchased in each annual teaching cycle. The majority, though, were dissected in the winter months because cadavers were preserved in the cold conditions for longer (see Chapter 3). Looking in more detail at the supply trends, seven key features of the body business are highlighted.

The anatomy trade was initially very successful when the Anatomy Act was introduced in 1832. Some 125 cadavers were acquired in the first 12 months. In the next two years, acquisition rates remained high at 108 (1833–1834) and 81 (1834–1835) bodies respectively. This was

132 *Dying for Victorian Medicine*

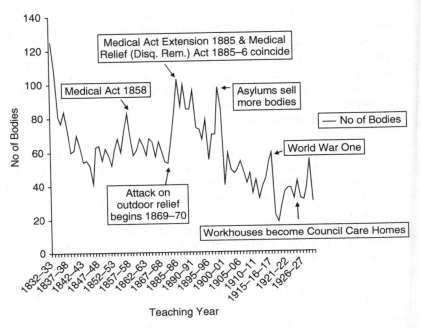

Figure 4.1 The Total Number of Bodies of the Poor Bought by St. Bartholomew's Hospital for Dissection, c. 1832–1929.

Source: St, Bartholomew's Hospital dissection registers, MS81/1-6.

during the first stages of the New Poor Law. Those figures support Ruth Richardson's important observation that it was a combination of the Anatomy Act and New Poor Law that together exacerbated pauperism when death overwhelmed meagre family economies.[56] That record, however, could not have been achieved without the concerted support of a wide range of Poor Law institutions prepared to be active participants in the body business. This is the first new finding. St. Bartholomew's established supply lines with a diverse range of sources and that helped hospital teaching staff pioneer new standards in medical education.

By 1840, the supply trends reveal that the body count dropped down to an average of 60 per year, with 1845–1846 being a low point of 41 corpses. Anatomists at that time had three major concerns. First, an Anti–Poor Law movement gained momentum and disrupted supplies everywhere.[57] Second, strong opposition in the national press, led by the *Times*, created negative publicity among ordinary people. Its impact meant that the poorest did everything they could to avoid dissection, including occasionally hiding bodies to give the bereaved more time to raise money for a pauper funeral by begging to bury their dead.[58] Third, guardians of the poor feared social unrest with just cause. Some experienced anti-anatomy riots in places like Cambridge, Liverpool, and York.[59] Most city authorities were worried about violent reactions on their doorstep. Guardians were thus more reluctant to cooperate. That was a worrying trend for anatomists. Numbers, however, soon stabilised again by 1850, when 62 bodies were acquired at St. Bartholomew's. The second new finding from this study is that good supply contacts were developed in the immediate vicinity of the medical school. Workhouse staff helped to source bodies provided they were paid for their trouble. In all the case studies in this book, doorstep deals are repeated everywhere.

Around 1858, new suppliers meant that anatomists were able to increase body-finding drives at a time when more medical students were coming into the profession. These body-supply schemes anticipated the greater demand for corpses around the time of the Medical Act when 83 cadavers were purchased in 1856–1857. Once the new medical legislation was enacted, however, the demand for bodies surged. The problem was that supply chains from prison ships were interrupted just when they were needed most in 1858. Dead criminals tended to be younger and therefore precious, adding to perennial supply problems.[60] In the 1860s acquisition rates were more erratic, though again they averaged about 60 a year. One reason for that trend was that the Poor Law was more generous by the 1860s, outdoor relief levels rising nationally by 16 per cent.[61] More people obtained poor relief outside than inside the workhouse, often in the form of pauper funerals paid from medical relief funds. That expenditure pattern did not favour anatomists. It also alarmed the Treasury (in charge of budgets), the Poor Law Board (responsible for local government costs), and an Anatomy Inspectorate (concerned about lower supply rates). A third new finding indicates that to describe the early and later body trade as alike is incorrect. The Anatomy Act, like the Poor Law, created trends arising out of policy developments. These can be traced in different phases across time. In many respects, the anatomy trade only really got underway in the later Victorian period. An austere anti-welfare climate was needed to expand the body trade exponentially.

Pressure was soon brought to bear on guardians of the poor to tighten their financial reins at the start of what became known as the crusade against outdoor relief.[62] Late-Victorian society was awash with harsher welfare policies that were to favour anatomists over the next three decades. A Poor Law directive known as the Goschen Minute (1869) proved to be a policy catalyst. It was written by a leading light in the Charity Organisation Society. Importantly, Goschen was a prominent civil servant who went on to play a leading role in the newly established Local Government Board (hereafter LGB) in 1871. He recommended that the disabled, infirm, sick, and vulnerable (orphans, widows, and insane) should come into the workhouse to get basic care. The Fleming Directive (1871) and the Longley Strategy (1873) then set out in quick succession how guardians should cut welfare bills across the country to a minimum.[63] The poor were encouraged to act independently and practice self-help. Posters were hung in workhouses, on church doors, and local public houses, setting out the harsh reality of a world-without-welfare. More of the sick poor were forced to seek health care from workhouse infirmaries. Out of fear, many came too late for treatment. At St. Bartholomew's this trend meant that body-supply rates nearly doubled. They rose sharply from 53 bodies in the 1869–1870 seasons to 103 corpses by 1885–1886. This 48.5 per cent increase was not a coincidence and was repeated everywhere, a fourth new finding. It is important to factor in two further legal changes – one medical, the other welfare-driven – that set the rise of 48.5 per cent in context.

In 1885, the Medical Act was extended.[64] More doctors meant more cadavers. The body trade expanded to meet higher demand. Medical schools were prepared to pay higher fees for dead bodies to workhouses. Many Poor Law unions were keen to find new ways to make savings in the underlying sub-structure of their welfare budgets. Their public pronouncements against the Anatomy Act have for too long been taken at face value in the historical literature. Most liberal, moderate, and strict Poor Law Unions sold bodies. That fourth finding is supported by evidence from the street geography in this case study that will be elaborated below. All medical schools found that there was a warm welcome when they knocked at the back-doors of workhouses that were always open for business at nightfall. The financial incentives might have been modest in some areas but, as we shall see, for many others it was a lucrative opportunity to offset welfare costs. It was therefore judged worth the risk of adverse publicity from time to time. A mentality of league tables, performance targets, and low taxation drove a harsh welfare culture. It was the poorest, most vulnerable who suffered most from it for the first time in Poor Law history. The ambit of the Anatomy Act was much wider than historians have hitherto appreciated.

Also in 1885, the Medical Relief (Disqualification Removal) Act was created (hereafter MRDRA).[65] This was another new policy initiative to affect the scope of the body trade. The significance of this minor Poor

Law change has often been overlooked. To recap briefly, the MRDRA stated that those in receipt of medical relief were now permitted to vote. Official reasoning was that to disenfranchise the poorest who needed basic health care for a short time was unfair. Yet this sea change ignored one of the foundation stones of the New Poor Law. Ratepayers in many areas were very disgruntled that non-taxpayers (those who claimed poor relief on a regular basis) were being given an equal say in how welfare budgets should be spent. Guardians of the poor in many areas of the country responded by stopping discretionary payments like alcohol for pain relief, food for nourishment, and pauper funerals on medical orders. Instead they insisted that the poorest come into the workhouse for health care, or go away. Central government did not anticipate the strength of bad feeling that the MRDRA would generate among ratepayers, but the strong reaction was not unwelcome for an Anatomy Inspectorate anxious to obtain more dead bodies. The Treasury likewise looked favourably on the unforeseen reduction in medical outdoor relief expenditure. Policy developments then in medicine and the Poor Law provided the perfect context to expand the anatomy trade. Figure 4.1 shows that St. Bartholomew's anatomists took full advantage of that trend, a fifth new finding. Throughout this book the chosen medical schools all benefited from the same changes. It is only by combining Poor Law and medical history expertise that a more nuanced picture of poverty emerges of the anatomy trade in the capital and the provinces.

Moving on to the 1890s, supply figures dropped down to an average of 40 bodies a year, reaching a low of 35 in 1895–1896. Again this was not a coincidence. Politically it was a challenging decade for the landed elite and big business in England. The passing of the Local Government Act (1888) heralded a new era in county politics, though arguably its consequences were not felt until the Local Government Act (1895) came into effect in April 1896. For the first time the Poor Law experienced democracy.[66] Property qualifications were abolished in guardian elections. Everyone could now vote to decide local welfare policies. The initial enthusiasm of the labouring poor pressured many older guardians to stop their body trade. On being elected to office for the first time, many working people challenged the right to sell on corpses. A successful campaign was waged in Nottingham for instance and supply chains stopped in the late 1890s (see Chapter 5). Across Northamptonshire it was also a prominent political issue.[67] The Honorable Charles Robert Spencer was unseated in 1895. On election day, he lost the Liberal stronghold of the Mid-Northamptonshire division to James Pender, the Conservative candidate. Pender was committed to old-age pensions and using local taxes to bury the poor, both popular local issues. Poor Law democracy and an expansion of the parliamentary franchise together at first promised to check the anatomy trade. Figures show, however, that over time the business of anatomy remained lucrative, a sixth new finding. The trade was muted when the Poor Law was democratised. It then expanded again because many staff in the capital's infirmaries agreed to be St. Bartholomew's keenest suppliers. They were soon joined by others.

Around 1900–1901, another new supply source opened up. Asylums had always been very reluctant in the past to sell their dead poor. Their trustees were keen to avoid bad publicity. Now, though, many were prepared to trade bodies, especially females. It is difficult to generalise about why this happened across the country. One key factor was a minor but significant rating trend that happened in 1890.[68] Most Victorian asylums were exempted from large property taxes. This was to encourage ratepayers to invest in capital-building projects for the insane. It was common to rate asylum property according to agricultural, rather than commercial, valuations. That policy was sensible since many asylums were built on the outskirts of towns and cities in the greenbelt. Over the course of the last decade of the nineteenth century, there was, however, a key policy change that had a positive impact on the anatomy trade. Under the Lunacy Act (1890), all asylums were re-rated by central government along commercial lines. In some areas, the difference between the agricultural and commercial values of asylums was four-fold by 1900. Many asylums were now located in suburbs that had been consumed by the pace and spread of urbanisation. This drove up their commercial values. The Treasury and local government saw an opportunity to make more money from Victorian property taxes. Selling the bodies of the insane poor offset those increased rate bills: each corpse could be sold for up to '£12' in places where supply was low and the dead plentiful. The trade had to be covert to avoid bad publicity but it was worth doing. The fact that body supplies leapt back up around 1900 to almost match the 1885–1886 levels stresses the importance of not overlooking this asylum rating-change in medical historiography. It is the seventh new finding of this study. At the time, dead women were in short supply. They commanded higher fees. More of the females categorised as insane thus entered the anatomy supply chain at St. Bartholomew's.

Supply levels dropped off again on the path to World War I. The passing of the Old Age Pension Act (1908) was one of the biggest contributory factors to a downward trend in the supply figures. Fewer older people needed to enter the workhouse and so less corpses were sold on for anatomy. Yet body supplies increased to nearly 60 per year once war broke out. St. Bartholomew's central location is one feasible explanation for why hospital staff could continue to trade in dead bodies during the war. Its wards contained casualties transferred from the Western Front who on return home were treated locally. Many died from their terrible injuries and were transferred over for dissection. This finding is atypical though compared to the other case studies in this book. Although it merits closer scrutiny it will require another separate study in the future. Meantime, by the 1920s, most workhouses were being converted into care homes run by local county councils. This was a time when the poor experienced the early stirrings of the welfare state. Local authorities started to take more responsibility for the nation's health care. The supply trend consequently rose slowly upwards again. The elderly had state pensions but many needed concentrated nursing care in the last year of their lives. Paradoxically, much feared welfare institutions run by the Poor Law had simply been re-vamped once the old workhouse became an infirmary and then the local hospital. All were run on the same basic management model. Their ethos did not alter significantly; the poor should repay their welfare debt to society in death.

Sampling dissection registers after 1930 shows that supply trends were now stable. Council care homes were the mainstay of the supply chain for most medical schools up until at least the 1960s. At present, the detailed analysis of that data collection is beyond the scope of this book for an important ethical reason. It is appropriate to retain sensitive case histories that may involve living relatives. The threshold of a 100-year rule should be respected, even if it is no longer a strict legal requirement for researchers today. These then were the general trends for the business of anatomy recorded in the hospital's dissection registers, with one key proviso. Earlier a gap in the primary evidence was mentioned that merits further investigation. Originally there were six dissection registers. One of the registers has been missing for 50 years. The body entries for those bought during the period 1873 to 1884 have gone astray. Since this marks the inception of the crusade against outdoor relief it is important to try to reconstruct what has been lost. We need to try to fill in the gap in what would have been 12 busy body-buying years. Those figures have important implications for the reach of the body trade across London. So in the next section the missing figures are extrapolated to complete the economy of supply picture.

Got two eyes... Got kidney and heart... Had offer of brain but declined!

It is fortuitous that there are two stray lists detached from the dissection registers that help to explain what has been lost from the missing book covering 1873 to 1884. A census point in 1872, and another in 1885, shows that 83 and 103 bodies respectively were traded. From these figures it is feasible to extrapolate average body-buying trends. Calculations show that not less than 83 corpses were acquired per year, over 12 annual teaching cycles. There are then a missing 996 cadavers.[69] We know that the official figures were 5,063 cadavers, plus 996 missing corpses, making a total of 6,059 bodies. St. Bartholomew's hospital may have had a crude Dead House and basic dissection room in Victorian times but it was a busy and crowded medical space. It also challenges conventional historical opinion about the scale of the body trade in London.

The *Times* listed leading medical schools in an editorial of 15 May 1885.[70] A General Medical Council (GMC) news bulletin was reproduced summarising 'an inspection of all medical examinations in all Universities in the United Kingdom'. The report concluded that, 'In London there were only 12 medical schools, with about 350 accredited teachers, in the provinces there were 9 schools with 250 teachers'. They all offered human anatomy courses that were officially recognised by the GMC, responsible for overseeing the licensing of new doctors. Of these, eight were large teaching hospitals. Even allowing for the fact that St. Bartholomew's was one of the largest medical schools, its body business suggests a minimum of 6,000 bodies were dissected in each location. To this we need to add the trade at the four smaller teaching hospitals. To be safe, halving the body count of the bigger medical schools gives an average of 3,000 bodies per smaller location. So, 48,000 bodies (eight major schools) and 12,000 bodies (four minor schools) produces an overall 60,000 body count. New research has thus produced a linear prediction which is hypothecated from reliable supply figures showing that the scale and scope of the anatomy trade in the capital was greater than the annual anatomy returns sent in to central government. The economy of supply indicates that a traditional estimate of 57,000 bodies should be revised upwards to 60,000 cadavers during the New Poor Law period.[71] And indeed that is a modest calculation because this study has erred on the side of caution for reasons of accuracy. That being the case, the upward trend suggests that some anatomists under-reported their body-buying activities and they did so consistently to central government. For professional reasons it was politically expedient to send in sketchy figures.

It is worth repeating that in 1886 the GMC recorded that there were '453 new entrants' registered for a medical degree, and of these '77 students' were signed up to study at St. Bartholomew's.[72] Significantly, the supply ratio was 77 students to 103 corpses at the hospital in 1886. This was why new recruits came to study at St. Bartholomew's. They were guaranteed at least one complete corpse per medical student. That said, this book has already recounted that in different phases of the anatomy trade the supply rates did vary considerably according to the background poverty conditions and underlying Poor Law policies. Later complaints about a lack of supply were not always disingenuous. Outside London, reports by smaller medical schools that inadequate body-supply rates were a constant headache were merited in some locations, notably at Oxford. It was self-evident that the medical fraternity had expanded considerably by the end of the nineteenth century.[73] Body-supply requirements had increased far beyond original expectations. It is important, however, not to get distracted by the figures for whole corpses. Anatomists after all bought and sold *complete* bodies and body *parts*. The dissection registers do not take this material fact into account. They provide an important but still partial picture of how the majority of students trained in human anatomy. It is not possible to reconstruct every body part sold across all of London. Records do, however, indicate that the trade was vibrant. Several representative accounts are explored here.

Henry Vandyke Carter, who produced the famous illustrations for *Gray's Anatomy* (1858), trained as a doctor at St. George's hospital. He noted in his personal diary that he often purchased body parts from the Dead House or coroner's cases after a basic post-mortem: 'Got two eyes. ... Got kidney and heart. ... Had offer of brain but declined'![74] Amputations from operative surgeries were likewise a lucrative source of anatomical supply. A visit, for example, by the Anatomy Inspectorate to St. Mary's hospital in 1905 revealed that bodies and body parts

> are stored in this [Dead House] room in wooden cupboards on wooden shelves. The bodies are not well preserved. The same chamber is also used as a furnace room having two heating furnaces in close proximity to the storage cupboards. ... The room is in fact partly a coal seller and is open to Porter's, coalmen, etc. At the time of the Inspector's visit the atmosphere of the room was very bad.[75]

Medical students bought body parts from the porters who ran a lucrative business from the premises. Bodies were often cut up and redistributed for bigger profits. The London Medical College on 6 May 1903 thus confirmed in a memo to Professor Alexander Macalister, Chair of Anatomy at Cambridge, that, 'My porter has a box of amputated parts which he has saved for Cambridge if they prove of use to you. There would be from 16 to 20 operative surgery bodies used every year in this school.'[76] At St. Bartholomew's the records show that the body parts from amputations usually came from Poor Law infirmaries. Trainee doctors were at the interface of two medical worlds, voluntary hospitals and Poor Law medical services. A guardian and medical officer of the Holborn Union made a request on 21 February 1871 for instance that read:

> Will you be so kind and give me an admission to St. Bartholomew's for a boy with serious disease of the femur. There may be a chance for him, and if not amputation will be better done at St. Bartholomew's than in the Holborn workhouse....
>
> I should like to take this opportunity of suggesting, as I have done already, to Mr White that a token understanding between the Parochial authorities and this Union and those of St. Bartholomew's would be of an advantage to both parties – we have now at work at St. Luke's a Capital Dispensary and I hope to get others in Holborn and Clerkenwell before long and I should like it to be understood that students at St. Bartholomew's should have the opportunity of attending the Dispensary and assisting the Medical Officers. They would have an admirable field of observations and would have a vast instruction which could not be obtained in Hospitals. In the Homes of the poor, all sorts of expedients have not been provided whereas in Hospital everything is to hand.
>
> In relation, the Medical Officers would have the great advantage of sending interesting cases for consultation and in any case...those likely to require an operation – I shall be glad if you will think the matter over and I have only to say it would qualify me to appoint in establishing a "rapport" which should be medically advantageous.[77]

The boy in question did require an amputation and his body part soon entered the anatomy business.

Regular entries in St. Bartholomew's Medical School Cash Registers attest to the profit-making side of the body-part trade. By the 1890s, for instance, students were paying fees of '£14 14s 0d' for 'entries to the dissection room' giving them permission to dissect 'bodies and body parts'.[78] Another petty cash entry a few pages on states that '£11 3s 0d' was paid in respect of 'addressing slips London clergy'.[79] These were payments for burial fees after dissection. They included both bodies and body parts. It is noteworthy that material transactions for body parts found in medical school petty cash books were never entered in anatomy registers or death certificates because it was not a legal requirement to do so. Surviving financial records are the only way to retrace this hidden side of the business of anatomy, a theme the Oxford case study explores in more detail. If a body came into a dissection room already dismembered into bits then it disappeared from official anatomy records. It is clear that this omission from standards accounts is important. For, as we shall see in Chapter 6, often the number of body parts *equalled* whole corpses. Breaking up a corpse for sale was more lucrative, ensuring that human material was distributed for profit widely. This trend is still very common on the internet throughout the poorest parts of the world today.[80] Until now, it has been the unwritten side of the anatomy trade and its continued presence in our lives makes it of some contemporary relevance in a biomedical age.[81] In Victorian times, brains, limbs, and torsos were all sold separately to medical students eager to qualify and start earning back their education costs. Historians have yet to appreciate the commercial scale, cash turnover, and the number of transactions involved in breaking up bodies along supply chains that ran across London and throughout provincial England.

The final point worth considering is the turnover of bodies in the dissection room at St. Bartholomew's because it was rapid and tells us about general trading conditions across London. Plotting the cadavers for each decade, as in Figure 4.2, highlights the scale of the business as it developed. Yet those crude numbers tell us very little about how long it took to agree to an average body deal and the time taken to dissect each corpse on the dissection table. Those calculations are significant because they reveal not only the physical pace of the body trade but the emotional impact of the speed of transactions for those micro-managing their grieving process. There would have been little time for consolation when deals had to be done very fast. Recently Julie-Marie Strange has encouraged historians to re-think the stoicism of the labouring poor.[82] She points out the importance of researching beneath a veneer of passive acceptance to touch the depths of unspoken human vulnerability. There exists a symbolic and verbal language that expressed the poor's sense of love and loss. It needs to be understood in its own cultural terms. The turnover of the body business can bring us closer to that lost emotional register when a loved one went speedily into and out of the dissection room at St. Bartholomew's. In the first trading

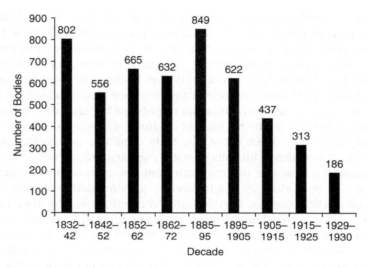

Figure 4.2 Number of Bodies per Decade Acquired by Anatomists at St. Bartholomew's Hospital, c. 1832–1929.

Source: St, Bartholomew's Hospital dissection registers, MS81/1-6.

period (1832–1872) no body was kept for longer than six weeks. Many cases were dissected within a maximum of 21 days. If the cadaver was 'prized' (an adjective sometimes pencilled into the dissection register) then it was dissected within five days. At the busiest times in the teaching cycle, a 72-hour turnaround was not uncommon. This meant that the body dealer had to be active on the street, develop good supply contacts, and know how to sell on the cadaver quickly to maximise profits. Suppliers were always commercially minded. One case history retrieved in the archives is symbolic of many thousands of prompt trade deals.

On 31 March, 1857, George Chapman, aged 75, a 'senile pauper' inmate died in the sick ward of Newington workhouse.[83] His body was transferred to the Dead House on site. The workhouse master, Mr. Albert Feist, then immediately contacted Mr. Robert Hogg, the undertaker.[84] In Chapter 1 this book encountered the inside dealings of this business partnership. Twelve months before Robert Hogg turned state witness at the Old Bailey in the case of *Rex versus Feist*, he was busy body-trading for St. Bartholomew's medical school. The anatomy registers show that Hogg was a paid intermediary in the hospital's supply chain. Beside the anatomy entry for George Chapman's cadaver, the demonstrator on duty in the Dead House wrote down 'paid up to this point'. He added up how much the hospital owed Hogg for being a regular supplier and then paid him in cash for negotiating body deals, transport costs, and burial fees. Across London, Hogg bought from a wide variety of health care and welfare institutions connected to the Poor Law. Reconstructing the paperwork attached to the anatomy registers and Poor Law accounts confirms that contracts were made with the County Lunatic Asylum (1854), Camberwell Lunatic Asylum (1855), London Fever Hospital (1855), Lambeth Workhouse (1853, 1857), and Newington workhouse (1857).[85] In the Old Bailey witness box, Hogg had revealed the details of his dealings with Guy's hospital during the *Rex versus Feist* hearing in return for immunity from prosecution in 1858.[86] He pointed the finger of blame at Feist, the master of Newington workhouse. This was a clever diversion tactic to distract official attention from Hogg's other body–outlets, like St. Bartholomew's. He was protecting his profitable anatomy dealings.

If one hospital was prepared to pay Robert Hogg more than another for a body, then he did a better deal across London. Guy's offered less for George Chapman's corpse than St. Bartholomew's. The latter lacked bodies in the busy teaching month of March and so paid a higher supply fee, even though the cadaver was a common case.[87] When George Chapman died on 31 March 1857, Hogg took a horse and cart to Newington workhouse. That night he collected the body and travelled through the streets some five miles to St. Bartholomew's Dead House under the cover of dark. Anatomists were waiting to receive the corpse. Over the next three days it was dissected and dismembered entirely. By '7pm on the 3rd April' the bits and pieces were left for burial. An entry in the anatomy register states the dissector's diagnosis was death from 'senile apoplexy'.[88] Poor Law records confirm that George Chapman was worn out by a lifetime of hard work, a spent force dying in poverty like many men of his time.[89] The likelihood is that his heart simply gave out and a minor stroke ended his life. Hogg gave him an anonymous burial in the hospital's common graveyard. In this case, the body deal was quick to ensure the decaying, aging corpse got to the dissection table as soon as was feasible.

Robert Hogg was kept on standby to serve the needs of medicine at St. Bartholomew's. The anatomy records show that on a regular monthly basis he brought cases of 'senile debility' and 'chronic bronchitis' to the hospital in the mid-1850s.[90] The turnaround rate for his body deals was always swift. On average he had obtained the cadaver within 24 hours of death. Once in his possession he then handed it over for a quick supply fee, generally from '£1-£12 a body'. Over the next 72 hours it was 'reduced to its extremities'. By midnight on the third day after death it was interred. Although Hogg stopped trading for a time after the *Rex versus Feist* scandal in 1858, he was actually lying low until the bad publicity abated. By the early 1870s, he was active again in the anatomy trade. Only detailed archive work brings to light the reach of his business dealings. Traditional histories tend to neglect the longevity of body traders and their wide-ranging income-generation activities.[91]

The anatomy records show that Hogg continued to buy bodies across London and from a wider variety of sources by the 1880s. In 1885—1886, for instance, these included 'unknown bodies found in the street', and from St. Saviour's Workhouse, Colney Hatch Asylum, lodging houses on George Place, Whitechapel, Greenwich Workhouse Infirmary, as well as Camberwell Workhouse Infirmary.[92] Hogg was also back trading from Newington Workhouse Infirmary. He had negotiated a new supply contract 30 years after his court appearance at the Old Bailey. It is not clear from the surviving record entries whether in the case of Newington it was Hogg in person, or a business partner, that picked up the actual bodies. The supply fees, though, seem to have seen him into retirement. He left a healthy family business thriving for the next generation of Hogg & c. undertakers. The speedy pace of his body deals is instructive and reveals a lot about what it must have felt like to be grieving and yet effectively excluded from the bereavement process. The poor could not afford either emotionally or economically to see a body to the grave.

In George Chapman's case, the records suggest that the bereaved were troubled by the anatomy trade. Cross-checking Poor Law and court records indicates that the grieving were still trying to come to terms with the final memories of the spirit of the person they had recently lost when the anatomy trade intervened. We saw ample evidence of this common situation in Chapter 1 at the *Rex versus Feist* case. Bodies had a fast turnaround and this did not change over time. Some of the bereaved were still in shock when they sold a loved one. 'Here one minute, gone the next' was how one pauper put it in his own eloquent words at the time. At most London workhouses it was not simply that the chalk on the coffin could be rubbed out with ease but that the corpses could disappear so rapidly without trace in the community. The deeper that we delve into the archives, making richer source connections, the missing parts of a mosaic of people that made up the body business starts to make more historical sense. The street geography of body transactions confirms the extent of trading deals, their rapid turnover, and true social cost across London.

Street geography

The anatomy registers facilitate an analysis of the street geography of the body-buying business. Again they have been split into the two trading periods, 1832–1872 (2,656) and 1885–1929 (2,407).[93] Table 4.1, shows the overall results for the entire period. It is clear from the data Poor Law infirmaries and workhouses were the chief source of cadavers at St. Bartholomew's hospital. Looking in more detail at the origins of the body deals reveals their geographic spread. These have been analysed in greater depth in Table 4.2.

The first new finding is that three infirmary-workhouses in the Holborn area were important sources of supply (see Table 4.2). Some 880 bodies or 35.03 per cent were traded in total. Of these, 400 cadavers were obtained from St. Giles Infirmary-Workhouse on Endell Street. Another 327 corpses were acquired from Holborn Union Workhouse located at Gray's Inn. The remaining 53 bodies were bought from St. Luke's Workhouse on Old Street, an old-style poorhouse that was

Table 4.1 Overview of the Body Trade by London Location, 1832–72 & 1885–1930

	Bodies 1832–72	Bodies 1884–90	Total
Poor Law Infirmaries & Workhouses	1,046	1,466	2,512
Hospitals	1,386	236	1,622
Home deaths (– sales made by intermediaries)	162	210	372
Asylums	6	276	282
Mental Hospitals after 1918	0	76	76
Care Homes (– workhouses converted after 1913)	0	62	62
Prisons	48	5	53
Abandoned bodies (– left on the hospital's doorstep)	0	35	35
Homeless (– found dead in the street)	6	28	34
Charity Homes	0	13	13
Sailor's Homes	2	0	2
Total*	2,656	2,407	5,063

Note: *There were a lot more bodies found homeless and bought on the street but they came via a Poor Law Infirmary or Workhouse body dealer in the archives and so are mixed up with those figures in the official records. They have been analysed exactly here.

Source: St. Bartholomew's Hospital Trust Archives, Dissection registers, MS/81-6.

Table 4.2 Infirmaries and Workhouses that Sold Bodies to St. Bartholomew's Hospital, 1832–72, 1885–1930

	1832–72	1885–1930	Total
St. Giles Infirmary & Workhouse, Endell Street, Holborn area	266	134	400
Holborn Union, Gray's Inn (later incl. St. Luke & City Road sites)	163	164	327
Whitechapel Infirmary & Workhouse, East End	174	96	270
St. Marylebone Workhouse, West End	38	221	259
St. George's Hannover Square (1870 became George's Poor Law Union, Fulham Road)	2	122	124
Southwark Union Infirmary & Workhouse (later with East Dulwich Grove Infirmary – East)	0	98	98
St. Margaret's Infirmary & Workhouse (from 1913 City of Westminster Infirmary – West End)	31	52	83
St. Giles & Bloomsbury Workhouse (incl. St. Giles-in-the Field, Bloomsbury)	4	69	73
Camberwell Infirmary & Workhouse, East End	9	61	70
Lambeth Workhouse, South East London	50	12	62
Kensington Infirmary & Workhouse (West)	1	61	62
Shoreditch Infirmary & Workhouse (East)	45	16	61
St. George-in-the East, Infirmary & Workhouse	6	50	56
Islington Infirmary and Workhouse (North)	6	48	54
St. Luke's Workhouse, Old Street (merged with Holborn Union after 1868)	53	0	53
Shoreditch Workhouse, East End	45	1	46
Wandsworth & Clapham Infirmary & Workhouse	0	35	35

Continued

Table 4.2 Continued

	1832–72	1885–1930	Total
Mile End Workhouse, East End	33	1	34
St. Mary's Infirmary & Workhouse, Newington	19	15	34
Chelsea Infirmary & Workhouse, West London	9	14	23
Southwark Infirmary, South East	0	22	22
St. Pancras Infirmary & Workhouse (North)	5	14	19
Strand Union Infirmary & Workhouse, Upper Edmonton, West End	14	5	19
West London Workhouse (joined City of London Union from 1869)	17	0	17
Spitalfields Workhouse, East End	14	0	14
St. Sepulchres, Workhouse, Smithfield, East End (West London Union & City of London by 1869)	12	0	12
Sub-Total	1016	1311	2327
Other location(s) selling under 10 bodies each			185
Total			2512

Source: St. Bartholomew's Hospital Trust Archives, Dissection registers, MS/81-6.

absorbed by Holborn Union in 1868. The close proximity of all the Holborn Poor Law institutions to the hospital made them ideal places to buy bodies. It was a very populous area. In 1835, the *Observer* reported on the area's population density: '20,000, 000 pedestrians; 870, 640 equestrians; 372,470 carts and wagons; 78,876 stages; 157,752 hackney coaches; 82,258 carriages; 135,842 omnibuses; 460,110 chaises & tax carts; 354,942 cabriolets'.[94] All of Victorian life passed through Holborn. Body-trading here kept any transportation costs to a minimum. Throughout this book that body-buying pattern will be repeated. Edward Stanley, hospital lecturer in anatomy and physiology, made the exclusive supply deals in the Holborn area. He was accused at the time of 'trafficking' in dead bodies on St. Bartholomew's wards, an allegation that proved to be false. Stanley had instead taken the strategic decision to secure a lucrative supply contract with Mr. Hewett, the master of St. Giles Workhouse.[95] It was no coincidence that he did so.

St. Giles parish had a long-standing reputation as one of the worst places to live in central London. The *Second Report of the Select Committee of Metropolis Improvements* (1838) stated that 'The Rookery on St. Giles Holborn' was one of the 'most troublesome' places to reside. It was described as 'a very old concentration of misery and crime, home especially to a ragged Irish population since the eighteenth century'.[96] Buckeridge Street stood at the centre of the parish, and 'an almost endless intricacy of courts and yards crossing each other, rendered the place like a rabbit warren...full of thieves and prostitutes and cadgers' lodging houses at fourpence a night or threepence in the cellars'.[97] Nearby Jones Court was where counterfeit money was produced by gangs. Nightly it was smuggled out through underground tunnels connecting the lodging houses in Jones Court to those on Buckeridge Street. Anything and everything was stolen to be traded for a few pennies in St. Giles parish, including dead bodies. Death, dearth, and disease were daily facts of life. The parish was filthy, full of excrement, and foul-smelling. The 1838 report reiterated that: 'In St. Giles one feels asphyxiated by the stench there is no air to breath [sic], nor daylight to find one's way'.[98] It estimated that 5,000 people lived in houses no more than ten feet square, each containing up to 40 residents a night. Accommodation was filled to capacity and contagion was rife. The grim reaper was everyone's neighbour in St. Giles Holborn.

St. Giles street scene did start to modernise between 1840 and 1870. Old buildings were knocked down. New housing had better sanitation. Notorious slums were replaced by smart new shops and the underground. St. Giles parish, though, retained its rough reputation. This was because the destitute forced out of one street never moved far away. When some of the most notorious lodging houses were demolished in 1845–1847, the poor moved sideways. Endemic poverty became denser, increasing mortality rates. There was a regular trade for instance in cadavers on Buckeridge Street that relocated a few streets along into Church Lane. In the 1841 census, 655 people stated they lived in Church Lane. By 1847 another census showed that 1,095 people had crowded into the same number of houses.[99] Victorian city planners ignored the problem of poverty by reshuffling social problems. After each slum clearance body dealers did not have to go far or wait long until the next trade deal. In fact mapping the street geography of the body trade is a useful way to visualise localised population movements among the poor. Trends show that the destitute moved from one slum clearance to the next, mortality rates becoming more and more concentrated by overcrowding.

Over time the poor were critical of how their dead were being treated in St. Giles and in particular Holborn Union. The *Times* on 6 September 1883, for instance, reported on events at the Holborn Board of Guardians, held at Clerkenwell vestry. The subject of the meeting was tawdry pauper funerals. Local people complained that a broken-down vehicle had hurled pauper bodies into the street. Their indecency was exposed for public consumption. 'Four mourners' had been ordered to sit on top of 'five coffins, three containing the bodies of adults and two the bodies of children' tied on with only a frayed 'clothes-line'. A subsequent investigation revealed that the conveyance crashed because the undertaker often stockpiled cadavers:

> The fact was they let a contract to Mr Hooper [undertaker] who in turn sub-let it to a man on the Old Kent-road, who in his turn, in order to make the thing pay, had to delay the burial until after there was an accumulation of bodies. There was no undertaker to accompany the corpses, but simply the driver who on arriving at his destination handed over the corpses to the gravedigger and when the corpses were taken off the coaches the relatives did not know one from the other. The names of the corpses, which had been written in the coffins with a piece of chalk, were by the shifting of the coffins indistinguishable when the cemetery was reached.[100]

There was a 'very animated discussion' among the guardians about this newspaper report. A spokesman for the poor claimed that the guardians 'were afraid to call independent witnesses'. Some of the destitute in Holborn did not know that their dead had been sold on and undergone dissection. The remains were being returned for burial in Colney Hatch Cemetery, when the accident happened in the street. Holborn Union was keen to avoid any further discussion of its profitable involvement in the anatomy trade. Mr. Hooper's official contract said he was 'an undertaker'. What this job title actually meant was that he was a body dealer. This was why he had no interest in burying the dead of Holborn. He leased out his so-called 'undertaking contract'. Another tradesman was prepared to stockpile bodies to make a profit from the meagre burials.[101] This type of evidence suggests that too many historians have taken at face value the actual meaning of undertaking contracts and their sleight-of-hand job descriptions in Poor Law accounts. Twenty-five years after the *Rex versus Feist* scandal not much had changed inside the body business. The poorest were still being exploited for a quick profit by those body dealers posing as undertakers.

The Holborn situation explains why St. Bartholomew's expanded its anatomy trade into the East End of London. Anatomists judged it prudent to recruit a range of suppliers. Scandals reported in the press made it necessary to juggle traders to avoid bad publicity. Whitechapel Union was thus the next important source of supply, selling a total of 270 bodies (10.74 per cent). Jerry White explains that: 'It was women in the cheap common lodging houses of Spitalfields and Whitechapel – the East End's west end, as it were – who provided commercial sex for dock and building labourers and market porters'.[102] Again these were the females at the roughest end of a spectrum of prostitution, staple dissection subjects. Anne Hardy also explains that the area was rife with disease:

> Certain parishes preferred to save the expense of treatment and deal with cases in their own workhouse infirmaries. Significantly, Whitechapel and St. Giles were both in this class. ...
>
> In the decade 1851–60 Whitechapel contributed to the largest number of metropolitan fever deaths, and the *Lancet* suspected it of playing 'too prominent' a part in fostering the great 1861–70 epidemic.
>
> In 1865 it was estimated that Whitechapel contained 9,000 houses, 5,000 of which were let out as separate lodgings, each house containing an average of three families. The total population in 1861 was 79,000; density in the parish was 195 persons per acre.[103]

The entries in the St. Bartholomew's anatomy registers confirm these general trends and later in this chapter the disease profiles of the destitute sold on will be examined.

St. Bartholomew's naturally looked towards the largest workhouses in the West End of London too. Marylebone sold on 259 bodies (10.31 per cent). St. George's Hanover Square likewise supplied 124 bodies (4.93 per cent).[104] In the West End, slums were demolished and domestic housing was replaced by new commercial districts along New Oxford Street by the mid-Victorian period. This forced the poorest to move west. The body traders were never far behind their supply lines. In 1866 a report compiled by the *Lancet*, concerned about overcrowding in London Poor Law infirmaries, set those mortality trends in context.

The medical officers of the metropolitan workhouses appear to me to do their duty to the best of their ability; but I am obliged to add that in many instances their duties are very onerous and their salaries inadequate.... We may add that for the most part he has to dispense all the medicine and in many instances to find the drugs at his own expense.... One surgeon assisted by one resident junior is expected to look after 300 acutely sick, and 600 chronic cases; or that (in another instance) is expected to attend 130 acute and about 300 chronic cases in the [work] house, in the intervals of private practice.... Taken as a body, the medical officers of the metropolitan workhouse infirmaries apply themselves with a zeal and an amount of success to their disproportionate tasks which are surprising, and it must not be forgotten that they have in most cases not only to perform most arduous professional duties and a large amount of desk work, but they have to fight the battle of the poor, with terrible earnestness, against the prejudices and the gross material interests of the worst members of their boards of guardians...and whose wilful neglect to build a properly isolated dead house for their parish has long exposed the poor of a crowded district to frightful sufferings and risk of disease.[105]

There are a number of ways to read this source. It was evidently in the *Lancet*'s interest to promote medical officers and their low-paid work. At the same time the report highlights that large infirmaries attached to big workhouses like Marylebone were often overcrowded. Medical treatment was sporadic. Contagion was rife because the living mixed with the dead. These dismal socio-medical conditions explain why the anatomy trade expanded into the West End. One noteworthy new finding is that all the body deals can be traced back to a main arterial route along which the dead were transported to St. Bartholomew's. Body dealers used one of the most ancient highways into London. It went from 'Oxford to Colchester' via London and ran 'through or around St. Giles to Holborn, Newgate, Cheapside, Cornhill, Leadenhall Street, Aldgate, and Whitechapel'.[106] At night, so-called undertakers, but really body dealers, travelled on a loop to and from St. Bartholomew's hospital. Along the Oxford to Colchester turnpike anatomists were waiting to receive the dead travelling to their final destination.

Another new finding of the geography of the anatomy trade is that body-buying schemes related closely to ethnicity patterns of settlement. The Irish lived in the centre of St. Giles parish along Holborn in an area dubbed 'Little Ireland'.[107] The Italians moved into Saffron Hill, along the streets that ran down to Leather Lane in the middle of Hatton Gardens and onto the Clerkenwell Road. Traditionally both areas were the main suppliers of St. Bartholomew's. The Italians came into the Holborn district to sell their cheap delicacies like ice cream during the day. Street-sellers mingled with servants and prostitutes. Organ-grinders entertained local people on every street corner. Standing by were the body dealers looking to make a profitable supply fee on the street.[108] The Irish and Italians were predominantly Roman Catholic. Those in poverty could not afford though to respect religious sensibilities. Ideally the body would be buried intact; practically it was a profitable anatomical commodity. Selling now saved the cost of a funeral later.

Many of the names listed in the dissection registers of St. Bartholomew's were of Irish and Italian origin. Representative cases include three Irish people that lay side by side in the dissection room in 1835–1836. These were 'Robert Morgan (aged 35)', 'Mary Leary (aged 49)', and an 'unknown Irishman (age unknown)'.[109] They were all sold by a body dealer working along Buckeridge Street. The cadavers were sold on to Mr. Hewett, the master of St. Giles workhouse on Endell Street in Holborn. Hewett brought them personally to the attention of the hospital for a supply fee. All were collected and dissected within three working days, having died of pneumonia, an asthma attack, and a fatal stroke, respectively. That season they were 3 out of a total of 19 bodies bought on the streets of St. Giles between October 1835 and February 1836. In 1854 the dissection registers likewise show that a young Italian man, 'aged twenty-six named Thomas Gasperini', was an emergency admission from 'Saffron Hill'.[110] He died on 20 July. Within 24 hours he was dissected entirely. He died from 'Phthisis' (tuberculosis) a common disease. His body lay alongside 'Elizabeth Pappelle aged 54', a middle-aged Italian woman of 'unknown address'. Sometimes the skin colour of the dead person was noted too. In February 1854 for instance, 'Ramsey Sammey, A Black, aged 32' was 'Destitute, in the Sailor's Asylum' and died. His cadaver was sold by 'Johns' (a body dealer at the asylum), dissected for three days, and 'Phthisis of the lungs' (wasting of the body caused by pulmonary tuberculosis) was recorded on the final death certificate. Occasionally a detailed account of an emigrant was also left by the anatomist. In 1839, 'George Yettett', described as 'Late from New York', collapsed at the docks having alighted from an emigrant ship. Twenty-four hours after emergency admission, he died from cabin 'fever' and was dissected for '23 days' before burial.[111] Other economic migrants had journeyed from the Midlands to London to make a new life. This was the fate of 'John Watkins' from 'Birmingham' who died from 'erysipelas on 9 February 1843'.[112] Lately arrived, he had

Dealing in the Dispossessed Poor 153

no friends to claim his corpse. For '22 days' medical students studied his anatomy. His human remains were deposited in a common grave in the early spring of 1843. In this way, ethnicity, migration, and race were key aspects of the anatomy trade throughout London.

Looking back to Table 4.1, and then forward to Table 4.3, research shows that a lot of cadavers were purchased on the wards of St. Bartholomew's hospital as expected.

Table 4.3 Bodies Bought from London Voluntary Hospitals, 1832–72, 1885–1930

	1832–72	1885–1930	Total
St. Bartholomew's Hospital Wards	1313	170	1483
London Fever Hospital	28	1	29
West Middlesex Hospital, Isleworth	0	21	21
Hospital Ship, Woolwich	11	0	11
Royal Free Hospital, Islington	9	0	9
Briton Hospital	6	0	6
Holborn and Finsbury Hospital	0	5	5
Bridewell Hospital	4	0	4
Cancer Hospital, Brompton	3	1	4
St. James Hospital Balham	0	4	4
Lambeth Hospital	0	3	3
St. Leonard Hospital Shoreditch	0	3	3
St. Marylebone Hospital	0	3	3
Charing Cross Hospital	2	0	2
Smallpox Hospital	2	0	2
Army Medical Hospital	0	2	2
Highgate Hospital Darth on South Park Hill	0	2	2
St. Giles Hospital	1	1	2
St. Peter's Hospital Whitechapel	0	2	2
St. Stephen's Hospital	0	2	2
West London Hospital	0	2	2
21 Hospitals with only 1 x body sale per location	7	14	21
Total*	1386	236	1622

Note: *Some hospitals kept all cadavers; others shared them: an important facet of supply
Source: St. Bartholomew's Hospital Trust Archives, Dissection registers, MS/81-6.

Nightly they were removed by the beadles and passed over in exchange for a small supply fee to anatomists. Susan Lawrence explains that this had been a common practice since the late eighteenth century.[113] Keir Waddington likewise states that the role of the beadles was to 'supervise all dissections' and make sure the 'body be sewn up and placed in a coffin'.[114] This standard practice 'guaranteed some lay supervision' of the body parts, a profitable sideline. It is important to stress, however, that the hospital was a more important source of supply in the earlier, rather than the later, period.

At St. Bartholomew's, 1,313 bodies (49.43 per cent of 2,656 bodies) were generated in the first phase of trading – compared to just 170 corpses (7.06 per cent of 2,407 cadavers) in the second period. Over time this was a significant decline of 87 per cent. The hospital subsequently had to rely on other sources to make up the shortfall. The number of bodies bought from Poor Law infirmaries and workhouses between the two trading periods increased from 1,046 to 1,466 bodies, a rise of 40.15 per cent (refer Table 4.1). That trend reflected the material realities and the true social cost of the crusade against outdoor relief from which anatomists benefited indirectly. It is evident that the 40.15 per cent increase, however, did not match the 87 per cent decrease from the hospital wards. What this meant is that the hospital needed Marylebone (221), St. George's Hanover Square (122), St. Margaret's Westminster (52), and St. Giles and Bloomsbury (69) to together supply some 464 corpses, or 18.47 per cent, in the later period. More and more infirmaries were drawn into the body trade to make up for the decrease on the hospital wards by the later Victorian era.

In absolute numbers street deals contributed 441 or 11.4 per cent to the total economy of supply. Of these, some 210 home deaths (8.7 per cent) were acquired by body dealers in the poorest neighbourhoods between 1885 and 1930 (see Table 4.4). Another 276 corpses (11.46 per cent) were purchased from asylums, 76 from mental hospitals (3.15 per cent), and 62 from new council-run care homes (2.57 per cent) refurbished from old workhouse premises (refer Table 4.5). Hospital staff evidently recognised the need to diversify body deals despite the background poverty conditions and New Poor Law serving their needs. What is significant about the second trading phase is that everyone in the supply chain devoted more energy and time to the business of anatomy, as we have seen already in this book. This was because the expectations of the poor had been raised by the expectation of Poor Law democracy. Civil rights and welfare entitlements were key election issues. Once recession hit in towns, cities, and the countryside, endemic economic problems

Table 4.4 Street Deals Sold to St. Bartholomew's Hospital, 1832–72, 1885–1930

	1832–72	1885–1930	Total
Home Deaths (– deals made with body traders on the street)	162	210	372
Homeless (– found dead in the street)	6	28	34
Abandoned bodies (– found on the hospital's doorstep)	0	35	35
Total*	168	273	441

Note: *Figures are conservative for reasons of accuracy. Only those that could be checked with other record linkage work have been identified here.
Source: St. Bartholomew's Hospital Trust Archives, Dissection registers, MS/81-6.

presaged further social changes. Those trends also coincided with the Third Reform Act (1885). The future of the anatomy trade was always going to become more, not less, complex given that context.

St. Bartholomew's had to respond to a more diverse cultural landscape as the century progressed. Notable changes included improving standards in education and literacy. Paupers had always written letters in protest about their common hardships and dire predicaments. Under the Old Poor Law they retained the right to appeal direct to a magistrate, in court if necessary.[115] A scribe would often write on behalf of the aggrieved. The New Poor Law tried to distance paupers from that decision-making process. Guardians generally made collective decisions behind closed doors at the workhouse. Paupers could turn up to make an appeal but often they were dismissed in a rudimentary manner. The problem for guardians was that despite this autocratic style the world outside was changing. Guardians had to live in the same parishes as the paupers on the workhouse doorstep. It was impossible to ignore deprivation for long in an era of expanding local political representation. The introduction of elementary education with the coming of the Education Act (1871); the growing popularity of national newspapers like the *Daily News* (the third most popular in England); and an extensive railway network connecting town to countryside broadened everyone's cultural horizons. The correspondence books of individual medical schools consequently attest to the determination of the poor to undermine the anatomy trade. One representative case sets in context why bodies from the hospital's ward fell by 87 per cent.

On 5 July 1872, a widower wrote to protest about the treatment of his dead wife on the hospital wards. He alleged that because her body was in a bad state and therefore unfit for dissection, nobody at either the hospital (she was a Poor Law referral) or Whitechapel workhouse (where she had been given rudimentary treatment) cared about what happened to her human remains. A series of letters between the Treasurer of the hospital and Board of Guardians of Whitechapel Union survive and indicate that the allegations appeared to be true.

> A woman named Catherine Reeves, whose husband resides at 10 Ely Place Mile End, New Town in the Parish of Whitechapel, died in this Hospital on the 28th (ultimo); after having been under treatment for eight weeks. The husband being too poor to defray the expense of her burial application was made by our Steward, in the ordinary course, to one of the Relieving Officers, to remove the body in order to its burial in your Parish.
>
> This your Relieving Officer declined to do so, stating that he had received positive orders from the Board of Guardians not to remove the bodies of any paupers who die in the Hospital [i.e. they were to be sold for anatomy].
>
> The body came in such an offensive state the Hospital had been obliged for health's sake to order its burial.
>
> Under these circumstances I am instructed to request you to bring this matter before your Board of Guardians; and I am directed to inform you that, whilst the Governors of this Hospital are anxious to extend the benefits of Charity to the Poor from all parts, they will most reluctantly be obliged to refuse admission into the Hospital to the Poor of your Parish if, besides treating and maintaining their illness, they are to be just to the further cost of burying them in case of their death [without dissection].[116]

In other words, unless a pauper body was in a suitable condition to be dissected the hospital would not pay the costs of burial, regardless of the pauper's circumstances. The Treasurer moreover explained that he did not want to get in the middle of a Poor Law fight. Reeves had made an official complaint and further bad publicity was unwelcome. The last thing that the hospital wanted was a whispering campaign by paupers about death and dissection on its wards. It was one thing to make a discreet body deal, quite another to open the trade up to public scrutiny. It took several threatening letters from the Treasurer to Whitechapel Union before the situation was resolved by guardians agreeing to fund a pauper funeral. In the meantime, the pauper exploited the situation. He put pressure on the

Poor Law to bury his dead wife, Catherine Reeves, with success. The files of the Local Government Board are filled with many similar cases.[117]

Table 4.5 and Illustration 4.1 exemplify that the body trade at St. Bartholomew's became more complicated as the century progressed.

Table 4.5 Bodies Bought from Prisons, Asylums and Mental Health Hospitals, 1832–72, 1885–1930

	1832–72	1885–1930	Total
Prisons			
Coldbath Fields House of Correction, Clerkenwell	30	0	30
Millbank Prison	11	0	11
Leviathan Hulk Prison Ship, Portsmouth Harbour	2	0	2
Holloway Prison	0	2	2
City Prison, Holloway	1	0	1
House of Correction, Wandsworth	1	1	2
New Bridewell, Westminster	1	0	1
New Prison, Clerkenwell	1	0	1
Pentonville Prison	0	1	1
Wandsworth Prison	0	1	1
Sub-Total (Prisons)	47	5	52
Asylums			
Claybury Asylum	0	53	53
Cleveland Street Asylum	0	53	53
Banstead Asylum	0	24	24
Leavesden Asylum	0	24	24
Bexley Heath Asylum	0	23	23
Poplar and Stepney Asylum	0	16	16
Central London Sick Asylum	0	14	14
Cane Hill Asylum	0	11	11
Tooting Bec Asylum	0	10	10
Hanwell Asylum	0	8	8
Metropolitan Asylum Caterham	0	7	7
Bromley Sick Asylum	0	7	7
Sub-Total	0	250	250
Horton Asylum	0	6	6
Manor Asylum Epsom	0	5	5
Asylums x 1 body each, early period	6	0	6
8 x Asylums x under 5 bodies each, later period	0	15	15
Sub-Total (Asylums)	6	276	282

Continued

158 *Dying for Victorian Medicine*

Table 4.5 Continued

	1832–72	1885–1930	Total
Mental Hospitals after 1913			
Leavesden Mental Hospital	0	22	22
Springfield Mental Hospital	0	17	17
Claybury Mental Hospital	0	7	7
West Park Mental Hospital	0	5	5
11 x Mental Hospitals under 5 bodies each	0	25	25
Sub-Total (Mental Hospitals)	0	76	76
Sub-Total (Asylums & Mental Health Institutions)	6	352	358
Total (Prisons, Asylums & Mental Health Institutions)*	53	357	410

Note: *These totals are the cadavers sold to St. Bartholomew's. The institutions were also supplying other medical schools too.
Source: St. Bartholomew's Hospital Trust Archives, Dissection registers, MS/81-6.

Even small trade deals were important and this extended the geographical reach of the economy of supply. It spread out from the hospital's doorstep because anatomists needed to replace those bodies that it lost from the wards in the later period. Increasingly anatomists were under more pressure to obtain female bodies once Old Age Pensions came into force. Corpses bought from asylums, mental hospitals, and prisons reflect this new reality. Gender re-shaped the anatomy trade in ways that merit closer scrutiny in the next section.

No man should marry until he has studied anatomy & dissected at least one woman[118]

The dissection registers recording 5,063 cadavers have been gender-profiled. They reveal that St. Bartholomew's demography is strikingly different from medical schools in provincial England. Sexual classification confounds the findings of the case studies that follow in the rest of this book. Elsewhere, regardless of the size of the medical school, for every three men acquired for dissection, just one woman was purchased. The opposite happened at St. Bartholomew's where men and women were bought in equal numbers until the 1860s. This new finding is explicable. In the capital there were more female servants, greater numbers of street prostitutes, and higher maternal mortality rates in

Dealing in the Dispossessed Poor 159

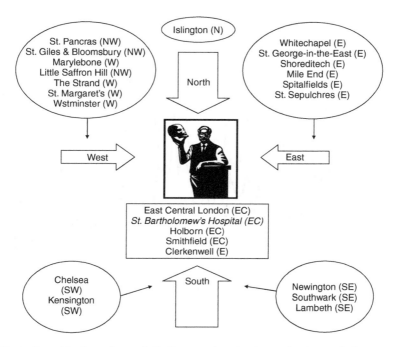

Illustration 4.1 Locations of Bodies coming in from the larger Infirmaries and Workhouses, purchased by St. Bartholomew's hospital, 1832–1930, author illustration.

the poorest neighbourhoods. That backdrop brought more female cadavers into the ambit of the anatomy trade, until the later Victorian period.

There has been considerable historical debate about the impact and timing of public health improvements on morbidity patterns during the Victorian era.[119] St. Bartholomew's dissection registers suggest that fewer women ended up on the dissection table after 1870. There are three reasons for this. As the century progressed better medical care did develop for the poorest in the capital's teaching hospital system. At the same time, there was a lower cost of living and this improved nutrition standards which improved maternal immunity. Public health measures then started to make an impact on everyday life by improving drinking water supplies and sewage schemes. Water-borne contagious diseases like cholera were better controlled. As a result of these medical trends, the body trade had to respond in kind. By the 1880s for every two men

160 *Dying for Victorian Medicine*

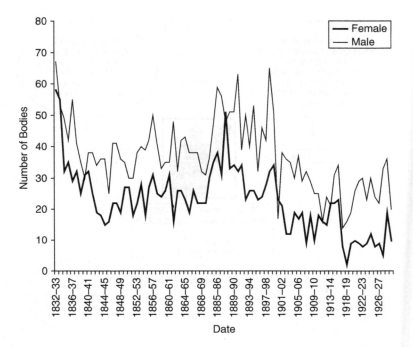

Figure 4.3 Number of Male and Female Bodies Sold to St. Bartholomew's Hospital, 1832 to 1929.
Source: St, Bartholomew's Hospital dissection registers, MS81/1-6..

purchased, one woman was sold for anatomy (see Table 4.6). Only at the end of the study period did the ratio increase to three men: one woman in the 1920s. There were specific socio-economic reasons that made the ratio of three men: one woman more common outside London.

The case studies that follow will show that normally a ratio of three men: one woman was the rule of thumb in provincial life. This reflected the socio-economic circumstances of those living in poverty. Since the 1980s, Poor Law research has shown that the workhouse was a care home for elderly males by the later Victorian period.[120] This was because men had fewer employment options compared to women as they got older. Women could contribute in many informal ways to family budgets. Traditional male labour skills were needed when couples were young but not when childcare, cleaning, clothes washing, and general household tasks had to be redistributed over the typical life cycle of an extended family economy. A multi-skilled grandmother who did household

Table 4.6 The Average Number of Bodies Dissected, Ratios of Male: Females Corpses by Decade at St. Bartholomew's Hospital, 1832-72, 1885-1930

Decade	Ratio Males: Females	Whole Cadavers Males: Females	Precise Figures Male: Females	Trends (Averages)
1832–42	448: 354	1:1	1.26: 1	1:1
1842–52	341:203	2:1	1.68: 1	2:1
1852–62	401: 248	2:1	1.62: 1	2:1
1862–72	377: 251	2:1	1.52: 1	2:1
1885–95	510: 328	2:1	1.55: 1	2:1
1895–1905	392: 228	2:1	1.72: 1	2:1
1905–1910	268: 166	2:1	1.66: 1	2:1
1915–1925	222: 89	2:1	2.49: 1	2:1
1925–1930	135: 51	3:1	2.64: 1	3:1

Source: St. Bartholomew's Hospital Trust Archives, Dissection registers, MS/81-6.

duties often freed up young parents to work full-time. This practical financial arrangement is common today in a world where childcare and mortgages are expensive. In the past, families relied on grandmothers, maiden aunts, and female family friends to bring up young children. This pattern of household economy is what social historians and behaviour psychologists call 'calculative reciprocity'.[121] Older women paid for their keep by providing domestic services. Men, by contrast, were an economic burden and this explains why families sold them on for anatomy. Poor labouring men often drained household economies once they could no longer labour in the fields.[122] Arthritis was a common medical complaint among agricultural labourers who had worked with damp crops. The late-Victorian flight from the land to avoid recession in the countryside exacerbated male unemployment trends. Their only choice was to enter the workhouse. In the provinces, the demography of the dissection room reflects this economic reality. On average men were worn out by hard manual labour by 50 years of age, whereas women could still earn their keep in their seventies. There were thus fewer female bodies in the workhouse to be sold on for dissection.

St. Bartholomew's dissection registers challenge these ideas. They suggest that the body trade in London was distinctive. That new finding reveals one key difference about medical training standards in the provinces compared to London. At St. Bartholomew's male medical students had ample opportunity to dissect male and female bodies. This is what made the hospital distinctive, even in the capital. Outside London, the case studies that follow in this book will show that male trainee doctors dissected male paupers predominately. They seldom saw, much less handled, a female body. Medical school records indicate that students who trained at Cambridge, Oxford, and Manchester tended to go into general practice, whereas those who qualified in London practiced hospital medicine or joined larger private practices. In the provinces, females reliant on Poor Law medical services or those who could afford to subscribe to a general practitioner were often treated by someone with a theoretical rather than practical knowledge of female anatomy. Learning about female anatomy from textbooks, lectures, and wax models was widespread in the regions because of body-buying trends.[123] A lack of hands-on human anatomy training using females is a finding that clearly deserves further consideration in women's studies of Victorian life. If, however, we delve briefly in more detail into the caseload, a selection of personal histories shows that myriad reasons brought men and women into the body trade in the capital. The year

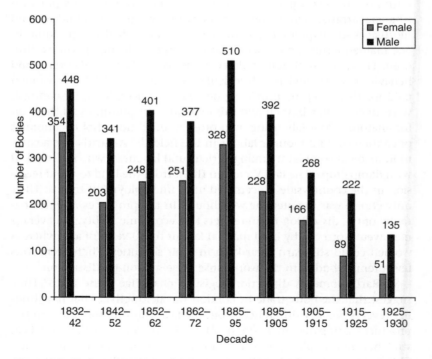

Figure 4.4 Bodies of Males and Females Acquired for Dissection per Decade at St. Bartholomew's Hospital, 1832 to 1929.

Source: St, Bartholomew's Hospital dissection registers, MS81/1-6.

1833 has been chosen as a representative sample because of the diverseness and richness of the case histories.

In January 1833, Shoreditch workhouse was rife with measles.[124] Over the next four weeks, 'Anne Blaine aged 3', 'William Hemmingway aged 5', 'Ebenezer Critcherf aged 3', 'Ann Debow aged 5', and 'William Storey aged 1 ½' became infected. The fact that they were lying side by side in the infirmary made the contagion worse. All died on consecutive nights from the fatal symptoms. Their poor parents had no money. Selling their children for dissection made financial sense. Mr. Teale, the body dealer and undertaker, made the supply deals. He buried them in a mass grave some 21 days later. The same year, 'Mary Lane', 'Sarah Tilby', and Ann Williams' became dissection cases too. They were all women committed to the young offenders' institute known as the House of Correction at Cold Bath Fields. The anatomy register says that they died 'diseased' but 'cause not stated' in April 1833. The records suggest that young men tended to get into more fights than women. Females died of physical problems compared to the fatal, violent episodes among men. Thus in 1833, 'George Greenaway' was described as 'a 50 year old male' of 'unknown address' and occupation. He died from 'delirium brought on by a leg injury' in the street. 'Daniel Haggarty', nickname 'Agarthy', likewise died in the street, in St. Andrews parish on 6 November 1833. His body was placed in the dissection room next to 'Frederick Starke', a '20 year old' young man found dead the same day. He died outside Smithfield market, a notorious meeting place for roughs. This case history contrasted with 'Catherine Stanton aged 33', resident of a rough lodging house at 'Essex Place' on the 'Hackney Road', East London. Like so many women in the sample, she 'died from childbirth' on 23 November 1833.

There were of course many cases when being 'worn out' with life was simply stated. This was the fate of 'Mary Carr', a pauper 'aged 80', who lived among thieves in 'Jones Court, St. Giles in Holborn'. Somehow she scratched together a living to survive outside the workhouse but died destitute at home in the bitter cold of January 1833. Whereas 'Gerard Helme', a lunatic 'aged 55', simply could not cope with life and died alone in 'Waistall Lunatic Asylum'. His body was dissected in March 1833 but he was not without a feeling family. The anatomist recorded that despite his dire situation and being isolated from his community, 'his friends on Sunday 10th March ... saw that his body [such as it was] was returned to them for burial'. Poor 'Thomas Haines', an elderly man 'aged 75', who died in Shoreditch Workhouse from 'decline' in April 1833 never got a pauper funeral. He was typical

164 *Dying for Victorian Medicine*

of many entries for men past their prime in the sample. After '25 days' there was in any case so very little left to bury of him. Across London, social deprivation and the problems of living in poverty affected both males and females of all ages and types. Given that fact of life, the next section explores aging in more detail to add historical detail to these gender traits.

Age

One of the striking findings in the St. Bartholomew's anatomy registers is that 20 per cent of the bodies sold for dissection were aged between 21 and 41. At least one fifth of the corpses were young, more than in the other education centres, and this explains again what attracted medical students to the hospital to train in human anatomy. Another 27 per cent were described as middle-aged, between 41 and 61. So trainee doctors

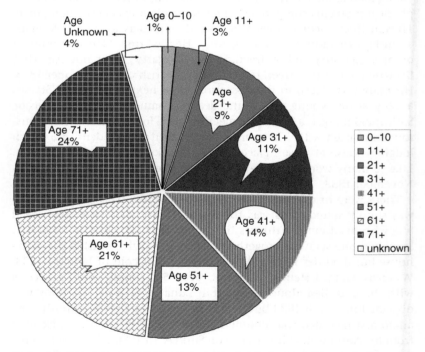

Figure 4.5 Age Range of Bodies Sold to St. Bartholomew's Hospital where n = 5062 bodies, 1832 to 1929.

Source: St, Bartholomew's Hospital dissection registers, MS81/1-6.

had a 47 per cent chance of dissecting at least one adult body below the age of 61. Although only 4 per cent of the sample size are listed as children aged below eleven, it is worth repeating that stillbirths did not have to be recorded officially in the dissection registers. Earlier in this chapter evidence was introduced that suggested anatomists often purchased dead infants and spontaneous miscarriages. Even though those figures cannot be accurately assessed given the gaps in the records, it is important to keep in mind that the official figure of 4 per cent is a conservative calculation. Overall, a medical student anxious to study the human body without the degenerative changes of old age obstructing his studies had a 50–50 chance – not less than 51 per cent – of dissecting corpses aged below 60. It is also striking that another 21 per cent of the sample were aged 61 to 71. These were once more predominately worn-out men in the later decades of the nineteenth century when work was scarce because adverse trade cycles exacerbated unemployment. Another 24 per cent were defined as over 71 years of age in the total sample. This finding adds further historical colour to the earlier observation concerning the differences in medical education in human anatomy in and outside London.

Recently some popular histories have suggested that 'male corpses were more highly prized than females, since they offered greater scope for the study of musculature'.[125] In fact anatomists generally took what they could get. Most wanted to dissect females but in provincial locations they were in shorter supply compared to the capital, a theme that later chapters will explore. There is no doubt that the study of childbirth excited medical science at major teaching hospitals throughout the Victorian era. There was a growing interest in the early stirrings of embryology, gynaecology, and paediatrics. Improved pathology started to link better health care interventions to treatments for malignant cysts and abnormal tissue, notably cancers of the breast and womb. There was also an increasing focus on hysteria and menstruation.[126] The development of well-woman medicine meant that the teaching of human anatomy at St. Bartholomew's was attractive. There were also numerous diseases that could be studied in the wider population and these are profiled in the penultimate section of this chapter.

Diseases of the destitute

One of the advantages of a study of this nature is that the dissection registers provide a unique profile of diseases of the destitute in London. Those that ended up on the dissection table lived on the bottom rung of

Victorian society. We can therefore be confident that the data produced provides a unique insight into the morbidity profiles of the sick poor, living in absolute poverty. It is fortunate that St. Bartholomew's anatomists recorded a diagnosis after each dissection in their registers. In this they followed the example of their counterparts at the Public Morgue in Paris. Often those records differed from hospital, infirmary, and workhouse death certificates. This means that a more accurate medical picture of destitution is being made available for the first time. It is evident in Table 4.7 that respiratory diseases were the biggest killers of the destitute for the entire study. In the first period, 1,127 cases, or 22.25 per cent, and in the second phase some 1,034, or 20.42 per cent, of the 5,063 bodies recorded were linked to fatal lung diseases. Broadly defined

Table 4.7 The Disease Categories of Corpses Dissected at St. Bartholomew's Hospital, 1832–72, 1885–1930

Disease Categories (Dissection Registers)	1832–72 Bodies	1885–1930 Bodies	Total Bodies	% of Sample
Respiratory Diseases	1127	1034	2161	42.68
Age-Related Diseases	61	362	423	8.35
Cardiac Diseases	140	241	381	7.53
Fever	263	0	263	5.19
Debility Diseases	181	28	209	4.13
Digestive Organs	114	68	182	3.59
Head/Brain Diseases	60	107	167	3.29
Nervous System/Motor Neurone Conditions	65	98	163	3.21
Cancer	34	121	155	3.06
Urinary Organs	33	86	119	2.35
Mental Diseases	44	44	88	1.74
Fractures/Amputations/ Broken Bones	43	25	68	1.34
Blood Diseases	42	19	61	1.20
Arteries/Vein Disease	0	46	46	0.91
Accidents	30	0	30	0.59
Lack of Nutrition	9	6	15	0.30
Others (– secondary diseases*)	305	112	417	8.24
Not stated (– unknown or illegible entry)	105	10	115	2.27
Total	2656	2407	5063	100

Note: *like sepsis or an incidental disease caused by a primary infection as originally described in the dissection registers.

Source: St. Bartholomew's Hospital Trust Archives, Dissection registers, MS/81-6.

this meant that the destitute had consistent shortness of breath. It was a fact of Victorian life that poor air quality polluted meagre lives. In both phases we can break down these figures to examine the epidemiology in closer detail. Before we do that, however, it is important to stress that the disease categories have been taken from contemporary sources and are analysed according to definitions stated in the primary records. This is to ensure accuracy and to be as historically sensitive as possible to the contemporary viewpoint of anatomists and their language of diagnosis. This method thus seeks to avoid the obvious demographic shortcoming of projecting back onto medicine in the past modern conceptions of disease classification taken from today's perspective.

In phase one, respiratory cases included in their order of ranking in first place some 416 (36.9 per cent) deaths from 'phthisis' or tuberculosis. This was not seen as a communicable disease but one that was common in families born into endemic poverty. It was believed to have had a wide variety of causes often related to one's constitution and general lifestyle. Then in second place there were 255 (22.62 per cent) fatalities suffering from 'bronchitis' of the lungs. This meant that the third most common condition with 145 cases (12.86 per cent) was described as 'those in the last stages of consumption'. Closely related, in fourth place, were a further 98 persons (8.69 per cent) who died from 'diseased, inflammation of the lungs'. Another 74 cases (6.56 per cent) in fifth place did not survive their last 'chronic asthma' attack. Broadly speaking that meant that in sixth place were another 57 (5.05 per cent) who expired from 'pneumonia'. Finally, in seventh place, the remaining 82 (7.27 per cent) cases had assorted breathing problems like 'emphysema of the lungs ... inflammation of the chest ... and pleurisy'. There was thus a spectrum of common respiratory diseases out of a total of 1,127 personal medical histories. It is worth stressing that these were painful, often prolonged medical conditions. Patients usually went into a steady decline over many years. Their employment and earning power would have been affected in severe winter weather. Environmental factors like crowded, damp, and sub-standard housing also exacerbated common respiratory conditions. Dense smog, smoked-filled streets, the foul smells of pollution were well-known contributing factors too. Today biomedicine is as easy as breathing. In the past, a baby's first breath was a gamble, a medical risk that congested with age.

In phase two, equivalent morbidity patterns were linked to the same types of respiratory diseases. Of the 1,034 cases (20.42 per cent overall), we find that ranking them produces approximately the same level of lung disease with similar diagnostic profiles. Again, in terms of ranking, we find that in first place there were 450 (43.52 per cent) 'chronic bronchitis' cases. The figures then show that in second place 'phthisis' or tuberculosis cases had exponentially declined to 231 cases (22.34 per cent). To this figure, however, we need to add another 85 cases (8.22 per cent) described as 'pulmonary-tuberculosis'. Better diagnostic tools and a more nuanced medical language had changed the pathology of dissection cases. Overall there were really 316 corpses (30.56 per cent) who died from tuberculosis-related conditions. This meant that in third place general 'pneumonia', with 144 or 13.92 per cent, was the next big killer. Again to this figure we need to factor in another 46 or 4.44 per cent 'bronchial pneumonia' cases, making a total for pneumonia-related fatalities of 190, or 18.3 per cent. There were 93 less lung disease cases in the later period, some 1.83 per cent, but it nonetheless remained a fatal condition for the poorest until at least 1930.

Recently demographers and historians of medicine have debated whether '33 per cent of deaths were from infectious diseases' or whether that figure was much higher 'nearer 40 per cent' in Victorian times.[127] If we stand back from the finer points of that debate and look instead through the historical prism of the diseases as described in the St. Bartholomew's anatomy registers, they strongly suggest that what mattered to the destitute was the high incidence of fatal respiratory diseases in their midst. Some were infectious; some were not. The key point is that being able to breathe easier was vital for the most vulnerable. The poorest did experience a better diet, improved public health care, a lower cost of living, and improved environmental factors. Over time these factors reduced morbidity in the Victorian era. Yet the diseases of the destitute also highlight that for much of the nineteenth century it was still vital to escape the physical community of absolute poverty. Moving from one slum to the next street did not alleviate health problems. In *Howard's End* by E. M. Foster, a popular novel written in 1910, the main character, Leonard Bast, is a lowly bank clerk. He wants to climb out of poverty but his live-in lover, a former prostitute, limits his potential in a world where income, morality, and social standing define a family's economic fate. One starry night, Bast plays out his fantasy of walking out of London by the light of the moon to wander around the bluebell woods of Kent. St. Bartholomew's anatomy registers highlight that Bast's quest was not just romantic. It had a practical, medical dimension too. Leaving behind his life would have literally filled his lungs with the oxygen of a better lifestyle. At St. Bartholomew's, a final aspect of the body business underscores this point. The records suggest that there is the possibility that the anatomy trade had links to the Jack-the-Ripper murder cases.

Jack the Ripper and the anatomy trade

There has been considerable debate in crime historiography about whether 'Jack the Ripper' was a medical man.[128] Recent scholarship has focused on a spectrum of potential killers. Potential candidates include those who worked in local slaughter houses, farrieries with veterinary skills, and medical professionals connected to the police (coroners and surgeons) with the right knife skills, all seen in the area of the murders. It is not the intention of this final section to engage extensively with this 'Ripperology', but rather to suggest that solving the conundrum of 'who was Jack the Ripper' could involve studying the intricate operation of the anatomy trade in the capital. To recap, in 1888, not less than five female victims, street prostitutes in most cases, were murdered in Whitechapel in the East End of London. Events took place close to St. Bartholomew's hospital. Anatomists transacted body deals on a regular basis with nearby Lambeth and Whitechapel workhouses, close to the murder scenes. One of the harrowing features of the 'Jack-the-Ripper' murders was the speed with which the victim's bodies were cut up. Extensive mutilation took just ten minutes. The method of killing was shocking. It resembled closely dissection techniques. Whoever was responsible, the killings were very rapid. That observation raises the intriguing possibility that there might have been a strong link to the anatomy trade operating in the vicinity.

Generally each victim's neck was slashed within seconds.[129] The murderer used a knife to make a sharp cut across the underside of the chin, a long incision from ear to ear. It is noteworthy that this first wound opened the carotid artery. A medical person would have known where to cut. Each victim fell to the ground. Forensic evidence suggests that the murderer grabbed the women from behind, cut their throats, and then eased the dying corpse onto the street. In this way, they attracted as little attention as possible to the crime. Fluids drained away fast, generally behind the victim's head. Again it is critical to appreciate that blood gushed out of the body from the neck area but did not spill onto the torso. The murderer was then free to dissect the corpse cleanly. Each torso was opened from the neck to the navel. In a frenzied but highly skilled attack the womb was cut open above the upper vagina area. This exposed the pectoral muscles. The organs were then taken out undamaged, including the womb itself. On several occasions limbs were disgorged too. What is intriguing about this method of killing is that similar bodies that had been mutilated in this way were sometimes brought into St. Bartholomew's dissection room. Occasionally entries read 'mutilated entirely', 'some mutilation', 'unsuitable for dissection, as mutilation', 'body had been opened' and so on.[130] Yet little historical link is made between the anatomy trade and the high-profile 'Jack-the-Ripper' victims.

The Home Office, Metropolitan Police, local coroners, and neighbours all speculated about the motive for such gruesome attacks.[131] Everyone agreed that it was the ferocity and speed of the murders that made them so distinctive and despicable. How could anyone walk into crowded neighbourhoods where the poor lived cheek-by-jowl and with such confidence murder quickly and then make a swift getaway? They had to have had an accomplice. There were certainly reports of someone carrying what looked like a medical bag in the vicinity of some of the victims. Speculation at the time was that the murderer was someone with basic dissection training. If so, then the anatomy trade could have been a factor in their murdering methods. The speed with which bodies were turned around at St. Bartholomew's explains where somebody attached to the body business might have learned to use the lancet with such lightning skill. How the murderer dealt with each victim's blood underscores this point. The crime scene was bloody but bodies were bloodless, a crucial forensic distinction.[132] Blood drained behind the head of each victim, and this left the corpse clean. Mutilation was methodical and therefore based on some form of anatomy skill performed at high speed.

To dissect a fresh body was a feat that could only be achieved with repeated practice.[133] There was no equipment to keep the corpse bloodfree. No plastic suction tube or latex gloves. In cold candlelight or by dim gaslight, it was hard to find the main organs much less cut them with anatomical precision. The carotid artery had to have a skilled incision so that blood drained away fast. Only in this way could the dissector see what he was doing on the dissection table. If he did not follow standard techniques, then he would be up to his armpits in plasma in minutes. Blood stopped coagulating after death but this did not mean that dissection was not in disarray. The stomach contents, festering food and blood, often flowed from fresh corpses cut by amateurs. The smell would have been overpowering and the corpse a bloody mess. If the body was kept for longer than two days, then preservation fluid had to be injected into the carotid artery. The blood was slowly replaced by chemicals. In cases where the body contained a baby, the dissector had to be highly skilled to be able to extract the dead infant undamaged from the deceased mother for teaching purposes. Whatever the circumstances in each case, the murderer who cut the victim's necks in the 'Jack-the-Ripper' incidents must have had a basic knowledge of what had to be done in the first hour on a dissection table when a fresh corpse was acquired for anatomy. They knew the importance of a fast and skilled turnaround, akin to standard dissection procedures in a Dead House like that at St. Bartholomew's.

Wynne Baxter, the coroner in Mary Ann Nichols' case, held an inquest that concluded on 26 September 1888 that,

> the injuries must have been made by someone who had considerable anatomical skill and knowledge. There were no meaningless cuts. The organ [upper pelvis area of the womb above the vagina] had been taken by one who knew where to find it, what difficulties he would have to contend against, and how he should use his knife so as to abstract the organ without injury to it. No unskilled person could have known where to find it or have recognised it when it was found.[134]

Paul Begg points out that the *Lancet* agreed with this conclusion. Its editorial leader stated that the speed 'pointed to the improbability of anyone except an expert performing the mutilations described in so apparently skilful a manner'.[135] Someone of a pathological frame of mind could then have been attracted to hospital Dead Houses where the work suited their disturbing personality traits. In the autumn of 1888, that person or someone associated with the anatomy trade perhaps took basic dissection skills out into the Whitechapel streets at night. Begg adds that the fact that most of the victims were street prostitutes, and their wombs subjected to a frenzied knifing attack, is not conclusive proof of a sex crime. This is an important observation if we factor in the anatomy trade and its fast turnover.

In the anatomy season of 1888, women were not in short supply at St. Bartholomew's but the turnaround rates had now changed in the dissection room. Preservation techniques had improved and most bodies were therefore being preserved with chemicals longer. On average they were cut up over 100 days as opposed to just seven days: a major change. Anyone who preferred fresh, female corpses would have needed to seek them on the street for the first time. Inside the dissection room they would have looked like wax specimens and less like human beings because of the chemicals in use. A basic alteration in preservation methods could have been behind the motive for murder. Longer use of human material was certainly a new style of dissection training in most London teaching hospitals. It coincided with the time when the 'Jack-the-Ripper' cases started in the East End. The medical profession stressed at the time that no doctor who upheld the Hippocratic Oath to care for the living would ever commit such a cruel and vicious crime. It nonetheless remains a possibility that a warped mind, with some form of human anatomy training, went in search of what they lacked to dissect.

Wombs were prized specimens and it was prostitutes who were the staple of dissection tables at St. Bartholomew's and elsewhere. In avoiding capture, the killer had no to time to ask whether the women were pregnant or not. In all likelihood, that opening question would have sent them screaming for help. A lack of oral contraception meant that abortions, miscarriages, stillbirths, and full-term pregnancies were a constituent feature of an average prostitute's street life. From a medical standpoint, the chosen women were more, not less likely, to be carrying a child in their womb when they were murdered. The killer would have expected the law of averages to work in his favour. The forensic evidence thus speaks of a clinician's mentality, not simple sexual perversion. He could reasonably expect to cut up some pregnant women. This was a person perhaps less interested in sex than in discovering the creation of life. It is a terrible irony that a twisted form of scientific enquiry might have involved turning wombs into a dissection trophy. What is also disturbing is that the murderer had to be a familiar face in the crowd, otherwise he could not have walked away so fast into the shadows. It was a fact of life in the East End that there were many intermediaries involved in the body business with a working knowledge of anatomy and dissection. Body dealers were familiar on the streets and included Dead House staff, infirmary nurses, workhouse porters, undertakers, coroners' mates, even midwives. 'Ms. Biers', for instance, of 'St. James' Workhouse' was listed as a female dealer from the anatomy season of 1840 onwards in St. Bartholomew's registers.[136] The potential list of people involved in the body sales is extensive and throws up all sorts of unproven but intriguing possibilities. Though why "Jack" murdered for such a short period remains a mystery.

Reviewing the evidence again for this book, one curious feature of the murder cases is that all of the female victims did not flinch at first when approached. Yet, as the murders went on, they would have been more wary of being solicited by a stranger. At the subsequent coroners' hearings, witnesses testified that some victims had to risk their personal safety for paid sex to survive outside the workhouse. Others needed a few shillings for a lodging room for the night, offering oral sex in an alleyway. Free drinks at the local public house motivated some to walk the streets. They all nonetheless were on the lookout for danger, more sensitive to unusual advances. Several women had also been in and out of Holborn, Lambeth, and Whitechapel workhouses. They were well known to the authorities, knew a wide cross-section of the population, and were streetwise. The forensic report in Mary Ann Nicols' case, for instance, noted that her petticoat was marked with 'Lambeth Workhouse P. R.', a pauper stamp.[137] She had been treated at the workhouse infirmary on a regular basis. This observation is noteworthy because all the female victims were left in a rather unusual physical pose of some relevance to the medical mystery surrounding events.

Begg explains that the women were found flat on their backs: 'the legs were drawn up, the feet were resting on the ground, and the knees turned outwards'.[138] In the past this was how anatomists dissected to expose the muscles. It is also the standard medical position when a woman is having a smear test, or internal physical examination to check progress in childbirth. Yet crime historiography seldom discusses the possibility that a woman made the first initial approach, not a man. This would explain a lack of circumspection in all the cases. Maybe an intermediary known to the victims, someone who traded for instance in their stillbirths, whom they were used to acting as a go-between, made the first fatal move. The actual attack had to be done by a man because of the physical strength needed to carry it out with such ferocity and quick skill. That, however, does not rule out the possibility that the women were groomed by a female accomplice, familiar to them from a workhouse infirmary or with whom they had done body deals on the street. Records confirm that women dealers and male traders like 'Ms. Briers' and 'Robert Hogg' walked the physical streets of Whitechapel in 1888.[139] The ending of a supply partnership between like-minded individuals prepared to murder might explain why the killings stopped as abruptly as they started.

Murdering to dissect is an unanswered question. Like many associated with the 'Jack-the-Ripper' story, and indeed other high profile murders, the gaps in the crime records are also the potential research leads of the future. It is appropriate and a mark of scholarship to leave ajar the dissection room door to fresh enquiry. The grim murder cases that were never solved could yet prove to have been part of the anatomy trade in London.

Conclusion

The anatomy trade at St. Bartholomew's was extensive. It stretched across London. Men and women were bought in equal numbers from a wide variety of suppliers. The lifestyle of servants caught up in failed courtships, disconnected from kin, made them natural targets for body dealers. Likewise a cross-section of asylums, infirmaries, and workhouses were also keen suppliers. In the early Victorian period, patients on St. Bartholomew's hospital wards were bought in high numbers. Meanwhile undertakers, but really body dealers, distributed available corpses. Cadavers were bought on the front doorstep of lodging houses, from the back of Poor Law institutions, and on the street. The Irish, Italians, and those economic migrants anxious to make a fresh start in the capital were purchased with regularity. Prostitution was an important source of supply; men and women got caught up on both sides of the deal. Sexually transmitted diseases were very common with the ill-fated repaying their health-care debt on the dissection table. Above all it was respiratory cases that served the needs of Victorian medicine. Lung-diseased beggars, thieves, and the dispossessed lay alongside young children who died from major child-killers like measles. Abortions, spontaneous miscarriages, and infant stillbirths were valuable anatomy material. In the most notorious neighbourhoods the body dealer on the street was a familiar face in many central London parishes. Most had long careers and made enough money to make it worth the time and trouble. Robert Hogg, the undertaker, exemplifies the longevity of traders everywhere. In 1858, as the Medical Act was being passed by Parliament, he exploited his court exposure at the Old Bailey to gain immunity from prosecution and thus protect his longer-term business interests at St. Bartholomew's. If one scandal came to light, then anatomists simply juggled their suppliers until the bad publicity abated.

The coming of Poor Law democracy meant that the poor protested louder about the anatomy trade. Anatomists were thus compelled to diversify their body dealings. Theirs was a complex, complicated, and increasingly covert business as the century progressed. It involved whole bodies and body parts. There was a fast turnaround to maximise profits. The poor had little time to stop to grieve. Coroners handed over, for a small supply fee, drowning cases, suicides, and corpses found dead in the street. Hospital porters traded amputated parts from operative surgeries. These were placed in wicker baskets or wrapped up in deal boxes, sold on to the highest bidder. Medical students paid for extra human material to train more quickly and start earning back the cost of their expensive education. Females were prized subjects because both the anatomy of childbirth and gynaecology could be studied. At St. Bartholomew's neither age nor gender nor family affection protected the pauper from an anatomy sale. That fact of life will be repeated throughout the rest of this book. For, as Bill Hayes reminds readers, 'a book ends, a story ends, a life ends. But the desire to study anatomy never will', and while there is anatomy, there is always an anatomy trade, at Cambridge in the next chapter, and elsewhere.[140]

5
Pauper Corpses: Cambridge and Its Provincial Trade

> We can offer you two methods of anatomical instruction: that of the dissecting-room and that of the lecture-room. In the first, you are to learn the art of the anatomist; in the second, I shall endeavour to teach you the science of anatomy. In the former, you learn to cultivate your powers of observation and manipulative skill; in the latter, your memory is to be impressed and your judgement educated. In both, I shall do what I can to help you to prosecute your studies.[1]

In early October 1883, on a bright autumnal morning, a tall, thin academic, dressed impeccably in a tweed hiking suit and sturdy leather walking boots, left London on foot.[2] Alexander Macalister was determined to walk in record time to Cambridge University. He had just been appointed to the Chair in Human Anatomy. His promotion was an important opportunity to raise his personal profile. Macalister put his best foot forward, displaying courage and endurance. It was a physical feat that he would repeat often throughout his prolific career to promote public engagement with the anatomical sciences. Macalister was the sort of character who made an immediate impact on his contemporaries. They recalled that he was always 'robust in constitution and energetic in temperament'.[3] In Victorian medical circles, he was soon regarded as a man of action and a spokesman for anatomists across the United Kingdom.

Macalister left the capital on the Great North Road. He hoped to reach his destination by nightfall 'in a little over twelve hours'.[4] On arrival in Cambridge, his 'celebrated walk' was achieved in set-time. It gained him notoriety and earned the respect of younger medical students, more of whom were eager to study human anatomy. News soon spread of his intellectual and physical prowess. The medical press recorded that

Macalister had in fact been breaking records throughout his working life.[5] It was reported that, aged just 14, he had gained admission to the Royal College of Surgeons of Ireland. This meant that he could register for a medical degree at Trinity College, Dublin. There he trained to become a demonstrator in anatomy. At 17 years old, he was licensed to practice medicine, earning a 'double qualification' to teach medical students. Six years later, he was offered the Chair in Zoology. He was just 23 years old. Few doubted that he was one of the most talented men of his generation.

In 1883, Cambridge University had recruited a man who believed passionately in his career choice.[6] Macalister would devote his working life to reforming provincial medical education. He thought that anatomy was not just an art form but a scientific subject that gave one strong 'mental discipline'.[7] He described it as 'a splendid introductory gymnasium' for young minds in need of exercise and training in medicine. Students, he advised, needed to enter the dissection room where 'you can find out facts for yourself, you can verify every single statement, you learn to take nothing for granted on mere authority'.[8] Macalister insisted that cutting up cadavers was of 'unspeakable value' compared to anatomical textbooks. Each medical student was 'brought face to face with tangible objective realities; easily grasped, easily understood'. Looking back 25 years later he remained convinced that:

> Nature is immeasurably greater than our most advanced conception of it, and it is Nature that we are ready to study, not merely the diagrammatic representation of it. We have to deal with facts, and it will be your business to make yourself familiar with them, and to remember that in your work you are really doing original research. That body which you are about to dissect you will find does not in all details conform to the book description. It has characters of its own that you ought to note; just as each patient that you will see in afterlife shows his own idiosyncrasies, of which you must take account.[9]

At the end of his career, Macalister regretted that 'the student is now spoon-fed by his teacher' talking from a medical manual.[10] In the mid-Victorian era, 'textbooks were poor and often with few, or no, figures and consequently we had either to make out all of the structures for ourselves with very little external help, or be satisfied with the unintelligent absorption of the cut-and-dried' specimens laid out by the demonstrator in anatomy. By contrast, hands-on anatomy was a slow learning process and yet dissecting on a regular basis encouraged 'spontaneous and original' work. Macalister lamented an over-reliance on rote learning and wax models too. Students invariably saw what they were supposed to see, rather than questioning their teachers' instructions. The best pupils were healthy sceptics but they did need an economy of supply in the dead poor. It was this problem that proved to be the greatest challenge of Macalister's career at Cambridge from his appointment in 1883 until his death in 1919. In this chapter his concealed struggle to establish that covert anatomy trade is scrutinised in depth. For the supply networks that he developed became a template for the business of anatomy in provincial life in Victorian times. The next section introduces new material from Northamptonshire because this was in the catchment area of Cambridge's body business and a key target. Sources explain why certain types of dead paupers entered the chain of supply. Newspapers and the private papers of anatomists then expand that discussion by highlighting some of the public and private problems that concerned Macalister. These set in context the work he did to try to develop better body-finding schemes.

The lamentation of human life

On 12 March 1839, the Reverend Pyrse Jones, workhouse chaplain, resigned from his clerical post before the board of guardians in Brackley Poor Law Union, an area of Northamptonshire targeted by Cambridge and Oxford anatomical schools. He told them that he was very unhappy about the mistreatment of the poor inside the workhouse. Doctors were able and willing to treat paupers at home. Too many were sent to ill-equipped medical wards. The accommodation was sparse and staff seldom trained properly. In a controversial resignation letter he claimed that the New Poor Law was failing the sick poor:

> I believe I am not mistaken, ever lament, the *lamentation of human life* [sic] in the aged, profussively [profusely], though unwittingly, super-induced [sic] by causes which might have removed and [caused] the *increase of human suffering* [sic] in the afflicted... which care and due application of proper skill might have been prevented... I have never observed in the last days of age out of the House that rigidity of Limb, especially in the lower extremities, accompanied as the nurse has told me, with an unusual change of colour; bodies [in the workhouse are] more macerated than commonly seen; countenance presenting one uniform impression and appearance of the successive steps of the cause of death. What I have remarked and here described, I have

submitted to my Medical friends, not connected with the Brackley Union, four of whom on different occasions [have] gone with me through the House and their opinion was the same, that the heated impure atmosphere which the Inmates breathed and the laxative Dietary combination would be the means of causing such Effects as I have described and they did not doubt from these appearances that a disordered state of health [has] prevailed.[11]

Pryse Jones's letter caused a local storm because the poor suspected that limiting medical care was a scheme to cut rates. Eight ratepayers, all agricultural workers, risked the wrath of their farmer employers to sign a petition, sent in protest to the Poor Law Board in London. A civil service investigation concluded that the chief source of the bad publicity was 'John Henry Parkins, publican of the village of Helmdon'.[12] This was no coincidence. The public house was a political hot spot in many English villages. The coroner's court, funeral raffles, and family wakes were often staged on the premises. Poor Law accounts show that often the publican was the ringleader of protests about the anatomy trade too. The Brackley petitioners alleged that 'Mr Gee, the surgeon' at the workhouse was to blame for the lack of medical care and higher mortality rates locally. He was, they claimed, typical of the sort of medical man contracted under the New Poor Law who ignored the psychological impact of workhouse admission. This, Gee denied strenuously, but 'Alice Rubra the Nurse' admitted privately to the petitioners that their medical grievances 'were merited'.[13]

By the 1840s, the issue of the proper care of the sick poor and medical relief had become a divisive political debate in local and central government. Concern was expressed that paupers had lost out in the transition from the Old to New Poor Law. So much so that a *Select Committee on Medical Poor Relief* (1844) was convened. It asked 'Mr. Rosebrook Morris, surgeon to the Brixworth Union, in Northamptonshire', to present his casebooks in evidence to a public enquiry. They had some telling revelations about the plight of those in need of medical relief in an area that lay within the catchment of Cambridge's body business.

Surgeon's cases:
Average salary £80 per year, paid per case

1837–8   170 cases, paid at 5 shillings per case
1838–9   66 cases, paid at 7s 6d per case
1839–40   105 cases, paid at 7s 6d per case

Fees changed to a new contract, on a paid fixed salary of £50 per annum

1840–1	156 cases, paid a fixed annual fee of £50 in total
1841–2	154 cases, paid a fixed annual fee of £50 in total
1842–1843	208 cases, paid a fixed annual fee of £50 in total
1843–1844	254 cases, paid a fixed annual fee of £50 in total

Conclusion: When paid a fixed salary average cases were 188 per year, when a per case fee, they fell to 80 cases per annum.[14]

The implication was self-evident. If the figures were accurate, then guardians encouraged the poor to see the doctor on a fixed-fee contract. The problem was that the doctor could not afford to devote the same amount of time to their care. On the face of it, paupers seemed better off. In fact the opposite was true. A senior civil servant asked Rosebrook Morris, 'Do you think that justice can be done to the sick poor under the present system of remuneration?' He replied:

> It cannot, unless the medical officer makes a sacrifice of his time, and I might also add of his money... I may instance one case which cost me in a fortnight 13 s[hillings] for one pauper, a poor woman, suffering under cancer, not with the view of a cure, but to do away with the offensive smell, and to deaden the pain.[15]

He was asked whether by treating the poor at cost it was at 'a loss to yourself?' This he affirmed and explained that the fixed salary did not even cover travel. For, he said, 'I must have a horse to do it; the £50 I receive from the district is the bare keep of the horse, with taxes, saddles, shoeing and wear of the horseflesh'.[16] He estimated that in the early 1840s the number that died alone, without adequate care, and no funds for a pauper funeral were 'for the last 3 or 4 years, one in every 10' in the Midlands. These families were the natural targets of anatomists. Having denied them medical care in life, many were being exploited in death. Given the seriousness of the allegations, Rosebrook Morris was pressed to elaborate on local medical conditions:

> From the 4th November 1840 to the present time (14th June 1844) one in every ten cases has died without medical assistance, not wholly dependent upon Medical Orders. In some cases they have told me that they had applied and could not get an order [for medicine or a pauper funeral], and they could not afford a doctor; in other

cases they have applied to old nurses and women in the parish, who have administered a little syrup or rhubarb or syrup of violets; in other cases, particularly with the bastards, they suffer the children to linger for a long time without any medical attendance whatever; where they could afford medical attendance, they will perhaps call in the clergyman of the parish or his lady, to administer medicine, and they will go off quietly [i.e. die] without any medical attendance whatsoever... Of those that died in a state of destitution without medical relief, some of them stated to me that they had applied for orders, and been refused them, and they could not afford a doctor themselves. When they came to register the death, they [bereaved families] stated that [i.e. death from medical neglect]. I kept a note of the cases at the back of the death register [i.e. dissection cases].[17]

There was a lot of disagreement between local guardians about Rosebrook Morris's statement. He therefore produced some statistics to prove his point, reconstructed in Table 5.1. The surgeon claimed these were bodies that became available for dissection due to inadequate medical care. In the Brixworth Union there were 33 parishes and a total population of 13,371.[18] Of these, 2,736 destitute persons were said to have died from medical neglect in the eight years between 1836 and 1844. The surgeon could not control for the aging profile of the population but he estimated that the death toll was 'thirty deaths a month for eight years' (that is, 91 deaths a quarter or 364 deaths a year). He thought that this was too high for the area. There were of course sound financial and professional reasons for a local surgeon to

Table 5.1 Mr Rosebrook Morris the Surgeon's Death Register: Deaths Due to a Lack of Medical Relief in the Brixworth Union, c. 1836–1844

Date	Indoor Deaths	Outdoor Deaths	Total Deaths
1836–7	13	117	130
1837–8	91	115	206
1838–9	156	87	243
1839–1840	209	159	368
1840–1	219	167	386
1841–2	252	181	433
1842–3	248	213	461
1843–4	260	249	509
Total	1,448	1,288	2,736

Source: Northamptonshire Record Office, ZA2037/8, 'Mr Rosebrook Morris, surgeon, medical case books and death statistics, 1836–1844'.

make these allegations. It was in his interests to stress his heavy workload and the need for better medical standards. Nonetheless the fact that Rosebrook Morris was prepared to be so outspoken, potentially damaging his client-relationships with guardians, and that he produced his case books for public inspection, suggests multiple readings of his true motivations. It is noteworthy that he was neither evasive nor closed medical ranks, and hinted heavily about the dissection of the destitute. And this at a time when the New Poor Law was still getting underway, and is generally regarded in standard historical accounts as more generous to the poor in the provinces because of a fear of political unrest against the Anatomy Act. As a medical man he had moreover every reason to treat dissection as a taboo subject. Yet he was living and working among the poor. He was troubled by the medical experiences of his pauper patients in a catchment area so close to Cambridge.

Rosebrook Morris was concerned that the progress of medicine was being impeded, not improved, by the Poor Law. Paupers were returning to quack remedies, cheap nurses, and cunning women for help.[19] However one chooses to interpret the evidence, the fact that the dispute was taking place in public reveals that there was genuine concern in this surgeon's medical circles about the plight of the sick poor in local life. Across the county, the destitute were felt to be more vulnerable in death under the combined pressure of the Anatomy Act and New Poor Law. It was perhaps, though, a small newspaper report in the local *Northampton Herald* that eloquently summarised the voice of the poor. The friends, kin, and neighbours of a Bozeat pauper who feared dissection held a funeral raffle to make sure John Partridge was buried intact. His passing, and that of his wife, Mary, two years earlier, was marked with a small headstone. It was carved with a biblical curse. The inscription was Partridge's favourite saying and a dire warning from beyond the grave. It threatened anyone who tried to dig up his human remains for dissection:

In memory of John Partridge, who died March 30th 1840 aged 67 years
Also Mary Partridge, wife of the above, who died October 1836 aged 68
May all the afflictions of Job be his lot
that disturbs the remains
of those that repose below,
J. P.[20]

In English provincial life, Poor Law records show that controversies were frequent in the Midlands throughout the Victorian period. Over time, the tensions between medical services and the anatomical needs of medicine often worsened. Coroners' cases are an important litmus test of local medical sentiment and yet, until recently, the social history of coroners has been neglected by historians of medicine.[21] This oversight means that the perspective of the poor has been understudied. Their witness statements and testimony in coroners' courts betray the true social cost of poverty. Throughout Northamptonshire and those places on the border with Huntingdon there were a lot of sad circumstances that brought bodies into the dissection room at Cambridge. One case in the records stands in for many at the time. It elaborates that Rosebrook's criticisms were sometimes merited and often based on tragic circumstances.

On 9 March 1841, Lucy Ladds of Peterborough died in childbirth, having been attended by a local woman. Like many paupers, Lucy had no money to pay for a proper doctor and did not want to enter the workhouse. A Surgeon-Apothecary, William Sadler Esq. of Peterborough, told the coroner that he was called to the house of Joseph Ladds 'about 40 minutes before eleven on Friday night last, 9th March'. There he found Lucy Ladds in labour 'in a deplorable state'. He explained:

> There was a great quantity of blood about the floor and the bedclothes were exceedingly bloody. Her countenance was exceedingly pallid and she was moaning very much – she exclaimed frequently – *I shall die – I shall die –* I immediately turned up the bedclothes and saw an inversion of the uterus ... the woman continued gravely – her pulse was then like a thread. She died a few minutes before 11 ... I have no doubt that the inversion was produced by an improper, forcible and untimely exclusion of the umbilical cord. I further think that if the case had been left to the natural efforts of the patient the afterbirth would have been expelled and the patient would now be living – a very skilful effort from a skilful hand would have removed it and left the uterus in the maternal position.[22]

Lucy Ladds' baby had a large head that got stuck in the birth canal during labour. The umbilical cord was also tied around its neck. Joseph had no money for a doctor and he begged his mother, Mary Ladds, to help. She was a local cunning woman who often earned extra birthing-money. Mary pushed down on Lucy's stomach with a lot of force to try to squeeze the newborn out. The infant was born, breathed, but the afterbirth did not expel. So Mary pushed and pulled at the umbilical cord hoping to get things moving but the uterus came away with it. Lucy haemorrhaged badly and bled out on the bed, some major blood vessels severed by the tearing. Mary Ladds told the coroner:

> I am in the habit of going out as a Nurse – have been for many years. I had attended the deceased in her confinement seven times. Had no premium but did it out of kinship. I have been a Nurse 20 years and have had nine children myself. I never met a case like this – I observed that the afterbirth was rather hard. She said *it was the last time*. She was very cheerful at the birth. I was nearly half an hour going there. I left my house at eight – I set her up again at quarter to nine – the child was born about half past nine. It is a fine girl – it was 10 minutes or a quarter of an hour after the birth of the child I removed the afterbirth – something I did not understand came at the same time...The afterbirth required more force in the removal than usual,
>
> The mark of Mary Ladds – X[23]

Lucy Ladds' womb had prolapsed with fatal consequences. A frightened Joseph Ladds ran for the surgeon but to no avail. The coroner asked Mr. Thomas Walker, another surgeon, for a second opinion. He did a post-mortem and concluded that 'the excessive loss of blood was the cause of death – it is hardly possible that the inversion could not have taken place without violence'. The verdict at the conclusion of the case was: 'That death was caused by haemorrhage produced by inversion of the womb which was occasioned by unskillfullness [sic] on the part of Mary Ladds midwife who attended the deceased in her confinement'.

Whatever the rights and wrongs of this case, it is clear that the physical evidence was damning and the medical men closed ranks. In mitigation, Mary did not set out to kill her daughter-in-law, Lucy. She was thus reprimanded but not prosecuted. Prior to the Medical Act it was difficult in any case to prove medical culpability. The coroner judged that the best way to deal with Mary Ladds was to give her a bad reputation. In the future most women would avoid her birthing services. That outcome did not though resolve the tragic plight of this poor family. A death in childbirth for a man with seven children to feed was serious. Lucy's corpse was, like so many, a commodity, a financial opportunity to make some money in dire circumstances. Peterborough was on the train route to Cambridge and as this chapter recounts childbirth fatalities like this populated the dissection table there. Bodies like Lucy's were often sold on. It really depended on a coroner's cooperation, the sentiment of the Poor Law, and the level of organised resentment locally. Anatomists had to be realists, which is why by the time Alexander Macalister came to Cambridge in 1883 he faced an uphill task. It was essential to undertake regular lecture tours to convince guardians that selling on cadavers was a national duty. In Cambridge's catchment (Northamptonshire, the Midlands, and East Anglia) the poorest had grown to resent body-buying schemes that targeted their loved ones. Macalister soon launched a strategic publicity campaign.

It was fortunate that Macalister enjoyed railway travel, as well as walking. For the next 37 years, he visited nearly every small town and village in East Anglia, journeying across the West Midlands, travelling up to Manchester and Hull, alighting as far down as Brighton and Southampton. Later this chapter will examine the geography of his supply relationships in more detail. In the meantime, his working life was a perpetual round of generating and regenerating anatomy supply deals. The master of one workhouse might help for a time but expected his supply contract to be renewed at a later date. Macalister's papers show that he had to tip porters, coroners' mates, Dead House staff, and undertakers. It was a small-minded business, funded by petty cash. Costs were constantly rising because of the complicated market legalised by the Anatomy Act. One of Macalister's first train journeys was to Chelmsford Union. A private letter to Macalister, dated 19 May 1885, confirmed their cooperation: 'I am now directed to inform you that at the meeting of the Guardians today it was resolved to comply with your request to send you the bodies of such persons as are unclaimed by friends or relatives after death in accordance with the Act of Parliament upon the matter' wrote the 'Clerk, W. W. Driffield Esq'.[24] Macalister had taken the wise decision to prepare a circular dated 9 October 1884, which he sent in advance of his visits around the country. In it, he justified the need for more dead bodies of the poor. There were seven points that he always stressed to guardians in his public engagement work.

1. We do not ask anything unusual...more than forty other towns send their unclaimed bodies to Medical Schools...
2. It is a statutory request...
3. It is the only way in which Candidates for Medical Degrees can learn Anatomy & Surgery...
4. It does not involve any disrespect to the dead...
5. It does not involve any expense; on the contrary, the University Anatomical Department pays all the incidental expenses, coffin

&c., carriage to Cambridge in a proper railway funeral wagon at the usual funeral rate (1 shilling per mile) and the Undertaker's & Cemetery expenses & fees here. Thereby the ratepayers are saved all funeral expenses in these cases.
6. The Medical School of Cambridge is the largest and is rapidly growing...It is reasonable to expect that more distant places should help in our efforts to make our education as practical & good as it is possible to make it.
7. It is for medical advancement...if students cannot practice on the dead they have to gain experience and dexterity from the living.[25]

For a time Chelmsford was susceptible to these arguments but later in October 1912 Macalister had to go back and start negotiations afresh. He wrote: 'Promised a long time ago (Master of the Workhouse does not answer letters)'.[26] The profile of the board of guardians had changed in the interim. Five guardians were unhappy about the anatomy business and the master exploited their misgiving to increase his supply fees. At the same time, Macalister also made some private notes about the nature of the typical supply problems he faced. Those covering the 1880s provide important insights into his working life and his struggles to expand the business of anatomy.

Macalister recorded that the supply of cadavers was not keeping pace with rising student demand after the Medical Act (1858) and its extension in 1885.[27] He was under a lot of pressure to increase his pupil numbers because during the agricultural recession of the mid-1880s the endowment income that most colleges got from their agricultural rents plummeted.[28] Increasing fee-paying students was a practical solution to cover costs. Macalister needed, however, a better economy of supply to balance the books. His estimates show that: 'For example, if there were 188 students, [they would need] 56 dissecting hands, 40 arms, 32 legs, 32 abdomens, and 8 Thoraxes'. The problem was that there were just '8 students per body' in 1883. Consequently, pupils could only dissect 'every second day' on the '45 cadavers' in each teaching cycle. He explained:

Directions in which reform is necessary[29]
1. The utilisation to the full of all material available
2. The systematisation of the practical teaching
3. Some method of checking the attendance & work

186 *Dying for Victorian Medicine*

Let us suppose of the	56 Heads,	24 wish to dissect in the morning
	40 Arms,	16 wish to dissect in the morning
	32 Legs	8 wish to dissect in the morning
	32 Abdomens	16 wish to dissect in the morning

No of classes:

10–12am	Groups	
3	A, B, C	Most of these would be every 2nd day
2	D, E	ditto
1	F	ditto
2	G, H	ditto

At 2–4pm	No of Classes	Groups
16 Heads	3	D. E,
16 Arms	2	A, B
16 Legs	2	C
26 Abdomens	2	A, D
8 Thoraxes	1	A

At 4–6pm	No of Classes	Groups
16 Heads	2	A, B
8 Arms	1	A
16 Legs	2	B, C
0 Abdomens		
0 Thoraxes		

These class rotations reveal three key difficulties that Macalister seldom admitted in public.

First, Macalister's body-supply scheme was a constant business headache.[30] He tended to underplay this problem for political reasons. At the time he was arguing for more resources from university and college funds. He did not therefore wish to reveal the scale of the problem that he faced. He needed to put a positive spin on his rising student numbers. Secondly, he relied a lot on body parts, not whole cadavers, to train students. Again he preferred not to discuss this in detail to avoid any bad publicity. Those conversations could not only exacerbate his local body-supply rates but also undermine his financial standing in the university. When he bought complete corpses he worked within the

Anatomy Act. Nonetheless, his staff for some time had also been acquiring human material from operative surgeries, paying porters for amputations. These were, for instance, bought from the back door of the Old Addenbrookes Hospital. They were never returned on the invoices sent into the Anatomy Inspectorate because it was not a legal requirement to do so. Strictly speaking, Macalister's annual supply figures were consistently under-recorded but he judged this a necessity in view of student expansion. In 1883, for example, he had '53 heads' but only '45' bodies. Macalister was very interested in brain research and therefore paid extra to improve supplies when a brain became available after post-mortem in the hospital (more of which later).[31] The dissection room was then thirdly a busy space, but training standards were questionable when he first arrived in Cambridge. In his private notes, Macalister regretted the 'wasteful use of human material'. He commented that 'more than half is sheer waste' and this had been the case for too long. The students drifted in and out of the dissection room. Corpses and body parts were not labelled properly and there was consequently 'no regular instruction in method'. He described the general teaching standards as 'desultory and unmethodical'. His greatest fear was that if these basic problems were not resolved then medical students would either go elsewhere or find conditions so repugnant that they would leave the profession before being qualified. Macalister knew he could expand student numbers but whether he could retain them all was another matter.

In Macalister's opinion, only concerted hard work would improve student retention. It was imperative, he wrote, that they avoid 'the effect of driving men into the hands of coaches'.[32] What he meant by this aside was that trainee doctors would sign up at night with private anatomy schools for extra lessons and his income would correspondingly drop off to the detriment of the department's long-term financial viability. Or some students might literally throw themselves under the stagecoach, driven to suicide by a bloody mess in the dissection room. Victorian students dubbed the dissection room 'The Meat Shop'.[33] Here demonstrators in anatomy were known as 'Meaters', working in the macerating room. This was where bodies were cut, parts weighed, and then immersed in carbolic acid tanks. That human material was then prepared for daily instruction. It was an untidy preparation area and off-putting for sensitive students. At the start of Macalister's career, he too was confronted by corporeal realities. There was always a lot of competition for corpses. Men ran down corridors when the latest body was delivered at the back door of the anatomy school. In the ensuing scrum, cadavers were often cut clumsily. This was why throughout Macalister's career he stressed that dissection should be an art form, a craft learned by repeated skilled use of the lancet on the dead body. Macalister during his 12 years of medical training calculated that he 'had taken part in the dissection of over 690 subjects, tedious and articulate'.[34] On average he studied the anatomy of some 58 bodies in an annual teaching cycle in Ireland. This was an exceptional amount for one medical student. It reveals not only the scale of regular supply and the background poverty conditions in Ireland, but, above all, Macalister's dexterity and determination to be the best in his field.

Macalister, like all anatomists of his generation, was conscious of the cultural divide between medicine and the poor in Victorian society. In his public pronouncements, he tended to use typical paternalistic statements that stressed the value of anatomy to the advancement of medicine.[35] In private, however, he often debated the unenviable position of anatomists engaged in a tug-of-war with the poor over their loved one's bodies. A representative letter of common attitudes survives among his confidential paperwork at Cambridge. It was penned because Macalister secured a liberal supply deal with Hull Poor Law Union (discussed later). He befriended an influential master, Mr. H. Jenkins, and they corresponded about the cultural dilemmas that dissection raised. Jenkins wrote in response to a personal visit by Macalister. In June 1895 he conceded that, 'I may say, I do not think that the bodies of the poor ought to be treated the same as the bodies of the rich'.[36] This attitude to pauper funerals reflected the fact that most ratepayers resented the regular claims on local taxes from non-taxpayers. In this, Jenkins' views were those that Macalister encountered everywhere and which he tried to exploit in the course of his anatomy business dealings. Nonetheless Jenkins revealed some more about a discussion that he and Macalister had enjoyed face-to-face. The difficulty was, he wrote, that it was hard to deny that 'the friendless ought to have as much consideration as those who have a circle of friends'. Jenkins had debated with Macalister the social cost of the anatomy trade and whether in fact medical students should donate their bodies as a matter of course. He expressed the view that it would counter the general criticism that medicine simply took from the poor without some equivalent gesture. As Jenkins summarised,

> And, if the students who claim these bodies in the interest of rightness and science, thoroughly believe what they preach, would it not be easy for the students to convince the public, by they themselves agreeing that, they – each and all students – shall be compelled,

to sign an agreement giving the School of Anatomy full Authority to use their own body in the like manner upon their death. This, I believe, would be just and satisfy the great majority of people that it is a necessity. Trusting you will see the justice of these ideas, etc.[37]

Macalister did see the social justice but he was a realist too. It would take a generation for enough medical students to die and donate their bodies to science. The poor were needed now and in more numbers. Before looking in more detail at the perspective of the poor, this chapter examines the actual economy of supply at Cambridge to set those pauper life stories in context. For Macalister and his colleagues established a business of anatomy that many regional medical schools copied in England during Victorian times.

The dead boxes on the dead train to Cambridge

Three times a week an express train left Liverpool Street station in London.[38] It travelled via Cambridge to Doncaster. On its return journey extra funeral wagons were attached discreetly by railways engineers to the rear carriages. They looked innocuous: simple wooden slats, straight sides, and windowless. Normally train passengers witnessed butter, milk, cattle, pigs, and vegetables being loaded up. The produce had come from the wholesale meat markets and market garden outlets in the North to be sold on in London. Often coal was transported too. Inside these funeral wagons were stacked 'dead boxes'. This was the 'dead train' that carried corpses to Cambridge.[39] A local undertaker brought the human cargo to the back of the station in a covered carriage. The design of each 'dead box' was the same for everyone on their way to be dissected, a basic double coffin enclosed in an elm or cheap pine chest. Few body fluids or noxious fumes leaked into the small cavity between the double layers of the wooden box, each containing one cadaver. In the early days of the anatomy trade, bodies were preserved in turpentine before chemistry improved. Macalister recalled in a lecture to medical students how basic standard preservation techniques had been in the early nineteenth century:

> Dissection...was usually hastily done, and restricted to the general topography of the viscera and to the cleaning of muscles and vessels. The latter could be injected easily, and the injected preparations could be rendered permanent by steeping in turpentine, drying, and varnishing. Preparations of this kind were kept in most medical schools,

and hence a fairly good knowledge of the courses and branchings of the arteries was easily acquired, and constituted a large part of the anatomy which students had to learn. Veins and lymphatics were little noticed, except such of the former as were superficial and the peripheral nervous system occupied but a small part of the course.[40]

In Macalister's time, it was cheaper and therefore common to immerse cut-up cadavers in carbolic acid tanks. The pungent smell soon spread through the dissection room. So much so that most medical students were chain-smokers.[41] The thick cigarette smoke and pipe tobacco was said to be eye-watering in the crowded anatomy department. Cambridge students recalled that it was very difficult to see through the dense atmosphere to the dissection table. It was a top priority therefore to transport cadavers speedily because students paid a premium for the freshest.[42] There was always a rush to get the latest bodies onto the fast train to Cambridge. 'The faster, the fresher, the better', went the saying in medical circles, before tissue decay rotted the corpse.

When Alexander Macalister arrived in 1883 he unpacked personally the dead boxes of 'Elizabeth Smith aged 47, Harriett Bubb aged 65, John Norman aged 32, and Peter Hogg aged 70', all despatched together on the train from Finchley in North London.[43] Later they would be buried in the same common grave in Mill Road cemetery in central Cambridge. There they repose still with some 2,953 bodies, acquired for dissection by the medical school between 1855 and 1920. Graph 5.1 shows the peaks and troughs of the anatomy trade at Cambridge.

It is important to appreciate that these supply rates were an impressive achievement on the part of leading anatomists like Macalister given just how basic the old dissection room at Cambridge University was (see Illustration 3.2). A late-Victorian description in the *Cambridge Review* described it as:

> A deal-lined shanty with corrugated top; the three cracked old stoves that, like veteran gossips, have brought the art of getting much smoke from little fire to five o'clock tea perfection; the row of basins, the tallow soap, and o'er hanging taps, flanked by the clammy towels; the coloured diagrams; the shelves and boxes; that high tier of window and the skylights that give copious access to draughty air; the skeleton gas jets in painted array; the stools; and then the stands whereon the dishevelled remains of our unpreened and ungarnished humanity.[44]

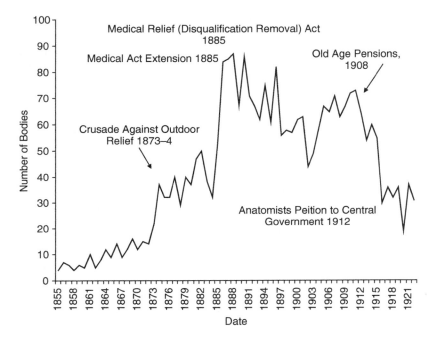

Figure 5.1 The Number of Bodies Sold for Dissection to Cambridge University, 1855 to 1920.
Source: Cambridgeshire Record Office, St. Benedicts burial records, PR25/21-23.

Mark Weatherall's meticulous work on medical student registers at the Old Addenbrookes hospital shows that 'over 3000 students...studied medicine in Cambridge in the nineteenth century'.[45] He has retraced the '42 students who registered with the GMC in 1873, the 70 in 1882, and the 125 in 1897'.[46] These figures reveal that so-called '*Medicals* comprised 1% of the annual intake of approximately 400 students in the 1850s; by 1900 they made up about 10% of the intake of over 1,000 students' at the University.[47] At least 5 per cent of these students needed access to cadavers over a two-year teaching cycle to qualify in human anatomy under the GMC regulations. This meant that in rotation every year 50 students needed 200 bodies to dissect for the next 24 months. Yet Cambridge acquired a total of 2,953 bodies between 1855 and 1920. There was always going to be shortfall in body supply given the expanding student numbers in Victorian times. As Weatherall points out, this was why 'the wit Francis Cornford' mocked human anatomy teaching as a science in which '*two* must share an arm or leg/ And *four* take head and neck'.⁴⁸ The actual economy of supply figures show that the business of anatomy moved in phases. It was connected intimately to the Poor Law. Four trends should be observed in the burial registers used to reconstruct the chain of supply of Cambridge anatomists.

At first glance the supply figures seem to suggest that anatomists had lower expectations when they were getting their business of anatomy underway. This outcome appears to confirm what for instance Ruth Richardson has observed in the patchy returns of provincial medical schools sent into the Anatomy Inspectorate in the 1840s.⁴⁹ Namely that supply was lower than expected, there was local resentment, and anatomists conducted themselves with decorum in provincial life. Yet this is a picture that should not be taken at face value. Delving deeper into the archives to reproduce a synthesis of the sources, it becomes apparent that behind the crude figures were more shady deals to get the business of anatomy established. In April 1847, Cambridge anatomy school, then in its infancy, had applied to the Anatomy Inspectorate for a license to authorise dissection.⁵⁰ The reason that it did so was because it was ambitious. Staff wanted some official backup in the event of its real supply chain coming to public attention. For some time bodies had been acquired from legal and illegal sources, sometimes publicly, more often by sleight of hand. Returns to central government were a masquerade.

The anatomical school at that time was led by Professor G. M. Humphry FRS (1866–1883, later of Chair of Surgery). He was assisted by T. H. Sims. Humphry recalled that Sims was 'the most truthful liar he had ever known'.⁵¹ Throughout the 1850s Sims had developed a network of street deals in the city that were not shared with central government.⁵² By way of example, when 'Edward Alexander Moyes' died at just 'two months old at the Lion Hotel', a lodging house in central Cambridge, on '7th May 1863', the anatomy school was on hand to make a discreet supply deal with the landlady and bereaved family in return for the infant's body.⁵³ Tragically, Sims came back to collect the cadaver of the child's mother, 'Mary Ann Moyes', who died seven weeks later, 'aged 27', having never recovered from her childbirth complications. She was buried on the order of Sims on 1 July 1863. Like Lucy Ladds, the Peterborough case earlier, her family had no money for a pauper funeral. In life and death, they had to be pragmatic. This meant that when 'Arthur Robert Moyes aged 32' died alone in the 'Bath Hotel', another lodging house, there was nobody left to pay for his pauper funeral. In the winter of 1893, Arthur was still living in the city close to where his mother (Mary Ann) and sibling (Edward Alexander) had died 30 years earlier. In the meantime his father seems to have disappeared leaving behind a grandfather, 'John Andrew Moyes'. He died, was dissected, and buried on '20th October 1888 aged 78'. Arthur was not a lucky man in family matters. Sims arranged for his interment after dismemberment on '22nd February, 1893'. He was laid to rest beside his near relations in an unmarked grave. The scattered fragments of his life in forgotten archives are now the only surviving evidence of the journey that the extended Moyes family made from the grip of pauperism through the dissection room door, and beyond into a history of silence. Sims likewise bought on the street the body of a typical tramp, 'James Bradbury', a man of unknown address, later buried on the '5th April 1862, aged 29' on behalf of the 'Anatomical Schools'. Meanwhile, the Anatomy Inspectorate returns show that Humphrys got the official agreement of Addenbrookes hospital to deal with Sims in 'December 1846' and then Cambridge Union made a similar commitment in 'July 1855'.[54] Nonetheless, street deals were an intrinsic part of the anatomical school's regular supply network that Sims pioneered. The official supply figures in Figure 5.1 for the 1850s and early 1860s should therefore be read with that caveat in mind. Official and unofficial deals were intrinsic parts of the plans for expansion at Cambridge.

Rumour soon spread that the anatomical school had obtained the agreement of Cambridge Union to exploit the poor in death. A cartoon was handed out in the poorest districts of the city. It depicted a butcher's shop sign: *"Purveyors of Paupers – Bodies to the Corporation of Cambridge'* (Illustration 5.1). It has been wrongly attributed to local protests about baby-farming.[55] The leaflet was designed instead to rouse the mob against the Anatomy Act in Cambridge.[56] Piecing together local evidence suggests that the cartoon can be dated to 1858. That summer the Medical Act was being debated in Parliament. Cambridge lost an important source of supply from the prison hulks floating on the River Thames because of the fear of social unrest in London. The cartoon anticipates both the greater demand for the bodies of the poor and the expansion of medical student numbers. Local people feared that the recent deal made by the anatomical school in 1855 with Cambridge workhouse would now come into force. In the caricature, Humphrys is depicted grinding down the bodies of children for mincemeat. Again this is a contemporary reference to the 'Meat-shop' (anatomy school) run by the 'Meaters' (demonstrators) of the macerating room.[57] It was well known that bodies were bought on Slaughterhouse Lane opposite a side entrance to Downing College

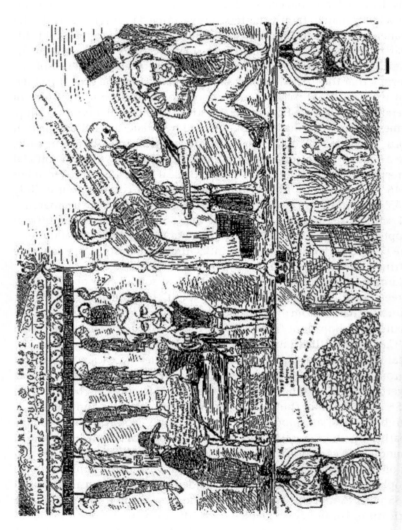

Illustration 5.1 ©Cambridge Library Local Studies Room, Cambridge cartoon, anti-anatomy, 1858.

where the anatomy school was located. There was a weekly meat market in the street thus its apt name. The poor were also angry that a local rector was helping Humphrys. A churchman is depicted saying, '*I say cocky, what a happy thought to 'convert' Pauper kids into Breakfast Dainties for the hot weather – I'd like one of those myself*'. It was common knowledge that Church of England clergymen profited from the Anatomy Act. After all, they buried what was left of the poor after dissection and dismemberment for regular interment fees. Sometimes a Roman Catholic priest conducted the basic service but this was less common in Cambridge. Looking up, above Humphrys' head, hang pauper bodies on meat hooks. Each has a label indicating their market price: a standard rate for a whole male body (£2 2s 0d), more for a woman (£3 3s 0d), less for a cheap old man (£1), and just '6d per leg'. These people are: '*To be Broken before Breakfast: Fraser's Patent for Reducing the Poor Rate*'. By the 1870s, the economy of supply figures shows that there was a lot of truth in this cartoon.

The second key observation in relation to Figure 5.1 is that Cambridge anatomists relied on the late-Victorian Poor Law.[58] At the start of the crusade against outdoor relief, when in most places small doles paid to paupers at home were cancelled, body-supply rates rose exponentially. Previous chapters have already introduced this common theme. This case study's findings reiterate those at St. Barthomolew's, namely that Cambridge's supply rates rose 100 per cent in 1873–1874. The needs of anatomy, medicine, and welfare converged. To put this trend in perspective, in 1877, of the 40 bodies bought from workhouses, almost half, 19 cadavers, came via Cambridge Union. These were pauper patients, emergency admissions in most cases, sent by the guardians to Addenbrookes Hospital for treatment. The elderly tended to die on the workhouse premises, whereas younger patients were transferred to hospital when very sick. Thus in a common pauper grave were buried together after dissection 'John Jones aged 62, John Christian aged 70, Jonas Collen aged 22, Mary Anne Browne aged 42, and William Wife aged 59' in early May 1877.[59] Multiple burials were the norm and this chapter will return to that theme later when it examines the geography of the economy of supply. Meanwhile the start of Alexander Macalister's tenure brings us to the third phase of the anatomical school's workload.

Returning to Figure 5.1 it is apparent that supply rates had become more sporadic when Macalister arrived in Cambridge. Figures show that between 1873–1874 and 1884 there were on average 44 cadavers bought annually for dissection. In 1884–1885 this fell to just 32 and that was why Macalister literally took to the road and railway to promote human anatomy teaching. He witnessed fundamental economic changes wherever he travelled. In most small market towns and cities there was a trade downturn that coincided with an agricultural recession.[60] Faster steamships brought cheaper grain and meat from the Americas, Australia, and New Zealand. Poor weather conditions at home produced bad harvests at a time when farmers found it harder to compete in markets flooded with international produce. The cost of living fell but then unemployment rose too. What this meant was that the poorest pressed guardians to relax the crusade by reintroducing outdoor relief. Many did for a short time and that trend accounts for an 18-month dip in supply at Cambridge.

Everywhere that Macalister went in the course of his public engagement work he kept stressing the need for bodies and new procedures that respected the dead at Cambridge. His private notes detail his draft speeches. They state that 'each student to have a dissecting card, written divisions for each day to be indicated by the Demonstrator – no one to be signed up for dissection when he has not attended for 3–4 available days'.[61] Rather than students working on random body parts, the 'teacher is to direct and demonstrate and explain the whole dissection and process'. To be more methodical, bodies that were bought were 'not to be touched for a week'.[62] He reassured guardians that they were kept 'without any injury except for a small incision in the neck to allow the tube [filled with preservation fluid] to go into the carotid artery'. If any mistakes were made bodies were sent back on the next train. These arguments were heard but it was the financial incentives that really interested guardians because the Poor Law was going through another policy review.

The fourth key trend in Figure 5.1 is that in 1885–1886 supply rates were 200 per cent higher than in 1873–1874 figures. Cambridge, like St. Bartholomew's, benefited from the passing of the Medical Relief (Disqualification Removal) Act (hereafter MRDRA) of 1885.[63] Again we have already encountered the significance of this minor but important administrative change to Poor Law rules. In summary, guardians were angry that the poor on medical outdoor relief were to be permitted to vote in parliamentary elections despite their non-taxpaying status. As a result, paupers were told they must come into the workhouse once more. Pauper funerals on medical orders were cancelled, and generally regulations were tightened up again. At the same time, Macalister stressed that the medical school would continue to cover all funeral costs and pay a supply fee for those bodies given up for dissection. The burial registers confirm that because of the MRDRA the geographical reach of the business of anatomy expanded significantly after 1886. And this was timely since the Medical Act had been extended that year driving up medical student numbers again. Macalister knew that the profession had become 'crowded' and was 'much-changed' and this would necessitate generating a lot more cadavers after 1885. In a speech to medical students in 1908, he remarked:

> Students of today have no such primrose path to a qualification. Each decade since [the Medical Act] has witnessed the lengthening of the course, an increase in the number of subjects of examination and a greater stringency in the standard required. The development of the curriculum is a Darwinian process of evolution and the result is the survival of the fittest as regards both the students themselves and the subjects of examination, in the more strenuous struggle for life.[64]

The assumption was that this struggle for a medical living would mean relying on more of the poor failing at life. The economy of supply figures show how medical reforms kept converging with New Poor Law orders. Together they brought more bodies by train from a greater variety of welfare institutions to Cambridge, notably after 1885. At one group burial in March 1887 – a typical example – three women and four men came from 'Luton, Whittlesey, Fulborn, Hitchin, Leyton, Doncaster, and Cambridge city-centre' respectively.[65] They were interred in a single pauper grave, stacked in parish coffins, one on top of another, and covered in quick lime. Sampling local newspapers adds further historical detail to this fourth trend.

The *Hull Packet and East Riding Times* in February 1885, for example, alerted readers to 'the question of sending unclaimed bodies to Cambridge' from Hull Poor Law Union.[66] Local people were angry that paupers were to be '*hacked and cut about by a set of sacrilegious medical students* [sic]'. It was said that guardians were 'befriending every friendless carcase which may be laid at the door of the union'. The newspaper editor recounted that the Sheffield school of anatomy had been run 'twenty-five years ago' by 'Dr. Craven' but was closed down after being attacked by the mob. There was evidently strong local feeling on the subject that had not abated. Nonetheless, the editor commented that ratepayers thought it was 'a pardonable action' to sell dead paupers 'on the part of Guardians of Hull'. It was no longer unreasonable to do so, given the downturn in the economic climate. The recent MRDRA meant that local people now had time, said the editorial, to 'reflect that the expenses of ... *the pauper's whom nobody owns* [sic] ... his burial will have to be borne by them. ... Until funeral reform steps in, our undertakers bills are not very *reasonable* [sic] either'. The following month at Luton in March 1885 similar views were expressed. The *Essex, West Suffolk Gazette and Eastern Counties' Advertiser* reported on 'The Dissection of Pauper Bodies' locally. Luton Poor Law Union had:

> decided by a majority of two to grant the application of Professor Alexander Macalister that unclaimed bodies of paupers should be sent to Cambridge University Medical School for anatomical examination before burial. The subject had already been twice discussed with an adverse result to the application; but the majority now pleaded that the concession was in the interests of medical science.[67]

The MRDRA had convinced Luton guardians to revisit the scientific arguments in the light of pauper funeral expenses, despite the fact that previous motions had been thrown out twice before. They now adopted the exact language of the original Anatomy Act – bodies being sent for 'anatomical examination' (see Chapter 1) – and closed ranks when questioned about what this really meant in terms of dignity. What happened in most places – liberal, moderate, and extreme Poor Law Unions – is that more bodies were sold. The economy of supply was now more diverse. Examining pauper histories uncovers how once more ethnicity played a factor in many anatomy sales. So, for example, two representative cases were 'Maria Assunta Salvatore, a 48 year old' Italian pauper from 'Finchley'. She was dissected and buried on '3rd June 1887'. Maria lay near 'Mary Sweeney aged 51', an Irish pauper, again from 'Finchley', interred on '15th June 1887'.[68] Another noteworthy finding is that the number of children sold on for dissection increased too. Naturally the anatomical school was always very sensitive about this aspect of its work.

In 1886, the anatomy school welcomed the purchase of three dead children because they had been difficult to procure. 'Richard George Campling', an infant from 'Biggleswade Poor Law Union', just 'one and a half years old', was sold in 'July' and arrived by train. 'Sarah Jane Parker Union' an infant of '8 months old' was bought from the premises of '110 East Street' in central Cambridge during late August. 'Bernie Prime' aged '5 months old' came via 'Addenbrookes hospital' in the autumn and was buried on '10th November' after dark.[69] All of the children were interred by the same clergyman to avoid any bad publicity and they were laid to rest in a separate burial plot from the adult bodies. The burial records reveal that Reverend J. T. Lang officiated at the burial of 80 per cent of the infant human material used for dissection between 1855 and 1920.[70] He was a senior clergyman and tutor at Corpus Christi College. Even after he retired from the living of St. Benedict's church in 1897 he continued to play an active role in anatomical interments. In Mill Road cemetery each parish church in central Cambridge was given a separate burial plot to inter the dead because local graveyards were overflowing by the mid-Victorian period. This meant that the anatomical school could redistribute the dissected bodies of children among three plots close to their main burial site for adult corpses. In Mill Road cemetery, children were thus laid to rest in the plots of St. Andrews the Great, St. Edward's, and St. Mary the Less.[71] Adult bodies were always interred in the area set aside for St. Benedict's, the local parish church closest to the anatomical school where Lang had his living. In the case of children, there was evidently need for continuity, discretion, and subterfuge.

The basic demography of the economy of supply shows that it was nonetheless always more difficult to procure children and young women for dissection. Between 1855 and 1920, the ratio of men to women was three to one (1,971 men, 978 females – four cases are recorded as 'unknown sex'). The majority of bodies were over 50 years of age and these findings evidently contrast with the St. Bartholomew's case study. The main difference between London and provincial medical schools was the age profile. Figure 5.2 gives an overview of pauper age at death in Cambridge.

Chapter 4 explained why men rather than women were sold on for dissection first by poorer families in provincial life. Men were often worn out by 50 after a lifetime of field work. They had fewer transferable household skills to contribute to family economies by doing childcare. Cambridge was surrounded by agricultural districts and this was why more men were available than women. Yet the raw data should not simply be taken at face value. In reality, the source material indicates that anatomists were always keen to buy children and women at a time when embryology, laboratory studies, and pathology were coming into vogue. This context explains why in the 1890s anatomists were so keen to take on 'Colney Hatch Asylum' as a new supplier. Four women – 'Mary Thomson (46), Catherine Ford (68), Bridget Mardin (58) and Mary Holden (81)' – were sold, dissected, and buried at the expense of the anatomical school in 'June 1892'.[72] Macalister paid up to '£12 a body' to get the new supply deal underway, securing female bodies on

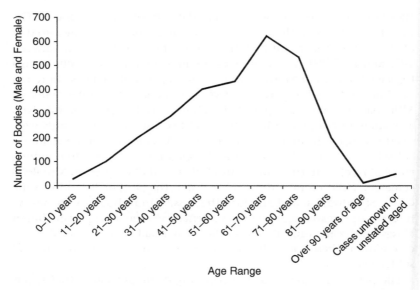

Figure 5.2 The Age Profile of the Bodies Sold to Cambridge Anatomy School, 1855–1920.
Source: Cambridgeshire Record Office, St. Benedicts' burial records, PR25/21-23.

a regular basis. He also favoured paying incentives for children so that students could study embryology without the degenerative changes of old age. A child like 'Evelyn Lace Stanley' who lived for just '15 hours' was usually entered in the *'Abnormalities and Deformities Register'*, setting out the origin of the deal – in this case a street transaction – she was bought from an intermediary on St. Benedict's Street in central Cambridge. During dissection, bones, tissue under the microscope, sections of the brain, and tiny sex organs were all studied by students, the demonstrator and other interested colleagues. Burial took place on the '28th April 1900'. The same was true of 'Gertrude May Firr' who died aged '5 months' from an 'unknown cause' and was buried on '30th August 1900'.[73] Figure 5.2 shows that age shaped the economy of supply. Nonetheless the private papers of anatomists also set out that they preferred to self-select females and children for dissection. In reality there were fewer opportunities to do so.

Returning briefly to Figure 5.1, the fifth and final trend worth observing in respect of the economy of supply is that the coming of democracy made it much more difficult to improve supply rates.[74] This happened a lot after the passing of the Local Government Act (1894) which abolished all property qualifications and brought full democracy to the Poor Law. Soon after, Macalister found himself embroiled in a number of unwelcome controversies. Of these, events at Leicester and Nottingham are instructive on wider body-sourcing problems. It happened that the centre of Leicester was populated by small artisans, shopkeepers, shoemakers, and the poor, who together had a tradition of being radical. They often defied national policy, notably resisting compulsory vaccination for smallpox.[75] In 1897 nonetheless a new resolution that had been rejected previously was put to the board of guardians asking them to supply Cambridge with pauper bodies. Macalister had pressed the need for their help during a personal visit and the Anatomy Inspectorate sent a follow-up letter stressing the problem of low supplies. 'Claude Douglas of New Walk, Leicester', a medical officer, replied that, 'As you are aware the Leicester people are a queer lot in all such matters. I will try and speak to some members of the Board on the subject. There are several lady members!'[76] Full Poor Law democracy and women having a voice in local government meant that Cambridge was rebuffed.[77] But again, it would be unwise to judge this exchange of letters in isolation. Leicester guardians were radical but they were also financial realists. It was later agreed that the bodies of the poor from the North Evington Infirmary on the outskirts of the town would be sold for a higher fee to Oxford anatomical school.[78] As medical education expanded it became a competitive field of study. The economy of supply was thus complicated by guardians trading one anatomist off against another. The next chapter will return to this theme. In the meantime the finding is another reminder of the importance of detailed reconstruction of the economic dynamics of the chain of supply. An equivalent financial dispute caused a lot more controversy at Nottingham the same year.

On 12 October 1897, the *Nottingham Evening Post* reported on a visit by Macalister to the board of guardians to discuss the 'Dissection of Unclaimed Bodies'.[79] During a heated and lengthy debate, Macalister was confronted by a new and rather aggressive guardian named Mr. Thraves. He questioned the social and scientific value of sending cadavers to Cambridge. In a private letter, Macalister had been forewarned by a medical man and guardian about Thraves's true motivations:

> I beg to inform that at the Workhouse Visiting Committee Meeting on Thursday morning the question of permitting the unclaimed bodies to be removed to Cambridge was brought up. Dr. Bolton proposed it and I seconded and it was carried by a large majority – 17 to 3, nearly all of the board. Now I have ascertained that since this, a

Mr Thraves, an undertaker, no doubt fearing that his trade is interfered with, is going around and doing his best to stir up bad feeling over this question.[80]

Among Macalister's private notes at Cambridge his jottings survive about this controversy. In a notebook he tallied up that: 'Old Board of 49 members, 17 voted for Cambridge, 15 voted against – New Board of 40 members, 7 old supporters gone, 4 new supporters gone – Of the new men several in their election address stated that they were opposed to sending bodies to Cambridge'.[81] Macalister therefore prepared a lengthy speech to be made at a public meeting, which the local press promised to cover in full. It soon became a heated discussion.

Macalister made his opening address to guardians and then Thraves went on the offensive. Thraves demanded to know why it was that 'anatomists wished to keep the bodies for a period of up to eight weeks before burial and how they would be preserved in Cambridge'.[82] Macalister was very unhappy about being asked such a delicate question. He instead kept stressing that 'anatomy did not involve any disrespect to the dead'. Why then, replied Thraves, keep cadavers for so long unless they were to be cut up extensively? Eventually Macalister's hand was forced and he elaborated that: 'The examinations come at fixed points in the year, and they could not arrange the mortality in the examination should have any relation to one another. They must have bodies for examination. It was a very slow business and the body was practically unaltered'. This was untrue. Thraves as an undertaker knew full well that after 'anatomical examination' there was not much left to bury. He had moreover expected after election as guardian to gain an exclusive contract for pauper burials in Nottingham. He was therefore angry about the loss of potential income and Macalister's slippery replies on the subject of complete dissection. He kept pressing Macalister to tell guardians more about the anatomy trade. The presence of the press and such a public confrontation were very uncomfortable for Macalister. Thraves nonetheless pressed home his case. By embarrassing guardians he hoped to keep up the public pressure to reverse the motion that opposed his financial interests. At the close of the discussion, Thraves proposed that no cadavers should be sent on the dead train to Cambridge.[83]

In the end, Thraves won the day because he adopted a very clever strategy. Guardians were so alarmed by the bad publicity that they agreed to suspend any further body sales until after the next election. The motion was therefore put in abeyance but it left a lot of bad feeling in Nottingham. As Dr. James Bolton, a guardian, scribbled in a note to

Macalister: 'I think the matter should have been carried yesterday if pushed to the vote but not by a large majority. Some guardians are afraid of the next election in April'.[84] Public opinion was a tricky thing to manage once the Poor Law had experienced full democracy. The fourth trend in Graph 5.1 should be seen in that light. Macalister had smaller teaching groups and therefore did not require as many bodies because the colleges rather than the central university now organised class sizes making them more manageable. Yet by 1908 Macalister had just 38 corpses. He had to make regular railway journeys, negotiating and re-negotiating with a wider variety of diverse interest groups. Leading guardians were often elected by the poor, now in the majority, to represent their viewpoint. This complicated the anatomy trade. The passing of the Old Age Pensions Act (1908) then further undermined those supply lines. Fewer paupers needed to come into the workhouse once they had a pension. By 1912, the lack of supply was serious. Macalister decided to act. His colleagues, like Dr. James Barclay Smith, were rather depressed by the turn of events as a letter from Dr. Thomson at Norfolk County Asylum confirms: 'So you too have the harmless, manic-depressive stampological [sic] mania; I have often thought we ought to have some Masonic sign whereby we recognise each other; one could beguile an hour or two on a railway journey or other occasion'. [85] Macalister tried to keep everyone's spirits up but losing cadavers from Leicester Union to Oxford anatomical school worried him. It also alerted Macalister to the need for anatomists to take a more unified approach.

In 1912 Macalister and Professor Arthur Thomson from Oxford agreed to work closely together to lobby the government about the need for a more organised, compulsory system of body distribution around the country. The idea was that if bodies were in short supply in one location but plentiful elsewhere then the Anatomy Inspectorate would authorise and coordinate the redistribution of corpses by railway. Inspectors of Anatomy had been reluctant to do this in person in the past to avoid any unwelcome publicity for central government. The workings of the Anatomical Supply Committee, sometimes called the Anatomical League in contemporary accounts, will be elaborated in the next chapter.[86] Essentially, Thomson at Oxford chaired the meetings, whereas Macalister collected the evidence on body-supply rates. The latter journeyed around most medical schools. Thomson in his private correspondence later recalled Macalister's hard work after his death in 1919:

> The late Professor Macalister is responsible for a statement made by Mr. McKenna, the then Secretary of State, to the effect that in

London in 1910 there were 700 such unclaimed bodies reported to the Poor Law institutions, of which only about half were available, whilst be believed that in the rest of the Kingdom the numbers considerably exceeded 3000, of which less than ¼ were sent to various teaching institutions... It appears that the total number [of bodies bought] for London in 1892–3 was 548 – in 1911–12 it fell to 280. Similarly in the provinces... the number of subjects fell from 274 in 1892–3 to 182 in 1911–2.[87]

Thomson did not admit that these figures had been manipulated, detailing average, not actual, body sales and omitting body parts purchased. He concentrated instead on the fact that Macalister blamed the coming of Poor Law democracy and reluctant guardians. Added to which was the problem that most hospitals were reluctant to share surplus supplies generated in the capital.[88] An Anatomical Committee – a cartel – prevented corpses going outside of London. During the First World War and thereafter supply levels never exceeded the 1873–1874 levels. The crusade against outdoor relief had been of vital importance to the expansion of medical education at Cambridge. Fewer corpses were sold on to the anatomy trade by the time that the New Poor Law was disbanded in 1929. The geography of supply chains and the reaction of the poor are analysed in the final section of this chapter, expanding on the five trends from Figure 5.1.

Sixpence a corpse: the demand for bodies

The Demand for Bodies –
The "Lancet" bemoans the dearth of subjects – not for discussion but dissection.
"A large number of the students are without parts", it says – not alluding to their want of intellectual capacity, but to the fact that bodies on the dissection table are too few to go round, so that every one may get a helping. Unclaimed paupers are it seems anatomised; but some guardians object to giving the bodies up – perhaps for fear that on their being opened it may be found how little there is in them. We do not know how the paupers view it. To a pauper of a well-regulated mind it should be of comfort to know that his death would benefit science as well as saving the rates. But consolation is hardly consonant with being "a good deal cut up" even after death and more of them are averse to becoming "anatomies"... with... " I do not wish to be regarded in that bony light"...[89]

This small newspaper extract, published on 27 November 1880, three years before Macalister came to Cambridge, indicated that the human face of poverty was no longer going to be so anonymous inside the anatomy trade. Throughout the 1880s and 1890s, newspapers, both national and local, began to focus more on the perspective of the poorest. Paupers became more visible and their voices started to come to the fore. This was because the Third Reform Act, the creation of County Councils, and the Local Government Act altered the fundamental fabric of local politics. Jon Lawrence points out that change was not immediate but once begun, it was unstoppable.[90] It created a 'politics of place' in local life that was irreversible. There was increasingly a lot of emotive discussion about what paupers felt about dissection. Richardson's excellent work on the cultural reception of the Anatomy Act in the early to mid-Victorian period first alerted medical and social historians to the importance of exploring pauper reactions.[91] Regrettably only a handful of historians have subsequently gone on to explore those views in depth. An important exception has been Julie Marie Strange's groundbreaking research on the cultural meaning of death and dying for the working classes.[92] The emotional range of their grieving process is now no longer taken as simple stoicism in the literature. Coroners' cases are an untapped source of this hidden world. One representative example is instructive.

In January 1890 a coroner's inquest was opened in Shoreditch, East London. It was convened to investigate why a widow was still 'sleeping with a corpse'. Her deceased husband had been Thomas Huckle, aged 60. Thomas was formerly employed by Messrs. Whitbread, a brewery, and he lived in a lodging house at Lyon's Buildings, Old-street, central Shoreditch, with his wife, Mary Ann Huckle. Sadly Thomas suffered from severe 'asthma and bronchitis'. After a painful night, he died in bed at '6.30am' on a Monday morning in early January. Under cross-examination, Mary told the court:

> *The Coroner*: Did you tell anyone that he was dead?
> *Witness*: No, sir, I slept with the corpse that night and the next day I told my brother and sister.
> *Coroner*: I hear you slept alone with the corpse till Friday?
> *Witness*: Oh no, not with the corpse, but alone in the room with it.
> *Coroner*: Why did you not tell some one?
> *Witness*: I went away during the day, and slept in the bed at night.[93]

The local sanitary officer explained that his enquiries found the house 'shut up' and 'it was not till Friday that he could get the body away'. A neighbour named Matilda Seeley claimed that Mary told her – *'My old man is gone'*. She elaborated that, 'In the evening she [Matilda] went into the room, and saw her [Mary] lying on the bed cuddling and kissing the corpse – She refused to let anyone touch it'. Mary did not confirm or deny these rumours in court. The coroner was perplexed. He said to the jury, 'it was a strange fancy of the wife's to keep the corpse in the house for four days and sleep with it for four nights'. In the end, a verdict of 'natural causes' matching the medical evidence was confirmed by the jury. The coroner, however, betrayed a lack of a common touch, concluding that the wife's behaviour 'could only be put down to eccentricity'. There is a lot to be learned from the subtext of this vignette.

Mary was evidently devoted to Thomas Huckle. She loved him and literally tried to hold on to his body for as long as possible when he died. Her sister and brother were alerted the next morning. This seems to have been a delaying tactic to give the family more time to find enough money to bury Thomas Huckle before the coroner took the body away and out of their control. It is likely that a range of normal human emotions – a mixture of shock, grief, and fear – motivated Mary to keep hold of a rotting corpse for four days. It was also the sort of strategy that a woman who suddenly found herself in poverty had to adopt to ensure that a loved one's corpse was not sold on to either St. Bartholomew's or Cambridge anatomy school (both bought bodies of the poor at the time from the East End). It is easy to misread the thinking behind Mary's actions. Logically she seemed to be holding on to a corpse for as long as possible. But acting as she did placed the case in the public domain of the coroner's court. By creating a sensation, Mary was actually putting pressure on the local community to help her pay for a pauper funeral and avoid dissection in a location where all bodies were normally sold on without exception.[94] The fragile line between someone's respectability and shame was sometimes sleeping with the corpse.

Local Poor Law records show that friends and neighbours often helped the bereaved to avoid the dissection of a loved one. In 1883, for instance, Macalister convinced Spalding Poor Law Union to sell pauper bodies on one of his first lecture tours. Three years later, the poor learned about the agreement because once the MDRDA came into force guardians were sending more cases to Cambridge. Two men, one 'named Bryan' and another called 'John Metcalf', were sold in June 1886.[95] This was done on the understanding that 'the expense of burial, coffin, and carriage' were paid by Macalister out of his petty cash expenses. Events then went awry. 'A day or two' after 'Bryan' was sold, 'his friends appeared' at Spalding workhouse. They were angry about dissection and told the local press about the scandal. Guardians debated what to do. The body of 'Bryan' was sent back by train but not 'John Metcalf'. He was dissected quickly and then buried on '16th June 1886'. Meantime, the adverse publicity meant that several guardians tabled a resolution now objecting to the supply agreement but to no avail. It was 'rejected by fifteen to nine votes' and supply to Cambridge continued. This episode contains, though, all the elements that made the anatomy trade sporadic. The importance of kinship, a scandal, guardians in disagreement, bad publicity, bodies shuffled back and forth by train, and supply disrupted.

In Croydon we see a similar chain of events but with a slightly different outcome. 'Joseph Piercy... an octogenarian inmate of the workhouse' fell ill in May 1890.[96] He was transferred to the sick ward and a notice put on the end of his bed. It read – 'None'. This meant that in the event of death he had no relatives and could be sold on for dissection. The sign was incorrect. There had been an administrative mix-up. His admission card said 'a son was in Croydon workhouse' but 'Mr Stevens the steward of the parish infirmary' did not double-check these crucial details against the bed notice. Joseph died and 'it was taken for granted that there were no relatives, and the body sent up to the School of Anatomy for dissection'. Later, the son in Croydon workhouse, and another pauper relative at 'Wood-green', tried together to claim the body. It was already dissected and dismembered. An added difficulty was that another near-relation and ratepayer from Mitcham 'had been contributing towards Peircy's support for some weeks after his death'. Croydon board of guardians took decisive action. They 'decided to refund the overpaid money and ordered the steward to send his resignation', but supply was not stopped. If anything, scandals like these made those in the supply chain more determined to act with duplicity and hide their dealings. When paupers did speak for themselves it could be disquieting.

In 1885, the editor of the *Essex Standard, West Suffolk Gazette and Eastern Counties' Advertiser* sent reporters to investigate workhouse conditions across East Anglia. This was the main catchment area of Cambridge anatomy school. Colchester workhouse featured as a main article in the March edition. By all accounts, it was clean and generally well run by caring staff. It had a reputation for being 'humane'. The reporter felt that the workhouse compared well to a hospital or large almshouse. It was filled with the aged poor but conditions were not a scandal, quite the reverse. The paupers were pleased to talk to the reporter and he learned a lot about their attitudes to poverty. Then the subject of dissection came up in conversation. The reporter stated:

> The majority of the old ladies in the House seem apprehensive lest their mortal remains should be sent to *"Cambridge for dissection"*, one old dame quaintly remarking that she would be of little use as her bones scarcely had any flesh on them. There was one exception however in an elderly inmate who joking remarked that she had already disposed of her body for the modestly low sum of six pence, and when asked where the purchase money had gone she pointed to her nose, indicating it had been spent in the much relished delicacy of snuff.[97]

The reporter was intrigued and keen to find out if this was a joke or not. He enquired whether other paupers had been encouraged to sell their corpses for sixpence too. Nearby sat an elderly ex-shoemaker and the man confirmed that the practice was common among the male inmates but he took a practical view:

> On being asked by a companion what he was going to do with his body, he exclaimed, *"Oh I don't care what becomes of me. I shall go to Cambridge"* [sic]. He also indulged in what appears to be a standing joke in the Workhouse stating that he had sold his body for 6d., and what was more, had *"got the money"* [sic].[98]

The reporter concluded that casual words spoken in jest really reflected the downside of a pauper's life. Most were physically content and well cared for in the workhouse. Yet the casual banter and practical turn of phrase also contained a world of meaning about the central part that dissection played in meagre death rites. A *Daily News* reporter covering a similar story in August 1891 stopped and asked a publican about the plight of aged men in the workhouse around Chelmsford. He enquired whether they 'find any fault with the workhouse?' The publican replied, 'No; he don't find no fault. He takes it all, quietly enough'.[99] Others were more outspoken. In October 1891, the same *Daily News* reporter met labourers in the fields around Oxfordshire in an area where anatomists at Oxford and Cambridge competed for bodies. One started up a conversation about pauper burials:

> *Ah*, broke in a rugged gaunt-looking fellow, taking out his short pipe from his mouth, and *'t'other day when a colt kicked my boy's brains out and I went to the parish to ask'em to bury'im, they told me as I ought*

to have money enough to do it myself [sic] ...These were just the same stories as I found everywhere else, and these men seemed eager to discuss things ...[100]

A poor labouring man told the *Daily News* reporter on 3 September 1891 why aged male paupers were admitted to the workhouse and then sold on:

> *I stuck to th' old man as long as I could,* said the labourer, *and I'ould do it. Th' old'un'ad a seat by my fireside and a bit 'some't to eat while I could gi't to'n. But when he went into the'ouse'cause I couldn't afford to keep'n at'ome and feed young'uns, they didn't aughter a come down on me for a shil'n a week and I told'em so* [sic].
> [Reporter] You went before the guardians?
> *Yes, and I d'ad to walk five mile through drenching rain and five mile back over it, and then they wouldn't let me off. They said I must pay, and I wa'n't working half time and they know'd it* [sic].[101]

The poor labouring man kept his father at home until his working hours were reduced in winter. There were too many young mouths to feed and so the old man had to be admitted to the workhouse but during the crusade decades there was a new Poor Law rule. It compelled labouring families to pay at least a shilling a week towards a parent's upkeep in the workhouse. This many could ill afford to do. Although the poor labouring man in question may have dramatised his predicament nonetheless an anatomy sale was a pragmatic calculation. If they had to pay a shilling a week to keep a grandparent in the workhouse then there was no money left to pay into a burial club. Aged paupers thus sometimes took a financial incentive, sixpence a corpse, to ensure they were not a financial burden to their loved ones after death. To put this evidence in context, anatomists normally paid sixpence for a foot or an arm. If what the paupers said was correct, and we have no reason to doubt their statements which the reporter corroborated, someone in the supply chain was making a healthy profit margin from their corpses. Macalister paid standard fees, usually '£3 per body' for an aged adult and up to '£12' for a young female or child.[102] It might have been practical but it was also illegal to pay incentives pre-death for human material. The body was the only commodity that a pauper in absolute poverty had left. The geography of that economy of supply is reconstructed in Table 5.2 and reveals the extent of Macalister's dealings around the country.

Table 5.2 Geography of Suppliers to Cambridge Anatomical School, 1855–1920

Original Location	Bodies Nos.	%	Original Location	Bodies Nos.	%
Cambridge City Centre	700	23.70	Hitchen	27	0.9
Hull Union	315	10.66	Leeds Union	21	0.7
Addenbrookes Hospital	305	10.32	Huntingdon Union	17	0.6
Cambridge Union	178	6.02	Whittlesey Union	13	0.4
Doncaster Union	146	4.94	Whitechapel Union	11	0.3
Finchley Union	138	4.67			
Leyton Union	129	4.36	Under 10 (% fractional)		
Brighton Union	104	3.52	Bishop's Stortford	9*	
Fulborn Asylum	100	3.38	Chelmsford Union	9	
Mildenhall Union	86	2.91	Hardingstone Union	8	
Wisbech Union	72	2.37	Kingston-Upon-Hull	8	
Biggleswade Union	70	1.99	Southampton Union	7	
Anatomical School	59	1.82	Basford Union	6	
Colney Hatch Asylum	58	1.79	Bury St. Edmunds	4*	
Hertford Union	54	1.45	Haverhill Union	4	
Luton Union	53	1.42	Reading Union	4	
Bedford Union	43	1.21	Thetford Union	3	
Three Counties Asylum	42	1.11	Saffron Walden Union	3	
Manchester Union	36	1.08	Ely Union	3	
Yarmouth Union	33	1.04	Nottingham Union	3	
Chesterton Union	32				
London County Asylum	31		Sub-Total	2944	
			Other single suppliers	9	
			Total	2953	

Note: *Poor Law Unions.

Source: Cambridgeshire Record Office, St. Benedicts burial registers, PR25/21-23.

It is evident that Cambridge anatomists relied predominately on a local supply. Of the total number of bodies, 2,953, not less than 1,183 cadavers, or 40 per cent, were generated within the city limits. These were street deals, hospital cadavers, and paupers supplied by Cambridge workhouse. This geographic trend applies to every anatomy school in the capital and the provinces. Anatomists always depended on the local dead to equip the dissection table. The anatomy trade was predominately a doorstep business. Linking census material to burial and surviving dissection registers is instructive on this point. The 1881 census for Cambridge workhouse reveals the typical life cycles of those traded.[103] Cases included 'James Brown, pauper, and farm labourer, born Cambridge' who simply lived too long. He died 'friendless aged 93' because he had outlived his kinship group. He was dissected and buried on '2nd April 1892'. 'John Chapman', a former 'farm servant' who had lived and worked at 'Fulborn' on the outskirts of Cambridge, had to come into the workhouse on his retirement too. He died, was dissected and buried on the '2nd February 1887 aged 77'. Chapman lay alongside a fellow pauper inmate named 'Thomas Leeding aged 70' who had been a 'college servant'. He fell on hard times and entered the workhouse where he was sold on for dissection. On '15th June 1887', Leeding was buried with nine other bodies that came by train into Cambridge. They were interred together in a large pauper grave in Mill Road cemetery near Cambridge workhouse. Close by, 'Sarah Perkins', her age estimated as '72', was buried two months later.[104] Sarah had been a 'domestic servant in Saffron Walden'. She had entered the workhouse, and was sold on, lacking a local family to pay for her pauper funeral. Her interment was on '30th August, 1887'. In the female wards of the workhouse in 1881 she worked in the kitchens with 'Mary Trayler', described as a 'pauper imbecile' from 'Thingoe, Suffolk'. Mary was typical of those paupers who fell into the 'imbecile' category and whose families could ill afford to keep a disabled relative once they aged. Mary's estimated age was '61' at death in the anatomy records.[105] She was dissected and dismembered, and buried two years before Sarah Perkins on the '9th July 1885'. A noteworthy finding that suggests most paupers knew the dissection fate of their fellow workhouse inmates.

It is worthwhile plotting the street names of the body deals done in central Cambridge.[106] The burial records show that the majority of cadavers were bought in the poorest parts of the city. Those who lived in either close proximity to the anatomy school or near the workhouse on Mill Road were the most vulnerable in death to dissection. The records suggest that bodies were also purchased by porters at the expanded Downing site where the anatomy school was situated. This would have avoided neighbourhood gossip. The alternative was to deal with a doctor or undertaker who acted as an intermediary when bodies became available in the city. Regrettably, these deals were seldom written down. The evidence that does survive is fragmentary. Of the '59' bodies described as being generated by the 'Anatomical Schools' in the burial registers, further analysis indicates that these were the early street deals of Sims who pioneered paying petty cash to set up local supply lines. Later, the way those transactions were described changed in the burial registers (they were incorporated into the general Cambridge city-centre figures). There must have been a system in place to expand the local trade, even if the specific evidence of it has been destroyed or misplaced. It is likely that Cambridge anatomists copied what their colleagues did in the capital, buying from a diverse range of people willing to be part of the chain of supply. Outside of Cambridge, this meant paying key personnel in asylums and workhouses, and targeting areas of the country with poor Public Health standards where mortality rates were higher.

The supply contract that Macalister negotiated with Hull Poor Law Union was important since it generated 315 bodies or 10.66 per cent. Likewise, Doncaster traded 146 bodies, almost 5 per cent, sent by train to Cambridge. Most anatomists were very focused on their teaching duties and scientific studies. Macalister's reading material was population statistics and railway timetables. Hull was one of five English cities with the highest mortality rates. It was a logical place for Macalister to alight from the train to obtain more cadavers. His body buying patterns match the north–south divide that demographic historians have noted for the period, excluding London, and with the exception of Brighton.[107] In fact, the campaign of Arthur Newsholme, a leading medical officer, to improve poor public health standards in Brighton, threatened the anatomy business.[108] Yet reducing mortality took time. The harsh attitude of local guardians, keen supporters of the crusade against outdoor relief, had a more immediate effect. It meant that Brighton was a mortality hot spot where 104 bodies, almost 4 per cent of the total number of bodies, were dispatched to Cambridge. Again the MDRDA was resented locally and this increased supply lines even though public health measures started to have an impact on background death rates.

The geographic spread shows that leading crusading unions like Brighton, Manchester, Reading, and Southampton, were keen to share their surplus supply with Cambridge. Bodies at Manchester, for example, were sold to the city's anatomy school first but then despatched to Cambridge when mortality rates peaked in the summer of 1906. Reading Union tended to trade with Oxford but again made other deals when they had extra cadavers to offer in the supply chain. Train lines were of vital importance. Smaller poor law unions, like Mildenhall, Wisbech and Biggleswade, had populations that shrank below 5,000 in the 1890s with the flight from the land by rural labourers. Yet they went on supplying a disproportionate amount of bodies because they were train stops on the branch railway line running into Cambridge. In the case of Leeds, it tended to supply its local anatomical school. Yet its surplus supply was sent on the coal train that went to Cambridge via the Leeds and Selby link. Likewise, Bedford, Hertford, Hitchin, Huntingdon, Luton, Stevenage, and the Three Counties asylum were all stations on the Great Eastern line going out of Cambridge. Meanwhile the daily train on the Northern and Midland mainline connected to Nottingham. The railway was a fundamental feature of the anatomy trade, rather than a symbol of modernity for the poor, and this finding will be reflected elsewhere in the other case studies that follow. Macalister meantime was a much-travelled man who spoke 'fourteen languages'. He was also a famous Egyptologist fascinated by forensic archaeology and anthropology. Everywhere that he went he collected skulls and brains. Cambridge soon had a famed collection in its basement and medical museum. Medical students purchased a skeleton or sets of bones through 'John Lane, the long-serving chief assistant in the Department of Anatomy'.[109] Lane's name appears numerous times in the burial registers of the anatomical school. He did the administrative work, signing off each body for interment, sending back and forth the 12 certificates that had to be filled in each time a body went from source to the dissection table and then into the ground. The choreography of this complex certification process was first introduced in Chapter 1. In the Cambridge papers and at the National Archives examples survive of this complicated and time-consuming bureaucracy. Most paperwork was destroyed after each quarter's audit. Since Macalister was travelling around so much, concentrating on his research when at the dissection table, Lane was in charge of processing bodies. In 1931, a celebration of his life in the *Medical Society Magazine* commented that 'if all the bodies which John [Lane] has injected were set end to end, they would stretch from the Anatomy School to just beyond Fulborn, in addition to causing a not inconsiderable sensation'.[110] It was five and half miles from central Cambridge to Fulborn village, a fair distance when measured in cadavers. Macalister was often prepared to pay more for brains and

Lane coordinated those purchases too. Two typical letters among his papers confirm that the railway was pivotal when moving around extra human material.

A former Cambridge medical student employed by London medical school sent a letter confirming, 'My porter has a box of amputated parts, which he has saved for Cambridge, if they prove of use to you'.[111] They came from '16 to 20 operative surgeries' and he was sending them in a deal box on the train from King's Cross station. Mr. W. H. Elkins, the medical superintendent of the Three Counties Asylum, confirmed a similar arrangement, and then revealed:

> In the ordinary course when a patient dies and is buried in the Asylum cemetery, we make a coffin and the Union to which the patient belongs pays 15/-[shillings] for the same.
>
> When we send to Cambridge, we make the coffin just the same and charge it to the Union, and as far as the latter is concerned, they know nothing as to the burial, and neither does the Auditor.
>
> When it goes to Cambridge, I pay the carriage [by railway] out of my own pocket, which is refunded by you, therefore it does not appear in the Asylum accounts and there is no reason why it should.[112]

Under the Anatomy Act, although all of the cadavers should have been declared, some were, but many others were not. Asylums were careful about their image in the local community since they already attracted negative publicity for locking up the vulnerable. This explains why Cambridge lacked the bodies of the insane poor. There was though no legal requirement for the trade in body parts to be disclosed. The business of anatomy in this respect carried on. An auditor did not have a right to know the details. Payment by results was therefore a regular occurrence. Cambridge knew this but ignored the social cost in the interests of its supply. Macalister did not want anything interfering with his brain research. After all, the only way to make scientific progress was to cut up human material. This meant that somewhere on the railway network each week was a pauper corpse and other body parts going up and down the track to the dissection room at Cambridge.

Conclusion

Alexander Macalister was a brilliant anatomist, a man of science, and someone who reformed medical education in human anatomy at Cambridge. He set new standards, introducing a *viva voce* to ensure that students had attended dissection demonstrations on a regular basis and were therefore equipped to enter hospital medicine or general practice. The template that he developed, expanding an anatomy trade on lecture tours and through widespread public engagement work, was to be copied everywhere. In subsequent chapters we will see the practical outcomes of his influence. The *Lancet* in 1889 praised his achievements. An editorial congratulated Cambridge 'on the fact that among medical schools in point of numbers it leads the list of entries' for the provinces.[113] But it had not been easy to expand or retain medical students because a lack of cadavers was an enduring supply problem. The broader lesson from this case study is consequently that the provincial anatomy trade was dependent on the phases of the New Poor Law. When rules and regulations tightened, anatomists benefited from parsimonious attitudes. The crusade against outdoor relief was not a friend to the poorest but it was a boon to Macalister at Cambridge. Supplies rose 100 per cent in 1873–1874, another 200 per cent in 1885–1886, and opened up many new sources across the country. In different places guardians reacted according to their local economic interests. It was the financial incentives involved that motivated a wide range of welfare institutions to become suppliers. The profits might have been small-minded but they were also judged worth the time and effort by those who claimed the petty cash incentives.

It helped a lot that Macalister was prepared to visit in person and persuade wavering guardians. Once contracts were agreed they had to be re-visited in subsequent years. Medical education was a competitive field financed by rising and falling student fees. It was also a business driven by profit margins made from paupers. Macalister for a time lost out to Oxford at Leicester because the former was prepared to pay more for bodies. He was determined not to repeat the mistake of paying less and won an important contract at Hull in 1885–1886. The railway was pivotal to his supply deals everywhere. If an asylum, infirmary, or workhouse was on a branch or main line into Cambridge, then Macalister was determined to secure the supply contract. Females and children were always in demand, with a three men to one woman ratio reflecting the realities of pauper lives and their family economies. The poorest did resent the anatomy trade but they reacted in diverse ways in a world-without-welfare. Some protested at the fortnightly meetings of local boards of guardians; others were resigned to selling on their corpse for sixpence. The coming of Poor Law democracy gave the poorest a voice and increasingly their vulnerability was a factor in the 'politics of place'. The negative side of the anatomy trade was 2,953 pauper histories lost. The positive outcome was a generation of medical students trained properly for the first time in human anatomy.

Along Free School Lane anatomists shared human research material with other dons in the scientific community at Cambridge during the late-Victorian era.[114] This little known fact is another unwritten aspect of the anatomy trade. Historians have yet to appreciate how many necropsies were shared and what sort of internal medical market was created in human material at a time when laboratory science was in its infancy. At Cambridge, scientists were renowned for developing some of the first skiagraphs in Britain.[115] This early form of X-ray produced a shadow image. Some of the first pictures were of dislocated joints, reported in the *British Medical Journal* (1896).[116] Those early reports stress that an accurate skiagraph took about eight minutes to complete.[117] It was a very shaky technology. Throughout the development process each patient had to be motionless when the image was taken. It soon proved much easier to record a small human trunk, compared to an adult. This was why the Royal College of Veterinary Surgeons tried to perfect early skiagraphs using small animals like cats. Yet, as the *British Medical Journal* pointed out, 'There was some difficulty keeping the cat at rest, and an anaesthetic was rendered impossible by the fact that during the trial with ether and another with chloroform the cat showed signs of danger [i.e. death]'.[118] Alternatively anatomists found that a tiny, dead infant before dissection was ideal for experimental skiagraph work. In any case, from its inception, skiagraphs were used a lot in child medicine. It therefore made sense to skiagraph dead children when they became available for human anatomy at Cambridge, St. Bartholomew's hospital, and elsewhere.[119]

Today, millions of people rely on X-rays in emergency medicine around the world. That technological leap is one part of a much larger untold, hidden history. There was a human chain of medical advancement that began with pauper children being used in early skiagraphs in 1896. Regrettably it is very difficult to match early images reported in medical journals to actual dead infants in the dissection records at Cambridge or St. Bartholomew's. The images survive but naming them is difficult. Most medical journals stress the use of living patients and omit references to the use of the dead for obvious reasons. The adverse publicity would have detracted from the positive news about a new medical advancement and instead opened up uncomfortable conversations about the anatomy trade. Besides, the Victorians had a very different mentality about photography of the dead. Michael Sappol's exemplary scholarship on the American anatomy trade shows that bereaved parents often took or were given a final photograph of their dead child to console them before it was sold on for dissection.[120] These post-mortem photo portraits were popular in England too. Examples survive in the Royal Collection; Queen Victoria, found photographs of Prince Albert on his deathbed consoling in her widowhood. More mainstream examples in a family setting survive in county archive repositories and these can now be viewed on open access.[121] They show that images, image-making, and preserving the identity of a deceased child were intrinsic aspects of the grieving ritual for many Victorian parents. This was though a private, not a public, act of consolation. The lack of reporting about unnamed dead children featuring in skiagraphs should not nevertheless cloud our historical focus. The fact is that living patients and dead children in the past helped to make fundamental medical progress for everyone. It is a curious medical fact that skeleton images, icons of modern medicine, were sometimes those of pauper children destined for the dissection table. This was one reason why Macalister, looking ahead at the turn of the century, commented:

> It has been said that human anatomy is a worked-out field; the comparison of old textbooks with those of the present day is the best refutation of this charge... It is by no means improbable that those who see the end of another century will be able to record advances both in knowledge and method even greater than those that have taken place in the last hundred years.[122]

Macalister was determined to press the case of science and its value to humanity. There was a major downside to the anatomy trade. Researchers had to ignore any call for consensual medical ethics. Throughout his long career, Macalister was a man of vision in an invidious position. He cared about medical science, better standards of education, and the proper burial of the poor. In Cambridge, research involving dissected paupers is one part of a mosaic of medical progress. In the next chapter we explore the contribution of Oxford. For this lost history was cumulative and involved many people's stories on both sides of the anatomy trade, at the threshold of the modern world.

6
Balancing the Books: The Business of Anatomy at Oxford University

Anatomy – An April Fool?
About 2 o'clock yesterday afternoon, I saw the remains of a child taken out of the water in this parish; there were a leg and thigh, and part of the spine and an arm. A fisherman by the name of Beesley found them when he was fishing; he said he had found another arm, and had thrown it again into the water. I examined the parts, and am of the opinion they were parts of a child of about 4 years of age. I think the body had been used for the purpose of anatomy for the arteries were filled with wax.[1]

On 3 April 1834, a controversial inquest opened at the Bull Public House in Oxford. The city coroner told the jury that an 'unknown' child's body was found on April Fool's Day. A local fisherman hooked a small torso floating downstream to 'Preacher's Pool' in St. Ebbe's parish. The decomposed cadaver was cut into parts. It was fished out of one of the deepest eddies of the River Cherwell. Coroner's records indicate that murder victims and infanticide cases were often discarded in pieces. Chapter 2 showed that corpses were thrown into fast-running, murky water late at night to avoid prosecution. In the watery grave forensic evidence became corrupted. The Oxford coroner asked a local surgeon, 'Mr Webb', to examine the 'drowning case' and rule out foul play. What he reported back caused an anatomy scandal. Local outrage was expressed that a lack of pauper funerals might have been behind a trade in body parts. The small corpse had been dismembered and sold to the highest bidder for dissection. This undignified end hinted at a complex economy of supply in Oxford.

An editorial in *Jackson's Oxford Journal* informed its readers that, 'There is an Act of Parliament for the regulation of anatomy; by that

Act persons are required, after dissection, to have the parts buried in consecrated ground'.² Anyone who ignored this legal statute was 'liable to imprisonment for three months, or a penalty of £50 at the discretion of the Court before which he shall be tried'.³ Despite extensive enquiries nobody could identify the child's torso and limbs. Neighbours could not recall a recent infant death. Newspapers tried in vain to make a link to the local medical school based at Christ's Church College. Police enquiries also drew a blank at the Old Ashmolean Museum on Broad Street where there was a small dissection theatre and collection of wax specimens. The jury foreman requested that the coroner report the case to the mayor and city magistrates: 'That measures may be taken to bring to justice such person or persons who have so shamefully and wantonly outraged the feelings of the public'. Yet like many breaches of the Anatomy Act, the case was unresolved. The medical fraternity remained tight-lipped. Local people were powerless. They lacked evidence of actual wrongdoing by anatomists and the anatomy trade in Oxford.

Sampling *Jackson's Oxford Journal* reveals that scandals about the anatomy trade tended to be brought to the attention of local coroners in central England.⁴ The press, like the community, suspected that local officials often colluded with medical men to acquire the bodies of the poorest by fair means or foul. Pauper funerals were under threat in the transition from the Old to New Poor Law. Customary burial rites differed a lot for economic reasons.⁵ The trend though was to be less generous as the nineteenth century progressed. In the case of Oxford, anatomists exploited parsimonious attitudes to establish their body business. The aim in this chapter is therefore to study in microcosm an anatomical school where the economic of supply was small-scale and the education of doctors was disputed. Oxford retained its reputation for being conservative in Victorian times.⁶ It was regarded as a provincial place to train in medicine despite its reputation for academic excellence in subjects like the Classics and Natural Sciences. Most medical students did pre-clinical training at Oxford. Students trained on the wards of the Radcliffe Infirmary and at Christ's Church College anatomy school. They then left Oxford to sign up for advanced human anatomy training at a London school. In Chapter 4 we have seen that there was a much larger economy of supply in the capital. By the late Victorian period, Oxford anatomists were determined to reform human anatomy teaching to try to attract back their medical students. This recruitment drive created practical problems and professional disputes that reveal the difficulties confronting a smaller school of anatomy when trying to sustain a profitable business of anatomy. The next section begins with the story of a pauper in the late-Victorian period. On first reading it seems an isolated case of a pauper being difficult. Later in this chapter the same story links up to a coroner's dispute. It was one of the key reasons that Oxford had such problems developing its body-finding system.

Pauper protests

> Daniel Hookham, an inmate of the Workhouse, and a frequent attendant at the Court, was charged with assaulting the Master, Mr. Stedham, on the 23rd March ... Mr Thompson clerk to the Guardians prosecuted, and the Master stated that just after Grace had been said the prisoner threw his bread and a pannikin of water over him, and followed that by throwing the pannikin also at him. A policeman was sent for, but the prisoner scaled a wall 9½ ft. high and got away.
>
> The defendant asserted that the Master had clandestinely had the body of a man named Taylor removed and sold from the workhouse for dissection, and all of Oxford would be up in arms against it.[7]

This 'City Police Report' was printed in *Jackson's Oxford Journal* in 1876. It alleged that dead paupers were being sold on a regular basis for dissection to anatomists at Oxford University. Its editorial line was neither a joking aside by the press, nor a student prank. It was a genuine claim made by Daniel Hookham, a pauper inmate of Oxford workhouse. This outspoken character was prepared to say in public what many gossiped about in private. The anatomy trade was active in some of the poorest parts of Oxford city-centre. Yet that fact of local life should not be taken at face value in the historical record. Anatomists had a great deal of difficulty improving supply chains despite paying generous incentives to suppliers in central Oxford. It was an uphill task to negotiate sales because of ingrained resentment in some of the poorest communities surrounding the University. The adverse publicity that Daniel Hookham generated was one indication of hypersensitive death customs. There was growing anger about dissection in a more democratic era. Reconstructing Hookham's personal history is instructive about what also motivated his attacks against the anatomy trade.

Hookham, like many casual labourers, relied on poor relief when work was scarce. He often witnessed the covert sale of pauper bodies. It was an unpalatable fact of life that a dead body was the one commodity that impoverished families had left to trade at times of financial crisis.

Hookham told the court that he was a 'plasterer' who picked up work as a 'painter and decorator' between building jobs.[8] He drifted from one short-term contract to the next, refurbishing College property. When times were good he could earn as much as '£1 a week'. Lately his wages had fallen to just '2s 6d a day'. Hookham complained to the court that piecework was paid below subsistence levels. Often it was impossible to feed his wife and children. Most labouring families lived on the precarious edge of pauperism. On a regular basis Hookham and his family had to enter the workhouse. In wintertime, when work was scarce, he was disillusioned and resentful about his limited employment opportunities. Out of frustration, Hookham admitted that he was sometimes violent. In the Police Court, his self-protective strategy was to expose the anatomy trade and lack of pauper funerals to public derision. Hookham hoped that this standard defence would justify his aggressive behaviour. It nonetheless generated undesirable publicity for anatomists in Oxford. Their counter-offensive was to ridicule Hookham's claims and demand in the press that law and order prevail in the local courts.

The local press reported that Hookham was 'an old offender' who tended to turn nasty when unemployed. He often threatened farmers that he would burn their hay ricks unless they paid him a small bribe of a few shillings to leave the area without causing trouble.[9] Passing through local towns, Hookham was seen walking the streets begging for food. The court was told that local tradesmen considered him to be a public nuisance.[10] He often threatened those who refused him charity. Hookham targeted undergraduate students too. During busy term time he demanded with menace a donation of a few shillings apiece. This soon brought him to the attention of the police who redirected him to the Poor Law for official help.[11] On entering Oxford workhouse, Hookham adopted a difficult and downbeat attitude. The master informed the magistrate's bench that Hookham was a 'foul-mouthed' pauper and an 'unsavoury' character. The guardians were fed up with his high-handed opinions and they wanted him incarcerated for being offensive, particularly about the anatomy trade. This only made Hookham more resentful. Ordinarily he tended to react badly when confronted by authority figures. Now he became even more offensive in court.

The labouring poor were often provocative in front of magistrates to whom they had a traditional right of appeal under the Old Poor Law.[12] It remained the case that even after the New Poor Law came into force in 1834 a standard pauper strategy was to be obdurate. Paupers still had agency and often exploited controversial issues like dissection to extract more money from guardians fearful of social unrest. In theory, aggressive behaviour could increase the chances of having an outdoor relief claim paid to remove a troublesome character from inside the workhouse. In practice, Hookham made a strategic mistake by threatening to rouse the mob against the anatomy trade. Local magistrates refused to tolerate such a blatant threat to law and order. They passed a sentence of 'two months hard labour' for being 'an incorrigible vagabond'.[13] If Hookham would not keep quiet about the anatomy trade then he would be locked up in prison for stirring up civil unrest about dissection in Oxford. This harsh reaction stemmed from historical precedents that made the city authorities wary of social disorder. Recounting that context briefly is important and so what follows returns chronologically to the eighteenth century.

John Bellars, a leading Quaker, first remarked in his *Essay Towards the Improvement of Physick* (1714) that 'it is not easy for the students to get a body to dissect in Oxford, for the mob being so mutinous to prevent their having one'.[14] In 1730, residents organised a riot to retrieve a body taken down from the gallows to be buried that was grabbed by medical students for dissection. A Victorian account recalled the struggle over the corpse. There was

> Hanged at Oxford, one Richard Fuller of Caversham in Oxfordshire, a young man of 26 years of age for murdering his wife – There was sad work on that occasion, the Scholars endeavouring to get the body, assisted by some Townesmen [sic] and others on the contrary hindering. The relations had provided a coffin to have it decently buried at Caversham but the scholars broke it all to pieces, the body being in it; after which those opposite to the Scholars had it again and so for several times, sometimes one side had it and sometimes the other, but the Proctors favouring the Relatives the body was at last delivered to them and brought to the Castle [prison]; about eleven at night when all was though still, it was taken to the water side to send away in a Boat but to their surprise, the Scholars were lying in Ambush and coffin and all was thrown into the water, but the Scholars soon went in great numbers and drew it out and carried it to Christ Church to dissect it. The Tumult was so extraordinary that the Town Clerk was forced to read the Proclamation [extolling law and order] but to no purpose, the Rioters crying they did not hear it.[15]

Against this backdrop, the Radcliffe Infirmary, the main voluntary hospital, established in 1770, was very careful to distance itself from the taint of dissection.[16] In the local press, a weekly list of fatalities was printed by the hospital governors. They always stressed that post-mortems were only permitted to resolve suspicious or unexplained deaths on the operation table. Surgeons were not allowed to run human anatomy classes for medical students from the mortuary room or dissect in the Dead House next door to the operating theatre. The hospital governors did permit medical students from Christ Church College to train on its wards in clinical practice provided they treated pauper patients under close surgical supervision.[17] These strict rules reflected strong criticism of the medical profession's attitude to the destitute. In Victorian Oxford they were charged with exploiting the poorest in the name of scientific progress at the University. There was some truth in this allegation. At Oxford's annual meeting of the Provincial and Medical Association in 1848 the assembled medical fraternity met to coordinate political pressure for the Medical Act. The main speaker referred to a popular satirical poem doing the rounds locally. In death, it was said, poor patients were often indebted to doctors. Their families were unable to pay off medical fees except by handing over a deceased loved one to the dissection table:

> Not for the sickly patient's sake,
> Not what to give – but what to *take*;
> To feel the pulses of their fees
> More wise than fumbling arteries;
> Prolong the lamp of life in pain,
> And from the grave, recover – gain!'[18]

It would seem that Daniel Hookham was not alone in ridiculing dissection in central Oxford in the Victorian era. It was the late-Victorian New Poor Law that brought these cultural tensions to the fore again in city life.

In Oxford, guardians adopted the crusade against outdoor relief with fervour in 1872–1873.[19] Over the next five years, new rules banned outdoor relief allowances except under very exceptional circumstances.[20] This about-face began to erode the makeshift economies of the poorest in the area. Daniel Hookham's father was at the forefront of a campaign to reverse the recent changes. He was aggrieved that not only outdoor relief payments but also pauper funerals were cancelled on medical orders. In February 1872 'William Hookham...60 years of age' wrote a letter of complaint to the Local Government Board (hereafter LGB) that was also published in the local press.[21] He claimed that he had been refused outdoor relief. This was a 'heartless' action that forced his 'wife and several children' to enter the workhouse. His immediate concern was a pending legal action about maintenance regulations. It stated that when William Hookham was out of work and in the workhouse, Daniel Hookham would have to pay towards his father's upkeep. William claimed that the Hookham family had no money for any extra welfare bills. The master, however, was determined to recover a few shillings a week from the children of worn-out labourers residing in the workhouse. In Oxford magistrate's court, standard prosecutions awarded a maintenance allowance of 2s 6d against each pauper family. William Hookham refused to accept that judgement. He claimed that guardians used 'improper language' in their court demands and he had 'a right to poor relief'. The relieving officer denied these charges. Central government closed ranks too. William was told to ask the newly formed Charity Organisation Society (hereafter COS) for help. He refused to accept charity. This context explains Daniel Hookham's obduracy. Delving deeper into the archives reveals another thorny issue. The issue of pauper burial was tense in the city-centre too.

Looking closer at the circumstances surrounding Daniel Hookham's court appearance, the evidence shows that it was prompted by a related controversy about a lack of burial space in Oxford. In 1876, the mayor of Oxford set up a Burial Board to find a practical solution to the problem of overcrowded churchyards. Ratepayers agreed that there was an 'urgent need for a general cemetery' in Oxford.[22] A survey was taken of empty landsites with adequate sewers above the water table outside the city limits. Oxford Colleges supported the scheme. They owned most of the proposed locations and would profit from an agreed sale. Poorer residents, however, feared that anatomists and local undertakers would exploit the burial review. If a large cemetery was purchased in the suburbs this would increase the distance people had to travel to bury their dead. Local undertakers would be able to charge more for transportation costs, the funeral cortège, and so on. That trend would compel many more to sell a loved one's body for dissection. The provost of Queen's College was concerned about the cultural divide between the poorest and the University over the burial issue. He informed Oxford Burial Board that he had spoken to a number of prominent clergymen about restricted burial space and the predicted increases in dissection sales. All agreed that 'on the ground of expense many... people would either be compelled to have a pauper's funeral or go round getting contributions'.[23] Begging for a burial would be a more common occurrence and those who failed to inter their loved ones 'decently' might generate social unrest.

Under these circumstances, Daniel Hookham would be able to rouse the mob against dissection without much effort in Oxford. He thus fell foul of full legal redress. Later this chapter will bring his story full circle (in its final section) when it connects Hookham's actions to the activities of a local coroner. Both acted in a confrontational manner over the anatomy trade, though for different motivations. Even so, anatomists decided to launch a new public engagement initiative spending the next 30 years trying to improve their supply. These coincided with the Medical Act extension and the arrival of a new anatomist, Professor Arthur Thomson, in 1885. His efforts were to shape the economy of supply in the foreseeable future.[24] Thomson's activities set Oxford's operational problems in context. His tenure in fact merits in-depth historical scrutiny about how he balanced his books.

Gentleman we are the geese – balancing the books

Anatomy may be likened to a harvest field.

First come the reapers, who, entering on trodden ground, cut down the great store of corn from all sides. These were the early anatomists of modern Europe, such as Vesalius, Fallopius, Malphigi and Harvey.

Then come the gleaners; all gather up ears enough from the bare ridges to make a few loaves of bread. Such were the anatomists of the last century – Winslow, Vic d'Azyr, Camper, Hunter and the two Monroe's.

Last of all come the geese – who still contrive to pick up a few grains scattered here and there amongst the stubble, and waddle home in the evening, poor things, cackling with joy because of their success.

Gentlemen, we are the geese.[25]

At Oxford University the latest intake of medical students was invited to an opening lecture on human anatomy at the start of the Michaelmas term in 1885. The lecturer commented on contemporary criticisms of the anatomical sciences in Victorian society. He did so because there was renewed criticism of the number of medical students and their reliance on the poor to train in dissection. The Medical Act (Extension) in 1885 was bound to exacerbate the anatomy trade. The lecturer admitted that human anatomy had a lack-lustre reputation in Victorian society. These 'geese' of the medical fraternity needed to improve their popular appeal.[26] During Queen Victoria's early reign, the discipline attracted a lot of public criticism.[27] The press highlighted the fact that the poorest saw little value in the 'few grains of medical knowledge 'gleaned' on the dissection table.[28] Although anatomists believed that dissection was central to the education of medical students, convincing local people that it improved their health care was an intractable problem. In Oxford, traditional death customs maintained a strong foothold in the public imagination. Religious sensibilities continued to stress the need to bury human remains intact.[29] A common attitude was that the cutting of the body on the dissection table was a sign of social failure.[30] To try to counteract these sentiments, anatomists pioneered public engagement work at Oxford.

Although the public engagement activities of leading scientific figures of national standing are well known, the popularisation of human anatomy in provincial university life has been neglected.[31] There were four key aspects of the outreach work pioneered at Oxford. Anatomists sought to generate fee-paying students to expand teaching provision; attract substantial university investment to build their disciplinary capacity; stimulate more government support for their research work; and counteract horror stories about dissection that tarnished their professional image in an era of widening democracy.[32] The public engagement reach of these provincial anatomists was not parochial but innovative and surprisingly modern for its time. The campaign was led by Professor Arthur Thomson. He was the man responsible for the economy of supply in dead bodies at Oxford University. His personal history is therefore of some relevance to this case study of the local anatomy trade.

Arthur Thomson was born in Edinburgh, on 21 March 1858, just four months before the Medical Act came into force.[33] His career path was to span the transition from irregular practitioners to professional medical men qualified for general practice. It was Arthur's father, James Thomson, who originally inspired his son's medical leanings. James had been a Royal Navy fleet surgeon and he encouraged his offspring to enter the profession.[34] In childhood, young Arthur had an inquiring mind, fascinated by anatomy and botany. That early promise flowered in his early twenties at Edinburgh University.[35] Thomson became a medical student, graduating with honours from one of Europe's leading medical institutions. He then chose an academic career, rather than entering general medical practice. After graduation, Thomson was appointed as a demonstrator in anatomy at Edinburgh. This position on the early career ladder was formative for talented anatomists in the late-Victorian era. It schooled Thomson in how to demonstrate the detailed anatomy of the human body to trainee doctors. He soon learned the art of clear communication and public speaking. Thomson, by now aged 27, decided to expand his career horizons. He applied for, and was appointed as, a lecturer in human anatomy at Oxford University. On his arrival in 1885, he joined the latest intake of medical students listening to the opening lecture on human anatomy staged at the Old Ashmolean Museum. He agreed that anatomists had an image problem and needed to redress their lack-lustre reputation. First, though, he had some immediate priorities.

On taking up his appointment, Thomson found that he had a major recruitment problem. Only three medical students wanted to specialise in human anatomy in 1885.[36] Thomson's solution was to recruit private donors from among leading dons of science to start a fighting-fund to create a central anatomy department in the University. He was a realist, admitting that he had three logistical problems. He first had to prove that human anatomy had academic merit and was financially viable.[37] It had to be capable of attracting and retaining regular student-fee income. His second challenge was to increase the department's visibility. Dissections were rare and not well advertised.[38] A third issue was poor demonstration space. It was difficult to watch dissections and so fewer medical students were encouraged to specialise in human anatomy. Thomson saw that the department needed to be housed in purpose-built facilities. Over time, his fund-raising efforts convinced the University Chest that he was committed to revitalizing the teaching of human anatomy. Convocation was impressed that Thomson increased his pupil numbers to 50 students per intake. By 1890, Thomson calculated that a minimum of 34 students were attending anatomy classes (most specialized in Preliminary Morphology).[39] The majority later signed up for medical degrees, which involved studying advanced human anatomy in their fifth year. Increases in annual student-fee income – 'from £18 in 1886 to £156 by 1890' – convinced senior dons to vote for building funds to improve facilities costing £7,250 in June 1891.[40] Thomson's hard work reinvigorated the Anatomy Department, housed in purpose-built premises by 1893. From its reconstitution, it became central to the division of medical sciences at Oxford.

On 21 October 1893, the day the new premises opened, Thomson recalled in a speech just how substandard the dissection room was on his arrival in Oxford.

> He was fortunate to be selected to fill the office of Lecturer in Human Anatomy in 1885. He looked back now with amusement at his earlier experiences. The great problem that he had set himself was whether

there was a future for the place. At times, things looked black enough, but he never lost hope, and his hopes were justified that day [i.e. today]. Coming as he did from one of the largest and best equipped schools in the world [at Edinburgh], the contrast was somewhat startling to find oneself appointed Lecturer in Human Anatomy with no place to lecture in, and possibly no one to lecture to [at Oxford]. He had begun his career as teacher in a shed containing himself, three students, a coal bin, and the "subject" and the number of students was now 30 ... recent history augured well for the future of medicine in Oxford.[41]

Contemporary accounts, like this, stressed that Thomson's early tenure was an unqualified success.[42] Yet he also paid a high personal price for prioritizing his management skills. The administrative burden of rescuing human anatomy from the doldrums meant that he had little time to fulfil his early promise as an anatomist. There were countless political fights to be won. After his appointment, Thomson was still fighting a rear-guard battle about how to teach medicine. In the mid-Victorian period, Oxford University sought to link its anatomical teaching to the rather cautious reform of the 'Natural Sciences'.[43] The aim was to develop a high standard of pre-clinical training for students who wanted to join the elite of the medical profession. This soon sparked a bitter debate about the quality of teaching in human anatomy at Oxford. In January 1878, the *British Medical Journal* accused Oxford of providing inferior medical training.[44] The editor bemoaned what became known as 'the lost medical school row'. In a series of bitter articles, dons were accused of lacking the basic skills in human anatomy to train '138 students'.[45] Thomson faced allegations that medical students still could not pass their licensing qualifications and were unfit for general medical practice after leaving Oxford. In 1885, the Medical Act extension added to his logistical problems. More medical students demanded bodies at a time when there was an under-supply of corpses. Against this backdrop, Thomson's early promotional work had a defensive tone and was time-consuming. His career at this juncture demonstrates the constraints that confronted anatomists in practical terms. Thomson thus concentrated on building his personal career profile in what little free time he had to spend in the dissection room. His petty cash records reveal his economy of supply. He was buying on a regular basis 'brains and heads' from porters at the back of the Radcliffe Infirmary for research purposes.

All his life Thomson was fascinated by the anatomy of the eye. His wife suffered from a number of troublesome eye complaints and her medical condition seems to have sparked his research interests in ophthalmology.[46] Thomson's meticulous research on 'the theory of the production of glaucoma' gained him respect.[47] His dissections established why glaucoma was worse at night. The eye, he explained, has a normal filtration system. Dirt is flushed away by a pumping action of small veins via Schlemm's canal in waking hours. At night, these veins in some patients relax when they sleep. Dirt can accumulate so that there is then a build-up of fluid contaminating the eye. The patient finds on waking that their eyes are red, irritated, and sore. It was well known that ophthalmic problems were common in pauper children.[48] An economy of supply would help to further his research work and be of national benefit. Thomson, however, did not just concentrate on this research. He was keen to diversify and reach out to other subject areas too. In recent histories his position as one of the foremost anthropologists, lecturing on medical and physical anthropology, has been recognised.[49] He is also praised for pioneering anatomy drawing at the Royal College of Art and becoming Professor of Anatomy to the Royal Academy in 1900.[50] Thomson was moreover an accomplished artist. He published the first life-drawing book in which anatomy photography illustrated the human body to art and art history students.[51] In 1885, however, those accolades were for the future. Most nights Thomson devoted his energies to calculating how to maximise his student-fee income and petty cash. His main concern was to resolve inadequate body-supply problems to allow the anatomy department to expand medical student numbers at Oxford. Ironically this delicate and sensitive issue that made other anatomists publicity-shy was to propel him into the public sphere. It is thus necessary to examine in detail the business of anatomy that he established and its practical operation since he found it hard to make the economy of supply pay its way.

In January 1908, Thomson wrote a frank letter summarising his financial situation over the last 20 years to the financial auditor of the central University Chest. He had been juggling competing economic pressures.

> Dear Fraser,
>
> The University is passing through a period of appeals. Everybody wants money. Now that it has been discovered that the University is in want of Funds many of the Colleges have likewise discovered that their purses are by no means burdened with cash. The result is a scramble for Gold and the whole matter has resolved itself in a case of "do take the hindmost" [i.e. any man for himself].

> Under the circumstances I am left no alternative but to follow the general example and beg for one's own department. I have less hesitation in doing this, for lately my energies have been mainly employed in trying to get money for other people's departments [in the medical sciences]. I trust that you will listen with patience if not with sympathy to what I have to say.
>
> I have long recognized that whilst the main function of this [anatomy] department is to teach, a no less important a duty is to turn out and encourage research work. To the utmost of my ability, I have endeavored to turn out original work.[52]

The double-dip recession of the 1880s and 1890s had reduced interest rates on endowment income.[53] Teaching budgets had been cut severely.[54] General research funds were depleted. It was nonetheless vital to remain competitive for cash-flow reasons. Bursars needed higher student fees to balance their books. It is therefore perhaps unsurprising to discover in Thomson's surviving archive common financial headaches very familiar to accountants in modern university life today. Thomson's public engagement work got underway because he urgently needed to acquire more cadavers.[55] At first he concentrated on making private transactions to try to improve body supplies. These were under constant negotiation. Suppliers had the economic advantage along a complex supply chain. An added difficulty was that medical schools had to compete with each other to improve their supply. Thomson's correspondence reveals that he thought this practical reality was divisive for the anatomical sciences. He soon approached his counterpart at Cambridge to coordinate their outreach activities, as the last and this chapter have already shown. As a first step, he helped to found the Anatomical Society of Great Britain and Ireland in 1886.[56] These professional contacts were to prove vital for pioneering public engagement activity in the ensuing years. Leading anatomists decided that they needed to convince the public that the anatomical sciences were not exploiting the poor to professionalise medicine. That public relations work, as we shall see, was concerted but had mixed success. It has consequently been neglected by historiography.

Thomson maintained that human anatomy still required sustained financial support. It was 'difficult without some pecuniary inducements to induce young fellows to engage in research in my subject from a purely scientific standpoint'. Thomson admitted that human anatomy was in competition with 'subjects such as Physical Anthropology or Human Embryology'. These 'new sciences' tended to attract the brightest students and he therefore appealed for an endowed 'Research Fellowship of the value of £100 to £200 a year' in his field. Thomson admitted that the discipline had its critics. It had been under pressure to be less parochial and more outward-looking since the Medical Act was extended in 1885. He admitted in a private letter to the Vice-Chancellor of Oxford University, 'It would no doubt be urged that Human Anatomy is played out. But take my word for it, this is not so, in the fields of Enquiry in which I have already referred, much yet remains to be done'.[57] Thomson emphasised that he had been working hard to develop body-finding schemes with Cambridge. He acknowledged that public sensibilities demanded practical resolutions. Together Macalister at Cambridge and Thomson at Oxford had adopted a concerted strategy to focus on the dignity of the human corpse on the dissection table to improve their professional image.[58] Twenty years later, Thomson explained that their efforts had 'been fraught with difficulties'. Only the economy of supply figures set those remarks in their proper historical context.

Thomson's petty cash records show that he travelled extensively to resolve his body-supply issues because he recognised the power of his personal touch to connect with the general public. He coordinated his activities with Cambridge, reconstructed in Table 6.1. In a private report, moreover, to Sir George Newman at the Ministry of Health, he recalled in 1921 that, 'As long as the results of my work depend on the voluntary attention of Medical Officers, Masters, and Stewards (particularly the latter two) in respect of the opportunities as they occur, there is always the fear of their efforts tending to slacken in the course of time unless restimulated at constant intervals by personal interview'.[59] Thomson complained that he could not, even if he had wanted to, just hide behind the high walls of academia. In his first year in his post (1885–1886) he journeyed by train around the Midlands (1885–6) to Birmingham, Leicester, Reading, and Wolverhampton. In Wolverhampton's case, to get supply deals underway he had to pay '£12 per body'.[60] As costs spiralled, he soon realised that a lack of visible central government support was serious. Civil servants refused to accept that what was needed was a government-run system of body redistribution to meet student demand. It was time to admit, Thomson felt, that the Anatomy Act had created a resentful political climate: 'the matter is largely a question of sentiment versus a distasteful duty'.[61] Of necessity, individual professors had to solicit in person public support for their work.

As Thomson travelled around the country and abroad his regular reading material was train timetables, weather reports, and shipping

Table 6.1 Travelling Anatomists: The Railway Journeys made for Public Engagement Reasons by Professor Alexander Macalister and Professor Arthur Thomson in Provincial England, c. 1880–1890

Railway Journeys Anatomist	England South	England Midlands	England East Anglia	England North
From Cambridge: Macalister	Bishops* Brighton Finchley Fulborn Hertford Hitchin Kingston Leyton Luton Reading Southampton Whitechapel	Bedford Birmingham Chesterton Derby Leicester Northampton Nottingham	Basford Biggleswade Bury* Chesterton Chelmsford Chelmsford Ely Huntingdon Mildenhall Saffron Walden Thetford Wisbech Whittlesey Yarmouth	Doncaster Hull Leeds Manchester
From Oxford: Thomson	Aylesbury Basingstoke Broadmoor Chichester Everton Wycombe* Newbury Reading Southampton Swindon Wells Wiltshire Whitney	Banbury Birmingham Chipping Norton Coventry Derby Faringdon Hereford Leicester Northampton Rutland Stone Warborough Wolverhampton	Q	Hull Leeds Manchester

Note: *Bishops Stortford, Bury St. Edmunds and High Wycombe, as cited in the original sources.

Source: Reconstructed from private papers and railways receipts in the Macalister Papers, Downing College, Cambridge and University Archives, Human Anatomy Department, Arthur's Thomson's files and petty cash books, HA1-3, 1885–1929. Bodleian Library, Oxford.

forecasts, not scholarly texts. His summer vacations (1885 to 1889) were spent overseas purchasing anatomical models for students to study when body supplies were low. Payments in his petty cash books confirm that purchases were made at renowned model-makers:

> Franz Joseph Sleger, Thalstrasse in Leipzig... Dr. Eger on Maximilian Strasse in Vienna... Dr. Ziegler at Freiberg in Baden... L. Cascian and

Sons 39 Wellington Way, Dublin...William Hume, 1 Lothian Street, Edinburgh...W. Barlow and Sons, 11 Forest Road, Edinburgh...K. Schall, 35 Wigmore Street, London...W. Wiltshire, 5 Pensions Gardens, St. Clements, Oxford.[62]

All the wax models were put in 'large show cases' costing '£5. 5. s[hillings]' and transported back to Oxford by sea and then railway.[63] At Freiberg in Baden it is noteworthy that he bought models of human embryos. Thomson's notes show that it was always very difficult to get the bodies of infants, children, or stillbirths to dissect.[64] Students also lacked dissection experience on young pregnant women. Wax models were therefore an essential aspect of medical training. The chief problem with these body-finding drives and model-buying expeditions is that they were onerous. Although they were not a new innovation – anatomists for centuries had purchased books and models[65] – the trips occupied most of Thomson's free time. Effectively Thomson had to become a travelling anatomist promoting his work by railway throughout England and Europe. The public engagement efforts of these anatomists were often arduous. Thomson also walked the streets of Oxford to buy bodies at nightfall.

Thomson's expenses reveal how he tried to expand the anatomy trade by being prepared to change his business style. His body-buying schemes are divided into two distinct phases in the historical record. In the first phase he made the deals in person and paid suppliers from his own pocket. By walking the streets on a regular basis, he tried to establish a local network of body dealers. Purchasing agreements facilitate a reconstruction of the local trading climate in Table 6.2. This primary data has been collected from burial records scattered around the city-centre before the Oxford Burial Board purchased land to set up a municipal cemetery in a suburb at Botley in 1894–1895.[66] That reconstruction shows the *minimum* number of whole bodies purchased for dissection that received a Christian burial. This distinction is important. Evidence from *Jackson's Oxford Journal* indicates that body parts were thrown into the river. It was also the case that the anatomy department tended to consign limbs to a clinical waste bin, rather than a multiple grave. Table 6.2 thus reconstructs conservative trading figures.

A total of '116 cadavers' were bought 'complete' by Arthur Thomson between 1885 and 1894.[67] He purchased them in the poorest parts of the city-centre. This was the fate of Mary Coolling, aged 25, described as a 'washerwoman from St. Thomas parish' who died from 'phthisis and exhaustion' in the Radcliffe Infirmary on 14 July 1886.[68] Her husband was out of work at the time and her body was traded on for dissection.

Table 6.2 Body Parts & Whole Cadavers Purchased for Dissection by Professor Arthur Thomson on behalf of Oxford University Anatomy School, 1885–94

Date	Number(s) and Type(s) Body Parts Purchased (Street & Hospital Deals)		Number(s) and Origin(s) Poor Law Whole Cadavers (Workhouse Sales)
1885	10 (arms, brains, legs and human material)	Witney	3 bodies
1886	14 (arms, brains, legs and human material)	Oxford	7 bodies
		Witney	2 bodies
		Wolverhampton	2 bodies
		Birmingham	3 bodies
1887	11 (arms, brains, legs and human material)	Oxford	13 bodies
		Wolverhampton	2 bodies
1888	10 (arms, brains, legs and human material)	Oxford	10 bodies
1889	10 (arms, brains, legs and human material)	Oxford	5 bodies
1890	10 (arms, brains, legs and human material)	Oxford	8 bodies
		Witney	3 bodies
1891	16 (arms, brains, legs and human material)	Oxford	10 bodies
		Witney	3 bodies
1892	12 (arms, brains, legs and human material)	Oxford	10 bodies
		Reading	5 bodies
		Witney	3 bodies
1893	16 (arms, brains, legs and human material)	Oxford	9 bodies
		Reading	2 bodies
1894	13 (arms, brains, legs and human material)	Oxford	16 bodies
Total	122 (arms, brains, legs and human material)		116 cadavers bought complete*

Note: *N = 116 (Oxford 88, Reading 7, Witney 14, Wolverhampton & Birmingham 7).
Source: Reconstructed from University Archives, Anatomy Department Records, Bodleian Library, HA1/1-3, record linkage work on daily financial records and petty cash expenses kept by Prof. Arthur Thomson, 1885–94.

Likewise, in December 1894, 'Joseph Jeffrey a 57 year old labourer' from 'St. Ebbes parish' was admitted to the Radcliffe Infirmary suffering from 'pneumonia'.[69] Three days later he died from an 'aneurism' and his body was sold on behalf of his poor family by the hospital porters. Thomson bought it in person at the back door of the infirmary. Similar deals were made with workhouses in nearby towns throughout the Midlands. Reading Poor Law Union, for example, was on the main railway line into Oxford. The master of Reading workhouse profited from a generous supply fee. Thomson paid him '£12' for each fresh cadaver sent by express train before nightfall.[70] The trade in body parts was vibrant too.

There are '122 transactions' in the petty cash records of the anatomy school for 'arms, brains, legs and human material'.[71] This finding shows what many historians of medicine have suspected but could never prove, that the trade deals for whole cadavers fell short of the body-parts business. It was more lucrative to cut up a body into parts and sell it on to medical students when complete cadavers were scarce in the supply chain. The business of anatomy was a breakers-yard for bodies. Human material was bought mainly from local suppliers to avoid adverse publicity. It could then be sold on to medical students in the locality with discretion, a recurring theme in previous chapters. Meantime Thomson's private notes indicate that, 'Demonstrations are also conducted with my assistant thrice weekly during term time... The dissection rooms are open daily from 10 to 4.30 and I supervise his practical instruction' on body parts.[72] He explained why these lessons were so important. He had 'soon reached a crisis, for in his lecture room it became physically impossible to squeeze more than twelve men together with his subject and Demonstrators'. Substandard accommodation and a lack of complete cadavers, these two problems made him reliant on body parts. The added advantage of buying parts was that they could be turned over faster in the department to maximise limited teaching space. On 20 July 1886 Thomson thus paid the 'burial expenses of S. Clayton £2 17s 1d [including] to porter [Radcliffe] Infirmary for brains, etc., 7s 6d'.[73] He again 'paid porter at Infirmary for '4 brains 10s on 31st August' the same year. Although he bought the body of 'Clayton the pauper', the '5 other brains' were much needed.[74] The human material came from unsuccessful operative surgeries at the Radcliffe Infirmary. An extract illustrating Thomson's petty cash payments in December 1886 shows this hands-on style of business and its black market in 'human material' transactions:

December 6th 1886	Tips to Porter at [Radcliffe] Infirmary	4s
December 6th 1886	Tips to Porter at [Oxford] Workhouse	1s
December 7th 1886	Tip to driver for Hearse	2s 6d
December 11th 1886	Burial expenses for 2 bodies [Oxford]	6s 2d

December 11th 1886	Expenses to Waterperry [Oxford] – W Henley supplier	4s 6d
December 13th 1886	Tip to [Radcliffe] Infirmary Porter for Brain & Leg	5s
December 20th 1886	Expenses to procure bodies – Wolverhampton & Birmingham	£18 4s 7d
December 24th 1886	Expenses Wolverhampton to Procure subjects[75]	£2 19s 3d

Even on Christmas Eve, Thomson was buying and selling the poor in person for the winter teaching session (cold weather preserved the fresh corpses and body parts for longer). By the 1887 session there was a complex array of people employed in the supply chain. On 25 April for instance Thomson paid for:

9 Brains [Radcliffe Infirmary]	£1 2s 6d
Witney [Dead House staff, supplier of body parts]	15s
Coroner's sub mate [for a body found dead in the street]	6s 6d
Washing [the body and parts by a local woman]	8s 9d
Tips to [Radcliffe] Infirmary porters & drivers of hearses	11s 6d
Sundries – cotton wool, chemicals & soap etc.[76]	18s 9d

The local voluntary hospital, coroner's office, washerwomen, hardware shops and workhouses were all in the unofficial employ of Oxford Anatomy School. Thomson must have carried around a heavy purse. He handed out a lot of petty cash on his city-centre walks. A few sixpences, a handful of precious shillings, sometimes as much as a guinea for a female body – these incentives ensured that Thomson was seen to be sensitive to cultural sensibilities. In the summer of 1887, for example, he paid funeral fees for three paupers who had been dissected. 'Matthew May, James Francis, and Anne Hargreaves' were all buried together at a cost of £2 12s 6d each'.[77] In the case of a fourth pauper called 'Margaret Nolan', a popular local figure, he expended '£9 19s'.[78] Thomson instructed a colleague 'John Milne Hastingdene' to make the funeral arrangements with decorum and discretion. The elaborate ceremony disguised the fact of dissection for a respected character in the community. Thomson was a pragmatist who knew the value of a good public relations exercise in Oxford.

Moving on to the second phase of the anatomy trade, surviving burial records provide a more complete historical picture because human material was buried together in Botley Cemetery to the northeast of Oxford city-centre.[79] A burial plot was set aside in this new municipal cemetery for the anatomical school. That move facilitates an accurate reconstruction of the body business between 1895 and 1929. Table 6.3 details cadaver numbers and their suppliers, indicating the distances each corpse travelled to reach the dissection table. The top three sources were Leicester Infirmary and its large city-centre workhouse (107 or

Table 6.3 Pauper Bodies Sold to Oxford Anatomy School, 1895–1929

In-coming Pauper Body Sales to Oxford	Numbers Sold, 1885–1929	Above 5%
Reading Union	102	25.25
North Evington Infirmary (– Leics. suburb)	87	21.53
Leicester Poor Law Union (– central)	30	7.43
Oxford Workhouse (– city centre)	29	7.18
Mental Hospital, Stone, Staffordshire	25	6.18
County Asylum (– Roundhay Street, Oxford)	23	5.69
Radcliffe Infirmary, Oxford	20	4.95
Crumpsall Workhouse, Manchester	9	
Hereford County & City Mental Asylum	9	
Oxford (– city centre street deals)	8	
Oxford Mental Hospital, Littlemore	8	
Knowles Mental Hospital (– Farringdon, Oxon.)	6	
Infirmary, Carlton Hayes, Warborough	5	
Leicestershire and Rutland Asylum	5	
Northampton Union, Northamptonshire	5	
Witney Union, Oxfordshire	4	
Wiltshire Mental Hospital	4	
Broadmoor Criminal Lunatic Asylum	4	
Poor Law Hospital Chipping Norton	4	
Graylingwell Hospital Chichester	3	
Wells County Asylum	2	
Southampton Infirmary	2	
Banbury Union, Oxfordshire	2	
Everton Infirmary, London	1	
Park Preweth House, Basingstoke	1	
Kidlington Union, Oxfordshire	1	
Camberwell Workhouse, London	1	
Unknown Sources	4	
Total	404	

Source: Oxfordshire Record Office, C/ENG/1/A1/2, Botley Cemetery records 1894–1929.

26.49 per cent); followed by Reading Union (102 or 25.25 per cent); and, in third place, the Radcliffe Infirmary and local welfare institutions situated in central Oxford (88 or 21.78 per cent). It is noteworthy that Leicester and Reading were stations on a main travel route on the Great Western Railway network. Thomson copied Cambridge's template and targeted sources of supply with fast transportation links. Thomson's transactions also reveal that his anatomy trade was small-scale compared to Cambridge – a total of 520 corpses as opposed to 2,986 bodies respectively (see Chapter 5). In round terms this meant that he lagged about ten years behind Cambridge; his supply figures were just 17.41 per cent of those of his Oxbridge rival.

Thomson knew his role was challenging when he arrived in Oxford in 1885. He had little control over curriculum disputes. Conservative dons were determined to promote the 'Natural Sciences'.[80] In this intellectual climate, it was always going to be an uphill task to kick-start the business of anatomy locally. Thomson's small-scale economy of supply should be assessed in the light of these background problems. In many respects the bare facts and figures obscure the true story of his tenure. Delving deeper into the archives reveals that Thomson worked very hard to make more supply deals. He was thwarted however by a combination of unfavourable circumstances. The next section explores Thomson's concerted efforts to resolve supply problems at a national level. That campaign was prompted by a professional dispute with a local coroner in central Oxford. This book reveals the importance of close professional ties between anatomy schools and coroners in the supply chain. When relations broke down they had a detrimental impact on the local business of anatomy. In what follows, the national and then local scene bring us back to Daniel Hookham, the pauper, by the final part this chapter.

The gravity of the present deplorable state of supply

> As anatomical teachers we hold that it is no part of our duty to provide subjects for dissection. The present system, whereby we are if necessary forced to appeal to the public bodies concerned often leads to a misunderstanding on their part that our applications are based on personal interests rather than on public grounds.[81]

Among Thomson's private papers is a dossier, marked 'confidential', of notes compiled between 1910 and 1921.[82] The file contains a great deal of information about his leadership of a new pressure group. In the provinces, prominent anatomists wanted the Liberal Government (1906–1914) to set up a National Anatomical Supply Committee System. The plan was to coordinate the redistribution of bodies on a pro-rata basis around the country. This would be calculated on the basis of actual supply lines and current medical student numbers. The Home Office on an annual basis would, via the Anatomy Inspectorate, apportion cadavers and then arrange to move them by railway from source to location after chemical preservation. Thomson hoped this would stop anatomists having to engage in black-market trading, a divisive modus operandi. Instead a concerted public relations offensive would involve anatomists working together to improve the anatomy trade on a national front. In 1987, Ruth Richardson touched on these plans briefly and outlined that central government did not lend their full support.[83] In 1912, just before the First World War and again after victory in 1919 (when the topic of death was a very sensitive one), politicians shied away from giving anatomists their public support. Complex behind-the-scenes negotiations, which Thomson coordinated, have recently been rediscovered. For nearly 100 years his private papers have lain undisturbed at the Bodleian Library in Oxford. This chapter opens the file on those secret negotiations for the first time.

In a sequence of private letters, penned over the winter of 1907, Thomson wrote frankly to the Home Office about supply problems between anatomy schools in England. The correspondence began when Thomson wrote a private letter to 'J. Pickering Pick, Inspector of Anatomy' requesting that he 'accompany him to Leicester Board of Guardians to persuade them to assist him in the supply of bodies'.[84] Leicester had refused to help Macalister at Cambridge in the mid-1890s, hiding the fact that he had been out-bid in a secret deal by Thomson at the time. Thomson felt unhappy about his conduct and regretted being in a similar situation again in 1907. New guardians now wanted higher fees and were threatening to trade one anatomy school against another. Thomson wanted nothing more to do with this type of sharp business practice and instead informed the Anatomy Inspectorate that it was time that they coordinated body-supply networks across the country. He explained that the previous arrangement had been that body-supply fees of '£10' per corpse were paid into a designated charity account called 'Leicester Poor Law Fund', held by local guardians at the local 'Westminster Bank' in the Midlands.[85] This ensured that the dissection profits remained 'off the books' of the Poor Law. They were always undeclared on the balance sheet at audit time. Pickering Pick disliked these financial schemes. He therefore agreed to help Thomson. The plan was that Pickering Pick would accompany Thomson on his return visit to Leicester to press for better supply in the national interest. He then had to cancel the appointment on the instruction of his senior colleague, 'Bennett at the Home Office', who inspected anatomical schools in London. Pickering Pick was told that Bennett had bowed to pressure from the 'Local Government Board'. Thomson made a private file note that civil servants

> were not sympathetic due partly to the attitude of Dr. Downs (Medical Inspector for Poor Law purposes) who seems to be almost antagonistic; for a memorandum from the Home Office asking for the Local Government Board to use their influence has apparently been shelved.[86]

Thomson had got caught in the middle of a complex political fight about the future of local government in a more democratic era.

There has been a lot of historical discussion in recent years about medicine and its professional standing in Victorian life. Two themes provide important context for this Oxford case study. Historians of welfare agree that inside the LGB there were important political factions among the civil service.[87] Their remit was to oversee the central administration of local public health and welfare standards. What this meant was that the Poor Law department dominated budgets and policy-making. It had the biggest tax revenues and therefore its work was audited carefully by the Treasury. Sometimes this practice operated in the LGB's favour. Civil servants could achieve major reforms, like the crusade against outdoor relief, on cost grounds with wider government support. At other times they could not carry out alterative proposals that might upset the social order in a democratic era. Nothing must be allowed to threaten the collection of tax revenues from an expanding electorate. Given that financial backdrop, the LGB favoured consensus in its working relationships with the general public. That style of management however worked against anatomists. Dissection was offensive to the majority of the population. Medical inspectors in the Poor Law department always urged caution about amendments to the Anatomy Act. The Home Department (later Home Office) had the authority to override the LGB on matters of national interest. Thomson pressed his case on this basis, but with no success. If reform threatened to upset the delicate balance of law and order, then it would have to be 'shelved' until a more politic date. A scheme to redistribute cadavers by railway around the country, argued Dr. Downs (medical inspector), could cause widespread public criticism at a time of prolonged recession. The Home Office concurred and withdrew their public support for Thomson's body-redistribution scheme.

Returning to Thomson's private dossier, his notes indicate that he thought the reasons for low supplies were a combination of: 'The operation of insurance and funeral societies; the claims on sentimental grounds of religion and other organisations; the operation of the Old Age Pension Act; and the recent decline in mortality rates.'[88] In a more democratic era, politicians had to appease the voting labouring poor with public health and welfare measures. These were beneficial to society as a whole but they also reduced the number of corpses in the supply chain. As Thomson explained: 'All supplies are a problem since no reliance can be placed on any individual source of supply, but that supply is liable to be interrupted if not permanently discontinued by the majority vote of those empowered to exercise a permissive control over it'. He admitted 'in confidence' that 'where bodies are in short supply, schools have offered burial inducements'.[89] Yet this created other problems since, 'The competition is so fierce that a black-market in burial is operating'. The only sensible solution was to create what he called 'a pooling-system' given 'the gravity of the present deplorable state of supply'. The stumbling block, however, remained, the LGB despite Thomson pushing for reform in 1908, 1912, and again in 1919.

The New Poor Law had been intrinsic to the operation of the Anatomy Act and so it was not easy to reverse its official reach in central administration. Besides, for 30 years the crusade against outdoor relief had revitalised the anatomy trade. The coming of local democracy was soon after to transform supply chains by making it much more difficult to purchase cadavers.[90] By then Thomson believed that the Poor Law now needed 'a new clause' that compelled guardians to supply dissection cases. Predictably politicians in areas with marginal seats were very anxious about bringing Thomson's new proposals before Parliament. He therefore suggested a compromise to the Home Office. Thomson was asked to design a pilot scheme that would be monitored by the Anatomy Inspectorate. If his redistribution model worked, it could then be authorised by a Select Committee rather than the full House of Commons. Thomson assured the Home Office and LGB that 'he was quite prepared to consult his legal advisers as to introducing the necessary clause [for compulsion] into a New Bill [Poor Law] and was quite prepared to explain quite frankly...what it was for' at a private meeting in front of a Select Committee.[91] Thomson admitted there were 'dangers attending to such a course of action' – it might be leaked to the press – but he thought a solution had to be found.

From as early as 1889 Thomson blamed the 'London schools of medicine entrusting too much descriptive aspects' when teaching human anatomy.[92] There was, he argued, too much emphasis on large lectures and not enough on hands-on dissection. Thomson regretted that metropolitan and provincial medical education systems had divided along two distinctive human anatomy lines. There was a two-track model of training in operation. Thomson stated in his private notes that lecturing to trainee doctors in a large lecture format functioned well since students were 'going on' with 'the ultimate object...to become surgeons or physicians [in] hospitals'.[93] At Oxford large lectures were unworkable because most medical students were destined for general practice. They needed a period of very concentrated and precise training in human anatomy to give them enough of a grounding in the basics. That compensated for the fact that they saw fewer patient case histories before qualification than on a busy London hospital ward. Thomson concluded that training on the human body needed to be appropriate to a medical student's career destination in local or city life.

Thomson recorded the job prospects of 471 students who registered for a medical degree at Oxford in the late-Victorian period (see Table 6.4).[94] Three trends are noteworthy from his calculations. Around 123 students or 26.11 per cent never completed their qualifications. Retention rates were low because of a lack of private funds to complete the training or because students tired of the arduous study hours. Of those who did complete their human anatomy training, 132 students or 28.03 per cent went into general practice after Oxford. Thomson saw these medical men as the stalwarts of the profession. Only 32 students (18 in London, 14 elsewhere), or 6.79 per cent, secured staff appointments in large hospital wards thus proving his point about the need for concentrated training in human anatomy. The profile of doctors registered with the General Medical Council had diversified after the extension of the Medical Act in 1885. In Thomson's opinion, the economy of supply and human anatomy training needed to reflect that changed reality. It was a national priority to supply the quantities of human material that matched rising medical student numbers and their future job prospects.

Table 6.4 Arthur Thomson's List of Medical Students and their Career Destinations after Oxford in the late-Victorian Era

No. of Oxford Students	Career Destination	%
123	Still Unqualified	26.11
132	In General Practice	28.03
18	Hold Staff Appointments in London	3.81
14	Hold Staff Appointments in Provinces	
17	Engaged in Science Teaching	
12	Are Abroad	
9	Are Museum Staff or Modelling	
4	Are in Indian Medicine	
2	Are in the Navy	
5	Hold Appointments in Public Life	
2	Asylum Physicians	
48	Have embalmed papers	10.19
17	To my Knowledge have Given Up	
56	I can get no records	11.89
12	To my knowledge are dead	
471 in Total*		

Note: *Reproduced as per the original source.
Source: University Archives, Anatomy Department Records, Bodleian Library, HA51, handwritten notes by Professor Arthur Thomson on his student's career destinations after Oxford in the late-Victorian era.

Thomson's work behind-the-scenes culminated in a pilot project that he submitted to central government in October 1919 when the Anatomy Inspectorate was merged into the new Ministry of Health. After World War I, this change of administration represented a new opportunity to press for reform. Thomson thus recommended:

A. Each school is to have first claim on sources of supply derived from towns within a reasonable area around it...More densely populated schools will not go far afield.
B. Therefore it will be necessary to consider the size of respective schools and their corresponding claims for proportional supply.
C. Any sources of supply opened up outside 'school areas' should be pooled in the interests of the ten provincial schools.
D. In the event of surplus...subjects to be added from the pool areas.
E. Assistance of government sought in order to reduce if possible the costs of transport.[95]

By way of example, Thomson outlined Oxford's pooled area over a 35-mile radius, with a total population of 1,082,078:

Oxford (52,979),
Northampton (90,064),
Banbury (3,118),
Witney (3,529),
Swindon (50,751),
Newbury (12,107),
Reading (87,693),
Thame (2,957),
Aylesbury (11,048),
High Wycombe (20,887).[96]

This was a wide geographic sweep across the south Midlands. Thomson also summarised the urgent situation in a private letter to Dr. Christopher Addison, MP, who was prepared to support a Health Bill to change the Anatomy Act, making it compulsory to supply bodies from welfare institutions funded by local taxes. He wrote:

Present Situation (1919)
A. 1913 conference lead to *no practical result* [sic]
B. 1914 war broke out, committee *could not press government so laden with problems.* [sic]
C. August 1914, mobilisation led to a fall in medical student numbers, reducing the levels of inadequacy of supply of subjects (to the extent that some supplies had to be refused)
D. Later in the War, the need for doctors led to a surge in the number of training causing the problems to arise again
E. Anatomical Committee convinced that *nothing short of an amendment of the Anatomy Act 1832 will surmount this great barrier* [sic].[97]

Behind closed doors, Thomson attended a secret meeting at the Ministry of Health on 26 November 1919. In his private notes made afterwards, he admitted that he said that 'the number would in fact satisfy the proportion of one and a half subjects per medical students for all purposes'.[98] It was 'not numbers but distribution' that was the key 'problem'. These private admissions are noteworthy since they further our historical appreciation of what was really happening inside the provincial business of anatomy. A wider variety of human material was being traded with regularity. These did supply medical students' needs because over the course of the New Poor Law the economy of supply had expanded into two businesses: one in body parts; the other in whole cadavers. Consequently the overall trading position was adequate by 1919. Human material was often, though, in the wrong place. Revealingly, Thomson admitted in his private notes that he was being disingenuous too. Thomson had carefully hidden his local difficulties from the Anatomy Inspectorate. This was, he wrote, the price of reform in the national interest.

This chapter now turns its attention to the complex relationships that developed between anatomists, the coroner's office, and main voluntary hospital in Oxford. A bitter professional dispute was what really undermined the local economy of supply. What follows emphasises the importance of collecting and collating archives when researching the history of anatomy and its chain of supply, and shows portable lessons that a study of this nature provides in terms of the operation of the Anatomy Act.

Corpses for the coroner

Oxford newspapers announced that a new surgeon had arrived from the capital in 1854. Edward Law Hussey Esq. had graduated in medicine at St. Bartholomew's hospital in 1845.[99] In the intervening nine years, he worked at several large city dispensaries where he gained a lot of experience caring for the sick poor. He also studied at the Lying-in Hospital in London, where he learned the art of midwifery. Hussey told reporters on alighting from the stagecoach at Oxford that his early career path reflected his interest in child medicine and treating the poorest in Victorian society. In his private notes, Hussey also wrote down that he left London for two personal reasons.[100] It would take many years to climb the ranks of a large hospital's surgical staff (he was impatient to make his medical reputation), and his uncle was Regis Professor of Ecclesiastical History at the University. He encouraged his nephew to set up a general practice at No. 2 St. Aldgate's in the city-centre. Local newspapers welcomed his surgical skills but little did their editors suspect just how many times they would feature Hussey in the press over the next 40 years. He soon proved to be an outspoken man of maverick opinions with a forthright professional style. Once Hussey was elected as city-coroner in 1877 he became a major stumbling block for the Anatomy School and its fragile economy of supply.

Returning to 1854, Hussey discovered on his arrival in Oxford a tense 'town and gown' atmosphere. Dons dominated local life from Oxford

University. They were guardians of the poor, members of the town council, sat on most local charities, and were board members of the main voluntary hospital. Civil society was thus run by dons with varying degrees of expertise. Hussey was an outsider and he would have to build professional alliances to make inroads into the medical marketplace, which had been expanding with some success. By 1875, Hussey noted that:

> The 'Means of Medical Relief' in the neighbourhood have been largely increased in late years. There are now 3 Hospitals with beds close to the town (in the Cowley road, at Bullingdon, and at Headington Quarry) maintained by the Local Authorities under the Poor Law. A dispensary in St. Clement's has been opened. The medical dispensary is now divided into two districts, with a separate Medical Officer to each; and a similar division into districts with separation has been made in the Lying-in Hospital ... There are many Sick Clubs and Benefit Societies – well managed and in active operation: the pecuniary subscription being small in amount and within the power of every working man in steady employment.[101]

Between the mid- and late-Victorian periods, the medical market in Oxford had become congested. Physicians, surgeons, apothecaries, homeopathic doctors, and quacks all advertised their services. In 1854, 'Nineteen surgeons and six physicians' listed in *Slatter's Oxford Trade Directory* charged sixpence for cheap medicine and up to a week's salary for personal consultations.[102] Hussey knew that provincial doctoring would be challenging but he did not anticipate just how difficult it would be to establish a private medical practice in a place where the University predominated. Once a rural recession got underway, the poorest outnumbered affluent residents. Numerous beggars, prostitutes, street-sellers, travelling show people, and vagrants pressed the charitable, health-care and welfare agencies in the city. Hussey had originally opened his business for private clients but as recession deepened he had to diversify his income streams. Hussey ran a cheap dispensary in a city-centre parish where the poor congregated. At St. Giles, he attracted plenty of new customers but most had no ready cash to pay for his cheap medicines. Hussey admitted in his private notes that the business climate was stressful and he often suffered from 'flagging industry and infirmity of temper.[103] For a time he tried working as a doctor of the sick poor in some of the villages surrounding Oxford. Again his business plan was to start with the lower orders and then widen his clientele. In

Stanton St John he set up a parish medicine chest and advised the local rector how to treat common complaints like constipation with castor oil.[104] This type of low-paid work was regular but not lucrative. Property owners admired his work for the poorest. Most, though, preferred to buy medical services from a more exclusive doctor. There was a lot of snobbery in Oxfordshire life. By late 1854, Hussey calculated that he needed a salaried position to supplement his struggling income. He applied for and was elected to the Radcliffe Infirmary as a House Surgeon.[105] It was a position he was to hold from 1854 until 1877. Meanwhile, he started a campaign to get elected as city-coroner – by 1877 he was successfully elected. This was the beginning of his confrontational relationship with the anatomical school at the University.

Hussey's tenure at the Radcliffe Infirmary was controversial from its inception but he saw his difficult situation as a career stepping-stone.[106] A synthesis of his private notes suggests that he was determined to realise his ambition of being elected to the city-coroner's office. By tradition, Oxford had three coroners who worked in close cooperation with the anatomical school at the University.[107] Two coroners were academic appointments with exclusive jurisdiction over deaths on University premises. A third position was appointed by the Town Council following election by property owners in the city-centre. The city-coroner was responsible for unnatural and unexplained deaths that occurred within a five-mile radius of the city walls. It was this role that Hussey sought because it had a high profile and placed him at the centre of the flow of local knowledge from the central Victorian information state. In the past, the person appointed was legally, rather than medically, qualified to avoid any potential conflicts of interest. The Radcliffe Infirmary had, moreover, a rule that House Surgeons should not carry out their own autopsies following deaths in the wards or on the operation table. If a member of staff became coroner then they had to keep their surgical work separate from any post-mortem duties.

Influential University medical men served as hospital governors. They took a great deal of interest in the appointment of the city-coroner in Oxford because several leading board members were also anatomists. Naturally they were always keen to buy the bodies of the dispossessed to dissect, those who underwent just the 'view' of the body by the city-coroner and a local jury drawn from local property owners.[108] Revealingly a report in the *Pall Mall Gazette* entitled 'After Death' explained that: 'The differences between a post-mortem examination for the purposes of scientific enquiry and a dissection for the purposes of anatomical practice, are very great'.[109] The major organs were looked at for 'signs of decay and the function and structure of disease', but coroner's cadavers were not sliced into pieces, cut into parts, nor decapitated, unless the crime committed or circumstances of the death were very exceptional. Corpses were displayed internally, the skin held back, but the skeleton, its skin, muscles, and tissue was left intact. Only dissection exposed the entire corpse to the cutting-up process. This confirms why coroner's cases were important sources of supply for a local anatomy school, until, that is, Hussey became city-coroner in Oxford.

Hussey had three major disagreements with those at the Radcliffe Infirmary.[110] At first he expressed the view that hospital staff had a disdainful attitude towards the poor. His colleagues, he claimed, lacked a proper understanding of the 'holistic' side of medicine. A better diet, clothing that fitted, and slum clearances would improve the material lives of those who came into the hospital to recover from the vagaries of living an impoverished lifestyle. A second point of contention was the Dead House at the Infirmary. Hussey told the hospital governors that porters were contaminated by rotten bodies, foul drains, and the stink of decomposition. He described the mortuary room as 'a Damp and unwholesome cell' where body parts 'were not safe from violence' because they were sold on for dissection.[111] Naturally this claim made in the local press was an embarrassment to the hospital and local anatomy school. A third disagreement arose over the right of the coroner to retain exclusive official jurisdiction over corpses generated on the wards of the hospital. There was an unofficial policy of selling the bodies of the poorest, with their family's permission and for a small fee, to the anatomical school. Hussey alleged that the poor resented having to sell corpses for dissection and that the covert trade could hold back the development of forensic medicine. He had, moreover, a low opinion of anatomists, describing them in derisory terms: 'The Teachers of Anatomy in Oxford are not the exalted persons of distinction whom you have found in other schools'.[112] A case in the winter of 1877 soon exposed these professional tensions.

In 1877, Hussey was elected to the city-coroner's office. He told an assembled crowd that as a popular candidate, respected for his work among the poor, he would work hard to win the electorate's respect and trust. He promised to do his best, 'by my constant care to show, by my future professional character, and whole conduct, that their good opinion had not been misplaced'.[113] His first coroner's case proved to be the most challenging of his career. A domestic violence incident was to have a significant and unintended impact on the anatomy school and its economy of supply. Hussey got involved in a medical case arising from a bitter dispute between a 'notorious' couple named James and Marion Grainger.[114] They lived in a lodging house opposite the entrance to the main city-workhouse on Randolph Terrace, just along Cowley Road, in one of the most deprived parishes in central Oxford. Neighbours told Hussey that the Grainger marriage was tempestuous. Marion Grainger was a drunk. She was often violent. Once sober, her temper was fierce. One Friday evening, 10 January 1877, James Grainger returned home after working all week in a local china shop. His abusive wife was waiting for him on the doorstep. Marion demanded that James hand over his wages. He refused. Marion became aggressive. She shook her fists. James was alarmed but determined to stand strong. Witnesses later recounted to Hussey that Marion demanded that her husband buy her some gin and stout at the nearest public house or else she would attack him with a kitchen knife. James held his nerve and his wages back. Marion's temper snapped and she became violent. James left the house, walked along the street, and fetched her some gin to calm the situation. Later that night Marion and James sat down to dinner together. By all accounts, another argument broke out about Marion's drinking problems. James told his wife that she must stop drinking hard spirits.

In a dramatic gesture, Marion Grainger lifted up a china plate that her husband had brought home from his work.[115] She threw it at James Grainger. It broke in two, pierced her husband's stomach, and he started to bleed badly from a large open wound. There was some dispute later in court about whether Marion also stabbed James with a large kitchen knife in the groin. Witnesses said that they saw James stagger out of the house. Robert Hawes, a neighbour living at 115 Cowley Road, found James Grainger standing at the door bleeding profusely. A local doctor, the constable, and a crowd soon gathered around the victim. Marion meanwhile was in a foul-mouthed drunken state, declaring: *'He has got no more than he asked for'* [sic]. Robert Hawes helped James Grainger into 'a privy' where he pulled down his 'blood-soaked trousers' to expose several 'deep wounds'. Robert tried to stop the blood flow with a tourniquet. James Grainger, now faint with pain, ignored pleas from his neighbours to press charges. Sensibly he did though agree not to return home that night. A local doctor consulted Hussey who recommended he come into the hospital. James declined until morning.

At first light Marion Grainger woke with a sore head and little recollection of the previous night's domestic violence. James stayed the night with a relative but this was not unusual in their stormy marriage. Marmaduke Pratt, son-in-law of James Grainger, was very concerned the next day when he saw the extent of the open wound in his father-in-law's groin. James was now in excruciating pain and the swelling had tripled in size. He was soon admitted as an emergency to the Radcliffe Infirmary. He was seen again by Hussey, the surgeon on duty and new city-coroner. Hussey operated on James to get rid of the infection in his groin. It was about to burst and cause fatal blood poisoning. Sadly James Grainger died from a haemorrhage in his stomach. Hussey now made a professional mistake.

Hussey elected to undertake a post-mortem examination of his hospital patient in the Dead House.[116] Yet hospital procedures stated that he should hand the case over to a local coroner for independent review. Hussey was though both acting city-coroner and surgeon at the time of the operation. The body could not be sent back home to place it outside the city limits in the county coroner's area because Marion Grainger had been arrested and charged with murder. Hussey decided therefore that he had the authority to perform the forensic examination and coroner's Inquest. This was not illegal but it was ill-advised under hospital regulations. Meanwhile the case was scheduled before a Grand Jury at the Quarter Sessions court. At that hearing, an Assizes Judge criticised the conduct of Hussey, calling his actions 'illegal and unprofessional'.[117] He expressed the view that a coroner should not conduct a post-mortem on patients in which he had acted as House Surgeon. The judge in his conclusion also remarked to the jury that a coroner should be a legally, not medically, qualified man to avoid any conflicts of interest in controversial cases. In fact there was no legal breach of the legislation governing coroners but there had been an ethical one. Marion Grainger was charged with 'wilful murder', but, because the medical evidence had been compromised by Hussey's actions, she was found 'not guilty' and released from Oxford prison.[118]

The governors of the Radcliffe Infirmary called a special meeting to review Hussey's conduct.[119] Some of those present were dons linked to the anatomical school. For some time, they had been unhappy about Hussey's outspoken opinions and criticism of medical colleagues. He was not a diplomatic man. This was their chance to get rid of a difficult character. They gave Hussey one professional choice, either he resign as House Surgeon or city-coroner, but he could no longer hold both offices. A furious Hussey tendered his resignation at the hospital and thereafter he became a bitter critic of his former employers and the anatomy trade. This professional bad feeling soon impinged on local supplies of cadavers. As city-coroner he now refused to supply any 'unclaimed' bodies to the anatomical school at the University.[120]

When Arthur Thomson came to Oxford in 1885 one of his first tasks was to try to re-negotiate a better supply deal with the city-coroner's office. Thomson found, however, that Hussey was not someone willing to compromise.[121] He felt professionally slighted by medical men at the University and was determined to oppose the business of anatomy at every opportunity. This was a serious logistical problem for Thomson because all anatomical schools relied on local supply networks for three key reasons. Purchases in the vicinity kept down transportation costs. Anatomists could move fresh bodies from source to dissection table quickly. On the doorstep bad publicity was avoided because deals were discreet. Hussey, by contrast, started a publicity campaign in the local press to try to mend his damaged reputation.[122] At its heart was a strategy to expose the dissection of pauper corpses and body parts sold to the anatomical school by the Radcliffe Infirmary. This meant withholding coroner's cases too. In this, Hussey hoped to raise his profile and establish the city-coroner's office as a popular place of local justice. This self-conscious attempt to re-fashion his professional credentials relied on exclusive access to, and jurisdiction over, any corpses that came under his official remit. As a result, there was a metaphorical 'tug of war' over dead bodies in Oxford. When, therefore, Daniel Hookham stood up in Oxford City Police Court in 1878 and accused guardians of selling on the poor for dissection, he was not only speaking from firsthand experience but also drawing on adverse publicity generated by the city-coroner's office. That context exemplifies the need to explore in greater historical depth the links between pauper agency, professional disputes among local officials, and the sensitive issue of legal access to corpses in the economy of supply chain.

In Hussey's first decade in office (1877 to 1888) he recorded that he held 605 official Inquests (see Table 6.5).[123] He also passed verdicts in 197 non-jury cases. Coroners had the discretion to record a verdict of a natural death when the cause of mortality was obvious (drowning, hypothermia, heart attacks, old age). In another 56 cases, Hussey complained to the Mayor of Oxford that he lacked the forensic evidence and proper coroner's facilities to reach a verdict. Rather than stage these inquests at the Radcliffe Infirmary Dead House (which he argued was substandard with unhelpful staff) the cases had an 'open verdict' recorded because of what he described as 'the want of proper accommodation'. Overall what these figures reveal in Table 6.5 is that some 253 bodies (197 non-jury cases and 56 open verdicts) could have been supplied to the anatomy school without public enquiry but most were not because of bitter professional rivalries. For reasons of accuracy these

Table 6.5 City-Coroner's Cases in Oxford, 1877–1888

Year	No. of Inquests Actually Held	Non-Jury Cases No Inquest Necessary	Accommodation Sub-standard*
1877	26	9	5
1878	55	18	6
1879	45	16	7
1880	55	20	3
1881	63	9	4
1882	66	24	10
1883	54	23	1
1884	45	18	7
1885	57	21	2
1886	49	14	4
1887	67	24	4
1888	23	1	3
Total	605	197	56

Note: *Original phrase was 'Want of Accommodation'.
Source: The National archives, MH12/9721, City-Coroner of Oxford's report to the Local Government Board, 1877–88.

figures have been cross-checked with anatomy department records. Counting up the bodies identified as lost sales reveals that in fact during Hussey's entire term of office he withheld 192 bodies (1877–1887) and another 91 cadavers (1888–1894), making 283 coroner's corpses; in round figures, of the 520 purchased in total from other suppliers (1885–1929) some 54.42 per cent of extra supply was lost to the anatomy school from the coroner's office. This was because even though Hussey retired in 1894, the under-supply issue with the city-coroner's office was never resolved until the demise of the New Poor Law.[124] It would appear that in terms of the economy of supply it was unwise for a struggling, smaller anatomy school to fall out with its local coroner's office. It unquestionably contributed to the failure of the business of anatomy at Oxford. The next section explores those pauper case histories that entered the supply chain and the broader historical lessons that they exemplify from new source material.

Pauper case histories: charity begins at home?

Throughout this book new evidence has been attempting to look up from, as well as down onto, and around, the dissection table. The survival of patient books and their death certificates at the Radcliffe

Infirmary makes it feasible to examine the burial records of the Anatomy School in more human detail.[125] Paperwork connected to pauper case histories can be used to reconstruct individual lives in reverse, from the point of death through to their living medical profiles. A selection of the best documented and representative cases sold for dissection will be set out below. First, however, an overview of the basic demography of the body supply chain sets those lost lives in context. Figure 6.1 reveals that Oxford Anatomy School bought three men to one woman, a ratio similar to Cambridge.[126] Females were always in very short supply, though, compared to all the other case studies presented in this book. Oxford medical students destined for general practice seldom dissected a female body. Some 95 bodies or 18.26 per cent of cadavers purchased were women traded between 1885 and 1929. This finding underscores that book-learning and wax models were essential educational tools. Children were likewise bought in smaller numbers. Only ten case his-

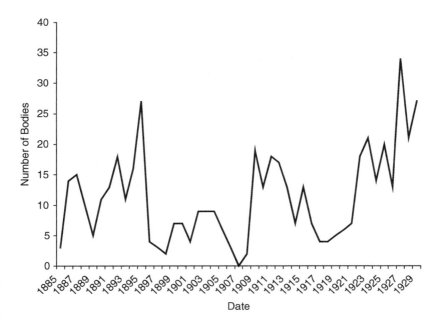

Figure 6.1 The Number of Bodies Sold to Oxford Anatomy School, 1885–1929.

Source: Reconstructed from record linkage work HA1/1-3, Human Anatomy department archives, 1885–1929, Bodleian library Oxford and Oxfordshire Record Office, C/ENG/1/A1/2, Botley cemetery records, 1894-1938, whole cadavers buried.

tories or 1.9 per cent can be accurately reconstructed from surviving records, a low number. Middle-aged men supplied almost two-thirds of the human material that was dissected. Human anatomy at Oxford was therefore a male preserve. Young male medical students cut up men over 50 years of age. Trainee doctors would have learned a lot about the musculature of the human body but much less about its different biological functions in the dissection room. Anatomy textbooks like *Gray's Anatomy* (1858) were a boon and yet leading anatomists regretted student reliance on rote learning. An intuitive feel, a hands-on sense, the ability to visualise the female body were beyond the grasp of those that attended some smaller provincial anatomy schools.

Returning to the human face of the anatomy trade, a female pauper like Martha Gates can be found in the historical record. At the Radcliffe Infirmary medical case notes describe Martha as '48' years old in 'July 1894'.[127] She had been a 'labourer' who worked in the fields but left the land to seek work in town. Rural work for females was in decline and once a recession took hold most women had to reinvent themselves as domestic servants or factory workers. Martha originally came from 'Turweston in Banbury'. She had a 'weak-knee' and a bad dose 'of influenza' which made her 'poorly' on admission to the hospital. For 21 days she was fed, cleaned up, and her condition treated. Sadly, on release she died soon afterwards from a second attack of 'influenza'. Her fresh body was sold quickly on to anatomists. The deal was made by a surgeon at the Radcliffe Infirmary, a go-between. Although Victorians believed that charity began at home, poorer families looked to the Anatomy School to fulfil those filial obligations.

Thomson kept detailed records of his burial expenses and undertaking fees. When he bought a body from across the county he paid extra to bury it back home decently. On 12 March 1886 an entry in his petty cash book states:

Body from Witney

To Evans (Master) order	£1 0s 2½d
Boy's Wages [to carry body]	14s
Material for Inspection [body parts]	4s 7d
Hearse to Witney	£1 17s 0d
Tip to Driver	2s 6d
Tip to Hine [body dealer]	5s
Paid Hine for carriage of goods	5s 10d
Turpentine [for the body]	6d
Total:	**£4 9s 7½d**[128]

If the body was sent on the railway from a town like Banbury then the fees, including its return for interment, were:

February 21st 1887
To coffin and removal from workhouse	£1 9s 6d
Railway Co. charge for carriage	£8 8s
Letters & telegrams re body	1s 6d
Sundries inclu. fee for body	£9 19s
Total:	**£19 18s 0d**[129]

To try to trim costs Thomson made an undertaking contract with J. Stroud Esq. in central Oxford and he sub-contracted coffin expenses to a local carpenter who was paid £1 10s each time.[130] By 1888, Thomson had managed to drive down costs to '11 shillings' per coffin. The reason that he did this was because burial fees in Oxford were high for the poor, especially after the new municipal cemetery was opened. Thomson explained in 1905 to the Cemetery Committee of the City of Oxford that at Cambridge each body cost £1 14s to inter compared to £4 5s locally.[131] This was because the city authorities insisted that only two bodies were placed in one pauper grave space, whereas on average at Cambridge six went into the ground together. If a body was brought into Oxford from the surrounding area and not returned home for burial by the anatomical school, then the city authorities levied extra fees to discourage medical men from contravening customary practices. In a private letter Thomson wrote that this meant:

Grave space [purchase] in perpetuity 9ft x 4ft [maximum 2 bodies]	£3 0s 0d
Interment of the person above 12 years of age	£1 0s 0d
Interment of the person below 12 years of age	7s 6d
Still-born infants	2s 6d
Fee of Minister	2s 6d

In the case of Non-Residents the above Fees for Interment will be doubled.[132]

So although it was very expensive to return bodies on the railway, Thomson sometimes calculated that if he could send back a lot together in large deal boxes to save costs, it could be less expensive than burial in Oxford. A summary of his financial records (Table 6.6) indicates that he was spending more than his student-fee income in the first decade of his tenure. This was the financial downside to the bitter dispute with

Table 6.6 Oxford Anatomy School's Expenditure on Human Material, 1885–95

Date	Estimated	Costs of Supplying Human Material			Total £ s d
		A* Tips	B** Transport	C*** Undertaking Cash	
1885–6	£12 18s 6d	£9 3s 3d	10s 9d	£3 4s 6d	£12 8s 6d
1886–7	£12 5s 3d	£18 19s 3d	£2 12s 2d	£36 18s 10d	£58 10s 3d
1887–8	£25 14s 11d	£24 11s 9d	£21 14s 8d	£43 13s 11d	£90 0s 4d
1888–9	£78 16s 9d	£21 0s 6d	£10 0s 0d	£48 18s 6d	£79 19s 0d
1889–90	£59 12s 10d	£26 8s 8d	£20 0s 0d	£13 7s 4d	£60 6s 0d
1890–1	£55 10s 0d	£34 17s 3d	£15 3s 5d	£33 8s 5d	£83 9s 1d
1891–2	£78 3s 4d	£65 10s 0d	£29 15s 9d	£24 7s 11d	£119 13s 8d
1892–3	£83 19s 1d	£56 13s 4d	£12 9s 0d	£23 17s 1d	£93 9s 5d
1893–4	£87 10s 9d	£49 19s 0d	£9 8s 6d	£41 16s 3d	£101 3s 9d
1894–5	£78 4s 7d	£50 11s 0d	£17 9s 7d	£30 6s 7d	£98 7s 2d
Total	£572 16s 0d	£357 14s 0d	£139 13s 10d	£300 9s 4d	£797 17s 2d

Notes: *A = supply costs, tips to masters, porters, and so on.
       **B = transports costs, road, horse or van hearse & railway carriage fees and so on.
       ***C = coffin, ceremonial hearse, and burial, and so on.
Source: University Archives, Anatomy Department Records, Bodleian Library, HA/1/1-3, reconstructed from petty cash books of Professor Arthur Thomson, 1885–1929.

the coroner's office. His estimated costs understated his actual expenditure, recorded faithfully in his petty cash records. This finding calls attention to why he relied on a body-parts business. It was cheaper, burials were multiple, and transportation costs minimal.

Returning to the pauper case histories is also instructive about the children Thomson managed to purchase from the Radcliffe Infirmary. On 6 July 1895, for instance, he bought the body of 'Florence Mary Abbott aged 3/[12] months [sic 12 weeks]', the daughter of a 'butcher from Lower Wolvercote', about two miles from the city-centre.[133] The three-month-old baby was 'an urgent' admission who died from 'pneumonia'. Her body was sold to porters at the hospital and dissected down to its extremities over the next six months. In the recession her father's business was precarious. She was thus buried by anatomists in early 1896. Likewise the male 'Child of Edith Mary Pick', described as a 'servant', was just 'eight weeks old' on admission. The infant was christened 'Reginald' and had severe 'gastric catarrh'. It had an inflammation of the lungs, breathed with pain, and 'died at 5.15am on 25th July 1901'. It too was sold for anatomy by the porters on behalf of its bereaved mother.

Alongside Reginald Pick in the dissection room lay 'Thomas Betteridge 2 days old from Oxford' who 'came with Imperfect Arms' into the

Radcliffe Infirmary on 26 March 1900. Medical records suggest that there was a digestive deformity at birth because the child died within 24 hours at '6.30am on March 27th 1900' of 'Complete and Imperfect Arms and the Colon'. This corpse was entered in an *Abnormalities and Deformities Register*, dissected down to its extremities, and the human material retained for almost a year before burial. Young people were used in this way too. Many accompanied travelling shows that came into Oxford in term time. Those who worked the fairs had no families and few social connections locally. This was the fate of 'Arthur Perry aged 24, travelling showman – no fixed abode – diagnosis Morbus Cordis [heart disease]'. He died of a fatal heart attack the day of admission and was 'given over to the anatomists for dissection'. There was no one to claim his body. Sometimes the person arrived in Oxford to sell their services to students. A man named 'Paul Hüber', described as 'aged 39, German tutor – no location', was an emergency case in late 1894. The diagnosis was 'Phthisis' and on his death certificate the surgeon on duty wrote 'tuberculosis of the lungs'. On this occasion a hospital 'post-mortem was undertaken' after his death on '30th December 1894'. Despite his education and middle-class appearance, nobody came forward to claim his body which was sold on for dissection and buried six months later in 1895. All of this anatomical material had a human face and yet got lost in the historical recordkeeping.

These bodies were very valuable given the local situation with disrupted supply. The chief problem that Thomson had was that the dispute with the city-coroner's office deterred local people from selling on their loved ones. Most were anxious to avoid the social stigma of being reported in the local press. Hussey was after all determined to object in public to bodies being moved out of his official jurisdiction. He did select cases for public discussion with care because he was trying to promote himself as a popular coroner. Nonetheless, the local poor were wary of their cooperation being exposed in an era when dissection was tainted with cultural failure. A controversial vaccination case in 1885 set the tone. A young baby was treated by Dr. Alan Thompson, Public Vaccinator against smallpox after an outbreak at Cowley, a poor city-centre neighbourhood.[134] At the St. John's Street Dispensary this was done in several stages: 'On 19th November 1885...James J Coolling (the child) was vaccinated from Chas. Boscott who was vaccinated from Rose A. Symonds who was vaccinated from N.V.E. lymph'. All the children were in perfect health and returned home. The vaccinator confirmed in his notes that 'I saw the child again on the 17th November...the vaccination was successful'. James however died after a severe rash, fever, and breathing problems on '22nd December 1885'. Hussey became involved as city-coroner because the child died in a local lodging house. In a succession of letters to the LGB, it became clear that the body was a valuable commodity and so there was a lot of argument about who it belonged to after death. The public vaccinator, Dr. Alan Thompson, was related to a respected surgeon in the city named Harold Thompson. He (Alan) asked his relative (Harold) to undertake an autopsy and complete the death certificate, to which Hussey as coroner objected on ethical and professional grounds. Harold Thompson thus reported to the LGB:

> I have examined the body of this child but have not opened it. The glands in all parts seem healthy and without enlargement. I have seen several cases of *Erysipelas* of late; and I have heard of several others. From the mother's description I think that the Child had *coetaneous erysipelas* [fatal skin infection].
> It might arise from cold, contagion or various other causes. Children will often die from convulsions in many weakening diseases.[135]

Three things are worth noting about this evidence. First, the surgeon carried out an autopsy, a 'view' of the body, but not a standard post-mortem. External physical evidence was used to establish the cause of death. Second, this made the child's body a commodity for dissection. It was uncut and therefore a prized teaching tool. Third, infant deaths were so common that the demise of a poor child living in a rough lodging house was treated as an everyday event rather than a suspect one. Under normal circumstances this type of case would have been an anatomy sale. Hussey, though, was determined to thwart that outcome. He called for another Inquest and made house-to-house inquiries which he subsequently reported privately to the LGB. His report included witness statements and revealed that the poor family might have been guilty of medical neglect. He thus justified his actions but at the same time he was preventing an anatomy sale. To civil servants this coroner's conclusion looked like an unintended outcome but not to those who knew the extent of local professional disputes. So alarmed were the labouring family by the gossip and rumours in circulation that they too sent in a letter to central government which read:

> Caroline Coolling, wife of Thomas Coolling, shoemaker, 16 English's Row saith:
> The deceased is the son of my daughter, Alice Coolling, needlewoman. His name is James Coolling, 3 months old tomorrow. I have

had the care of the Child from birth. It took to the breast well & was also fed from a bottle. The Child had good health. The crusts after the vaccination all came off, about a week after I took the Child the 2nd time. There was nothing wrong in the arm after the crusts came off. The Child was taken out frequently & he did not suffer from being taken out: he seemed more lively. The first sign of illness was he began to get very red around the neck. I saw it when I went to wash the Child. I think it was about the middle of the week after Mr Thompson saw the child the 2nd time. This redness spread a little one day & a little another. It was in patches as if scalded. It was *not* [sic] moist only redness. It went down to the ankles. It did not go over the shoulders to either arms. The left thigh was a little red on the Sunday morning before it died. It did not burst until after the Child's death. Last Monday week, I took it to Dr. Guinness. I took it again the following Monday. I got some medicine for it. The skin all peeled off, so the redness passed off. The Child got gradually weaker and died on Tuesday morning about a ¼ to 7. The Child had convulsions on Monday night and went off in a convulsion.[136]

What alarmed Hussey was that 'Dr. Guinness' was a 'homeopathic' practitioner who claimed to have medical qualifications – his business card said 'M.D. & F.R. C. S. I.,'[137] – but the coroner disputed those credentials. The Coolling family meanwhile were artisans that had slipped into absolute poverty. Alice Coolling was unmarried and so her child was illegitimate. Her mother, Caroline, cared for the baby so that the extended family could make ends meets – calculative reciprocity in action. When the baby boy fell ill, his grandmother bought a cheap skin remedy from a herbalist. The child should have been taken to a Poor Law doctor at the dispensary. Most labouring families though self-dosed to save money. The Coollings did not have any spare cash. Only a further archive search reveals the real depths of their hardship. Petty cash books kept at the anatomy school show that just six months later another close female relative would be sold on for dissection. This chapter has already made reference to 'Mary Coolling', aged '25', described as a 'washerwoman from St. Thomas parish' who died from 'phthisis and exhaustion' in the Radcliffe Infirmary on 14 July 1886. She was the cousin of Alice Coolling and her body was sent to the dissection room. There were few positive emotional choices for those whose makeshift economies collapsed in a prolonged recession. There was though one upside to this human story. The controversy resulted in friends, neighbours, and a local charity providing the Coolling family with enough money to fund a basic pauper funeral for baby James. The infant never entered the economy of supply, even though six months later the extended family had to sell Mary Coolling, a valuable young woman. The key point about this family history is that it reveals a local trend – for every one body gained, another was being lost. Once more a coroner's interference had disrupted the anatomical school's supply chain in the city-centre.

A final case in the records illustrates just what a 'tug of war' there was over bodies and why it had a detrimental impact on the economy of supply. A history of the Radcliffe Infirmary illuminates what happened:

> One Sunday morning, an old man, giddy, was brought to the Hospital unconscious suffering from hemipelgia. Four staff saw the man and agreed on a diagnosis of cerebral haemorrhage [a fatal blood clot on the brain]. Hussey demanded an Inquest must be held, that the Hospital post-mortem was incomplete and the stomach contents had not been taken for analysis, and persisted in his desire to hold another Inquest. The relatives were very much adverse to any publicity and made haste to secure the body, which was taken out of Oxford to be buried. Hussey posted officers at the confines of his jurisdiction on Headington Hill with orders to seize the body, but the relatives being warned went up another route and evaded them.[138]

The account does not state that normally this corpse would have been sold for dissection but it is the sort of typical situation that had developed at the hospital. Whereas bodies had been traded from the infirmary's back door with regularity, the city-coroner's interference made death a more public and stressful situation. Some poor families were doing all they could to avoid the hospital, coroner, and anatomy school in death. In another typical case where evidence does survive of the actual dissection sale, the labouring poor came to rescue the body out of the hands of anatomists to avoid controversy. An '84 year old woman' named 'Jane Burley' from 'Thomas Parish' died in Oxford Workhouse having been treated by the Poor Law and Radcliffe Infirmary.[139] The master sold her body because she was female and being elderly did not need a post-mortem. The records show that the city-coroner's office, even after Hussey had retired from the post, still wanted to interfere with that outcome. Jane Burley's corpse was thus 'rescued by family to save her from dissection' and buried in Botley cemetery in 1896. Although these examples are not statistically significant, they do illustrate common dilemmas and the reluctance of the labouring poor to be part of the economy of supply.

The outcome of this complex situation was that Thomson had to rely on cadavers brought in by railway. Yet this was an expensive business practice. Thomson's letter book indicates that he tried to secure cheaper railway contracts for moving bodies from source to dissection table. Most railway companies charged high fees to move corpses because of costly insurance. Professor Alexander Macalister at Cambridge did not suffer from this problem because he negotiated his contracts at a time when fewer medical schools were moving bodies by rail. He thus got a better deal to carry corpses at a rate of '2s 4d per 1 cwt' of body weight on the Great Eastern Railway line.[140] Thomson, by contrast, was paying 'about £12 to transport' each body. Correspondence indicates that this was an ongoing financial headache. In February 1909, for instance, he wrote to inform a University bursar auditing his books: 'My Undertaking bills, Railway fares i.e. amount to £164 2s 6d for the past year: during most times I had [whole] bodies. The cost per body works out at £11 14s. It seems a lot more than I thought but that is absolutely list price.'[141] The Great Central and LNWR based at Paddington station, Great Central Railway Office at Marylebone and the LNWR, located at Euston station, all refused to lower tariffs for taking bodies from source to Oxford. Thomson even tried threatening to move the bodies by road in a 'van' at a cost of '£10 per journey' and to share the expenses with Cambridge, but to no avail. The railway companies refused to budge. The freight was a lucrative source of income but it was also risky. If passengers saw the corpses then an unpleasant cargo would generate adverse publicity and might result in expensive compensation claims. Thomson reluctantly accepted the status quo because the railway was efficient, fast, and, discreet. It was also the preferred option of the Anatomy Inspectorate and since Thomson was keen to have their support for his body-redistribution scheme he remained committed to the dead train on the railway.

Conclusion

In February 1920, Thomson wrote confidentially to central government that: 'It is pitiable to think that a matter so entirely in the Public interest, and so obviously a case for State control, should be made a question of political prejudice and intrigue'. By return post, Alexander McPhial, Anatomy Inspector, wrote back:

> I have heard nothing *definite* [sic] about Government action. But I heard that Poor Law Inspectors after a canvas of a large number of

Guardians reported unanimously that only *legislation* [sic] will secure available sources of supply and in consequence of these reports, the Minister of Health had put the matter of the *least* [author's emphasis] form of legislation into the hands of legal experts.[142]

Thomson wrote again suggesting an interim solution:

An extensive list of further sources of supply which might be made available had been drawn up for central government's approval ... The list might be considerably increased if County Asylums were included and if also cases coming under the jurisdiction of Coroners were added.[143]

Yet Thomson also regretted his underhand actions, describing his daily body deals as 'a distasteful duty'.[144] In his detailed recordkeeping he left a fascinating historical account of how and in what ways the business of anatomy traded on the misfortune of the poorest.

One of the most striking features of this case study is that transactions that made up the body trade in whole cadavers (116 in total) often equalled and sometimes exceeded that of body parts (122 in total). In the early phase of the department's expansion plans, the survival of petty cash books has facilitated a reconstruction of the facts and figures over almost a nine-year period, 1885 to 1894, when Thomson was spending more on supply fees, undertaking, and burial than his student income was generating. His estimated costs were £572 16s 0d set against actual expenditure of £797 17s 2d. The basic demography of that workload indicates that anatomists were seldom able to self-select their preferred human material. Children and women were continually in short supply compared to middle-aged men. A ratio of three men to one woman seems to have been common throughout the provinces, reflecting family ties, as well as background economic and poverty conditions. That said, at Cambridge there were a lot more women compared to Oxford. This made human anatomy a male sphere: male students training on men in term-time. Corpses costing on average £12 came in by railway. The study shows just how much Thomson needed it for his public engagement work and trading networks. In every sense he was a travelling anatomist, much more than historical accounts have documented for the English provinces in Victorian times.

Another noteworthy finding is just how strained relations became between the Anatomy School and the city-centre coroner in Oxford, and what a detrimental impact that professional breakdown had on the local economy of supply. Thomson never admitted this problem in print to central government. It was nonetheless his chief operating headache. Despite inheriting the situation as a newcomer, Thomson could not re-negotiate a better deal with the coroner's office. This meant that although he generated 550 cadavers, he also lost another 283 corpses, or 54.42 per cent of the total supply, which in the past had been guaranteed.[145] The three lessons he learned were that it was ill-advised for any anatomy school to cross a local coroner; the home market in cadavers was important for the smooth functioning of the economy of supply; and close medico-legal cooperation was an important facet of effective public engagement work in the Victorian community. It has been estimated that there were 'roughly 330 coroners for England and Wales', generating 'around 5–7% of all deaths annually'.[146] This was a critical margin for somewhere like Oxford struggling to keep its medical students from drifting down to London.

The final finding is that Thomson's business was petty. He carried around a lot of small cash. His purse must have been very heavy with small coins spent on the street. Later he too employed go-betweens, such as hospital porters, workhouse masters, and so-called undertakers who were really body dealers. Everyone involved was making something, a pattern repeated everywhere. The fees must have been worth the time and trouble given how frequently they appear in the financial records. To the poorest, however meagre the amounts and scale of pauper funerals offered, those material benefits were precious in a world without welfare. Families did club together to avoid anatomy sales. Residents were charitable when a case like baby James Coolling's death reached the local press. Yet the same family that managed to refuse to sell his corpse did not have enough combined resources to bury another young relative, Mary Coolling, six months later. For every success story there was often an unspoken failure. This human side still lies neglected in local records.

Only by breaking down the anatomy trade on a case-by-case basis can lost lives be brought into sharper historical focus. This Oxford account has drawn out portable findings that deserve to be explored elsewhere. In the North of England, the focus of the final chapter, this book examines another important medical education centre at Manchester. Here anatomists traded across the Pennines, down as far as Cambridge and Oxford. Those North and South networks show that the business of anatomy was interwoven in the English provinces.

7
'Better a third of a loaf than no bread': Manchester's Human Material

There is one subject, however, which has given to our profession more odium than any other and it is singular that it should be one from which it is estimated surgical science has derived the greatest amount of advantage, – I mean the dissection of the dead. For the manner in which some men talk of dissectors and dissections, it would really seem that the anatomical labourer derived a gratification from his pursuit, beyond that which a conviction of utility as the basis of medical science affords; but I would say to such persons, *"Follow us into our secret researches, and you will not fail to be convinced that nothing but the love of science, the cause of humanity, the wish to prevent the immolation of our fellow-creatures, could incline us to exchange the pure air of heaven for the noxious effluvia of the dissecting room, or the tranquil security of other employments for that occupation which is pregnant with danger from contagion and wounds; it is seldom that a winter passes without some sacrifice of life to the perilous cause which we espouse."* [sic]...So long, therefore, as our researches are directed to the benefit of mankind, we have a right to claim all the protection that society and the legislature bestow...[1]

In the middle of the lecture theatre stood a 'small oval table painted green and set on small wooden wheels'.[2] Around it, the above address was made by Thomas Turner, leading Manchester anatomist, to new medical students at the start of the winter dissection season in 1840–1841. It took place at the Royal School of Medicine and Surgery, located on Pine Street, in the city-centre. Turner told students that he was very proud of his latest purchase, for this was no ordinary dissection table. It was an object of awe and veneration, symbolising the advancement of the anatomical sciences. Turner explained that 'on that very table John Hunter prepared some of those splendid specimens which now enrich the walls of the museum of the College of Surgeons'.[3] Its acquisition confirmed that medical education was in the ascendancy in the North of England.

Turner's lecture developed this theme by stressing that dissection was not a dirty science but a worthy endeavour. It was dangerous, difficult work, and only for the dedicated. Anatomists deserved the support of central government. The Anatomy Act was not perfect but it did need to protect the expanding medical profession from intrusive public criticism. It was this paternalistic view that predominated in provincial medical circles at the time. Turner was evidently proud of what had been achieved by Victorian anatomists. Manchester was improving its standards of Public Health and health care. The expansion of hospital sites and the growth of medical education seemed praiseworthy. More progress was sure to follow. However, a shady history of body supply – what Turner called obliquely '*our secret researches*' – lay behind this story of medical professionalisation in Northern towns and cities too.[4] This final chapter thus focuses on the human face of the anatomy trade in nineteenth-century Manchester.

Manchester and anatomy

Fiona Hutton's recent research has established that 'the growth of Manchester as an influential urban centre' began 'in the later eighteenth century' and 'was reflected in the proliferation of medical institutions' that served local health-care needs.[5] Medical education was located in private anatomy schools in central Manchester at the start of the Victorian era. It is important to appreciate that this type of human anatomy teaching was limited. It tended to be taken up by those surgeon-apothecaries who entered the profession at the lower ranks as emergent general practitioners. Two local anatomy schools attracted the majority of medical men from the Northwest of England. Joseph Jordan opened a small school of anatomy on Bridge Street. In 1814 he was advertising courses in human anatomy in the local press. Jordan's chief competitor was Thomas Turner who established his private anatomy school on Pine Street, close by, in 1824.[6] Jordan and Turner had very different teaching styles. The former advertised a practical and theoretical course in human anatomy; the latter offered a more hands-on experience of dissection. Through Turner's concerted efforts, Bridge Street and Pine

Street merged with the Royal Manchester School of Medicine in 1836.[7] Medical students meanwhile walked the wards of the Manchester Royal Infirmary (established in 1752). They could also attend 'a children's hospital, the Ardwick and Ancoats Dispensary, the Salford and Pendleton Dispensary, the Lock Hospital, the Eye Hospital, the Manchester and Salford Lying-in-Charity', and join 'the Manchester Medical Society'.[8] Though extensive, this list of medical facilities was not comprehensive. Fee-paying medical students were still ambivalent about whether Manchester offered the equivalent of a solid London education. The Royal College of Surgeons was likewise sceptical that a provincial training matched that of the metropolis.[9]

Chapter 3 explained that in the capital, pupils gained a certificate of medical attendance by walking the wards of a large hospital over a two-year cycle. They also had to prove they had attended two human anatomy lectures. It was a further requirement that they complete two dissection sessions at a recognised school of anatomy in London. In the provinces qualification times were doubled by the Royal College of Surgeons. This was because of concern about the lack of medical training facilities and lower standards of health care. John Pickstone's excellent history of health care in Manchester stresses that these criticisms were merited given how inconsistent medical education was at the start of the Victorian era.[10] At the Royal Infirmary there were no formal clinical classes, nobody could agree on a system of medical instruction, and there was little post-mortem work. Medical students first encountered patients walking the general wards but the training was not thorough. The passing of the Anatomy Act in 1832 would consequently prove to be an important catalyst for human anatomy teaching in Manchester.

The Anatomy Inspectorate was pleased to see the decline and merger of smaller anatomical schools into the Royal Manchester School of Medicine and Surgery between 1836 and 1856. Thomas Turner meanwhile was determined to get a proper economy of supply underway. He was delighted to report that anatomy was not only in the ascendancy but its body trade was vibrant too. In January 1849 he wrote to assure civil servants in a private note that:

> There exists no difficulty or opposition whatever on the Part of the Poor Law Guardians or Public Authorities and the sole reason for our not having an abundant supply, is literally that there have been comparatively speaking no deaths among the inmates of our workhouses...never did any public Bill act better than the Anatomy Bill does with us. We have no opposition, no annoyance in any way. We

are careful not to offend the public eye, and our removals and interments are conducted with great decency and care.[11]

This evidence indicates that the corpses of the destitute were being handled by anatomists rather than the city's Poor Law authorities, a theme discussed later.

Meanwhile to maximise that human material for teaching it had been essential to improve basic preservation skills. By 1871, new chemical techniques were being pioneered at the Royal Manchester School. Traditionally most corpses were preserved in alcohol for immediate use or set in wax for display in a museum collection.[12] During the Victorian period, anatomists started to experiment with 'picric, osmic, chromic and carbolic acids'.[13] The medical press reported that the problem with many acid compounds was that they 'make the fingers sore sometimes giving rise to small abscesses beneath the nails, and it is apt to soften the tissue of the subjects'. At Manchester, anatomists were forward-thinking, keen to use carbolic acid. Each new purchase was sent to a preparation room. Here a demonstrator assisted by a medical student stood on a table above the corpse. Carbolic acid was injected using 'hydrostatic pressure'. A solution of '1 in 40' was diluted into the 'thoracic aorta'. This 'hardened the tissues, both muscular and nervous, as well as preserving them from putrefaction for a considerable period of time'. Then 'three days afterwards' another solution was used, this time a 'coloured injection for the arteries... [a] cold red paint injection, propelled by a hand syringe'. The dye was a mixture of 'red and white lead, dryers, turpentine and oil'. The *British Medical Journal* praised Manchester's innovation, describing it as: 'a very beautiful mode of injecting the arteries, generally filling vessels only a few degrees removed from capillaries, such as the ciliary [sic] arteries of the eye and the minute vessels upon the intestines and in the skin'.[14] It was this dual preservation method that lengthened dissection times – from several days to up to six months. Surviving records suggest that bodies and body parts were in plentiful supply, even if anatomists were shy of admitting those financial details in print.[15] This practice was kept out of the public gaze. Thomas Turner did not, though, have everything his own way according to central government files.

Although smaller private anatomy schools had merged with the Royal Manchester School, it was significant that they operated on split sites in the city-centre. This gave entrepreneurs the impression that there was still scope for a new private anatomy school. Word soon spread that there was a large pool of fee-paying students studying at home. Arthur

Dumville (1813–1871) and George Southam (1815–1876) thus decided to start the Chatham Street School of anatomy in 1850, to Turner's chagrin.[16] For the next six years, it proved to be a thorn in his side. Until, that is, he persuaded them to merge with his Pine Street enterprise and thus amalgamate with the Royal Manchester School of Medicine & Surgery which was fully established by 1856. Meanwhile, a competition for corpses developed in the city-centre. The Anatomy Inspectorate was not happy about this rivalry. They warned Dumville that the economy of supply must not become divisive:

> It will be the cause of much upset to me in every way should anything occur to interrupt its quiet working. Will you bear in mind what I said about the windows in your dissecting room and have them made not to open at the bottom and altogether so arranged that you will not be overlooked by your neighbours.[17]

It soon proved difficult for the Anatomy Inspectorate to monitor the working relationship between Turner and Dumville. The rivalry was antagonistic and petty. Dumville complained about the quality of the human remains he was able to buy from the main workhouse in Manchester on New Bridge Street. The corpses they sold to him were often 'mutilated' and were poor teaching specimens for dissection.[18] These disputes culminated in Dumville's business partner, Southam, speaking about the situation to the local press. It was reported in the *Manchester Guardian* that Chatham Street admitted to medical students at a prize-giving in October 1852 that: 'One of the difficulties we have had to contend against last year, was the unfair and partial distribution of subjects for dissection; but this injustice has been effectively remedied, and the school now possesses all the privileges granted to similar provincial institutions'.[19] The Anatomy Inspector was not pleased about this public discussion. In a private letter he told Turner that he considered it to be 'the height of imprudence to allude to these matters at all in a report which no doubt was intended to appear in the public papers'.[20] The civil servant added that 'the statement with reference to their supply is not consistent with the truth' and he praised anatomists attached to the Royal Manchester School of Medicine for 'taking no notice of it'. They ought to get on with the task of building up the business of anatomy and ignoring their rivals. The reason for doing so was that by the 1860s similar medical schools were expanding their economy of supplies around the country. That trend threatened to challenge Manchester's predominance in Northern England.

Although the entire economy of supply up North is beyond the remit of this book, data collection indicates that there were extensive supply networks in operation at Birmingham, Leeds, Newcastle, and Sheffield.[21] These were growing exponentially with harsher Poor Law policies. It was noteworthy that many of the largest cities were also leading supporters of the crusade against outdoor relief. In Manchester a majority of guardians were keen to save rates. They compelled the poorest to come into the workhouse and withdrew pauper funerals paid on medical orders. The so-called *Manchester rules* were a financial trend that created a business opportunity to trade corpses. Alan Kidd points out that the 'campaign against outdoor relief was remarkably successful' in Manchester'.[22] Chapters 4 to 6 have illustrated that a lot of workhouses which denied this practice *were* selling on bodies throughout England. New evidence for Birmingham, another crusader, provides one of the best examples from the 1850s. The Anatomy Inspectorate wrote in a private memo for internal distribution how, 'At Birmingham the Guardians have methinks been well disposed to the working of the Act [sic].'[23] The destitute were bought on a sliding scale of fees, dissected and dismembered, and then buried by anatomists 'in Witton Cemetery – in a multiple grave'.[24] This was one of the largest pauper graveyards in England. Meanwhile, further north in Newcastle-upon-Tyne, the medical school was 'asking burial clubs for bodies' with some success where claims exceeded contributions.[25] Manchester had good reasons to be wary of its closest competitors and needed to respond in kind. Further changes in medical education were therefore imminent, as the *British Medical Journal* announced in December 1871:

> The Manchester school is now the largest in the provinces, and is indeed larger than many London schools. There are at present 106 registered students in the anatomy class alone, 65 of whom have taken out dissection cards. One of the great advantages of the Manchester school has always been the abundance of subjects e.g. in 1869–70 there were 43 distributions, and in 1870–1 there were 37. This abundance of the raw material is so constant that the authorities took no thought for the morrow, by preserving bodies prior to the opening of the school, a proceeding which the recent [Amendment to the Anatomy] Act allows. This regret is somewhat to be regretted, as for the first time for many years, there is a dearth of subjects at the School.[26]

By the 1860s, Manchester had lost its competitive edge in an expanding field of medical education. This was why anatomists were keen to pioneer new preservation, maximising the use of teaching material. It was also necessary to embrace further teaching reform or risk losing fee-paying students from their traditional recruitment areas in the Midlands and North of England.

Returning to Fiona Hutton's research, the establishment of Owens College in 1851 has been seen as a formative moment in scientific circles in Manchester.[27] Over the next 20 years this was the place where human anatomy teaching became civic, scientific, and non-sectarian. Owens College offered a broad curriculum in the Natural Sciences, classics, mathematics, and pre-clinical medical training.[28] Courses had a secular ethos, attracting the sons of businessmen, usually those keen to study anatomy, chemistry, natural history, and physiology. John Owens, a self-made textile merchant endowed the institution with £96,942 in 1846. New premises were soon added in 1851. This large amount of start-up money was not enough though to resolve cash-flow problems created by the Lancashire cotton famine in the mid-1860s. It was therefore the creation of a single Manchester School of Medicine by 1874 that persuaded Owens College that another future merger made sense.[29] Owens' men excelled at science teaching and they had the facilities to attract larger numbers of medical students to strengthen the academic standing of future collaborations. By aligning with a medical faculty they could together apply for university status.[30] By the 1880s, Owens College had a 'demonstration theatre, with accommodation for 100 students'.[31] The dissection room where practical work was done was large, '78 ½ feet x 32 feet' and 'lit by electricity'. This meant that in winter students could cut corpses 'three days weekly during the winter session and two days weekly during the summer season'. The dissection room was open 'from 9.30am to 4.30pm' each day. Each session was always supervised. Importantly dissections were conducted in rooms 'in a respectable and noiseless street'. These were behind the buildings of the medical school where there was 'no carriage thoroughfare' to disturb students studying. The design also discouraged nosy people, curious about what really went on inside the premises.[32] These arrangements helped Owens College gain a Royal Charter with its degrees overseen by the University of London.[33] Then, in 1882, a Royal Commission recognised the contribution of Owens College to general university education and gave it independent degree-awarding powers. This led to the creation of Victoria University, an umbrella organisation that comprised Owens College, University College of Liverpool, and the Yorkshire College of Leeds by 1887.[34] It was not, however, until 1904 that they were permitted to award medical degrees independent of the Royal College of Surgeons.[35] Permission was granted because of high academic standards and a plentiful supply of corpses to dissect throughout the late-Victorian era. A traditional history confirmed in 1908 that, 'One of the greatest difficulties of medical schools is to obtain a sufficient supply of subjects for dissection, but this has never been felt at Manchester...There is accommodation in the dissecting room for 250 students'.[36] From small beginnings in private schools the business of anatomy had burgeoned into a large-scale educational enterprise. It was dependent, though, on the removal of customary pauper funeral rites to facilitate the expansion of an anatomy trade in central Manchester. The next section analyses representative examples from new source material on local funeral customs. These explain what was under threat when dissection became more common and why what was lost became so contentious throughout the Victorian period.

Old and new poor law payments for pauper funerals[37]

Sir, the Ovorjeer of Hulme will pay you for the interment and funeral of Elizabeth Ross under Mrs Cropper midwife, Hewitt Street, Knott Mills to be interred at 4 o' clock tomorrow. 5th March 1833.

Sir, I will pay you the interment dues for Sophia aged 9 daughter of Hannah Tattersall in Jackson Lane Hollow this is to be buried in a grave. 3rd May 1833.

Sir, I will pay you the interment dues for Adelaide Landon aged 1 year 10 months to be buried tomorrow interred at Christ Church Hulme. 16th May 1833 interred And, daughter of Thomas and Sarah Landon, died May 13th of inflammation in the head. 15th May 1833.

Sir, I will pay you the interment dues for Geo Hughes aged 15 months. 31st May 1833.

Buried Geo, son of Josh and Alice Hughes aged 15 months. Died May 30th of inflammation. Interred, ground and dues at 6s. 10d. signed by the officiating minister at Christ Church Hulme, Geo Edwards. 2nd June, 1833.

Printed bill from St George's Church, Hulme for the funeral expenses of Esta Burgess, 12th January, 1833.
Burial   5s.
Dues   1s. 4d.

Service	1s.
Sexton	2s. 6d.
Bier	0
Bell	1s.
Entertainment	0
Total	10s. 10d.

Invoice for Edmund Waterworth making parish coffins. 17th January 1834

Large coffin Samuel Somerton	10s.
Small coffin for John Milner's	1s. 10d.
Large coffin for Jas Broadhurst	10s.
Small coffin for Bentley child	3s. 9d.
Small coffin for Catherine Ellen's child	2s. 6d.
Large coffin for John Tayor	10s.
Coffin for Thomas Landon	10s.

The above funeral payments were generated by the Overseer of the Poor for Hulme Township, at the time the New Poor Law was coming into force in 1834. Hulme was a suburb that got swallowed up by Manchester as the early Industrial Revolution gathered pace. These sorts of undertaking contracts have been used as a yardstick for sentiment towards the poorest.[38] The Hulme bills show that pauper funeral rites could be extensive before 1834. Three things are noticeable about these representative examples. First, the New Poor Law did not resolve the problem of what to do with the poor in death after 1834.[39] Someone still had to pay for their interment at a time when pauper funeral rites were an important symbol of a family's standing in the local community. Secondly, burials were frequent because during the early Industrial Revolution mortality rates were high in the North of England. On average 10,000 people died every year in penury in Manchester during the mid-Victorian period.[40] Of those, 3,500, never less than a third, applied for a pauper funeral. Some were granted, others refused and re-directed to Manchester medical school. Thirdly, all the Hulme funeral bills quoted were paid on medical outdoor relief orders. Throughout this book evidence has shown that any change to that traditional payment had an adverse effect on the bereaved and a positive impact on the anatomy trade. Hulme remained committed to pauper funerals but increasingly other areas of Greater Manchester did not.[41] This trend was what helped to expand the business of anatomy in Northern towns and cities. The context of these funeral bills can also be explored further via other pauper letters in the Hulme Collection. One well-documented medical case exemplifies many in this considerable archive.

In the 1830s, the Gunell family moved from Leicestershire to Manchester. For a time they got regular employment in warehouses but as they aged it became harder to make ends meet. Sickness became more frequent and ill health started to undermine their makeshift economy. Eventually they had to apply for a regular medical outdoor relief allowance. John Gunell was refused a poor relief claim in central Manchester and thus had to write back for help to his parish of settlement at Great Oxendon near Market Harborough in Leicestershire.[42] John Gunnell wrote:

Manchester August 19th 1832...

Sirs

I wrote a considerable time ago stating the depressed state of trade and Consequently our want of employ. I am sorry to say that after our waiting so long in expectation of some change for the better taking place things are Daily getting worse so that our situation is now become such that it is totally impossible for us to do any longer without Present Relief and not that only but we have been so far back in rent that our Landlady's patience is wholly tired out. In fact to be candid and plain with you (which I hope you will excuse as I consider such conduct to be most improper) I have to say that if you do not remit me to the amount of 2 Pounds my goods must be sacrificed and we must come to you altogether and if that has to be the case it must be under a Sick order as neither myself nor my Wife are able through bad health and infirmities to come any other way.

Manchester is now become a Scene of Distress through the Violent raging of the Cholera it is this day stated that there has been 100 deaths this last week (God knows how soon it may be our lot) and it seems everyday to gain strength. A Neighbour 6 Doors below us was seized last Wednesday about 4 o'clock in the morning – Visited in the forenoon took to the Cholera Hospital about 2 in the Afternoon and died before 4 the same afternoon...[43]

A follow-up letter is then instructive about the pace and scale of urbanisation and its impact on this poor family trying to survive absolute pauperism caused by cholera in 1833–1834. It is quoted in full because his address provides important context.

Sir

I am sorry to have to write to you upon the present occasion of Distress I had the English Cholera in July last and in consequence of age and infirmity I have not been able to do anything since. I therefore hope you will have the Goodness to Relieve my Necessitous situation, I am considerably in Arrears of Rent, I hope Sir that what is said is sufficient and shall therefore humbly wait for your answer by return of post and
    Am your most obliged
    Humble servant

John Gunnell
117 Portugal Street
Oldham Road
Manchester

I am still in the same house but the Street is much enlarged and the number augmented from 60 to 117.[44]

John Gunnell and his family lived in a location with one of the fastest growing population densities in Manchester centre. City census records show that 71 per cent more people came into the area between 1831 and 1841.[45] Portugal Street had lengthened, almost doubled, and so the Gunnell lodging-house number kept changing. The new premises were being built cheek by jowl, with the poorest crammed inside smaller social spaces. Public Health measures could not keep pace with the lethal cocktail of human excrement polluting the streets and wells. It was only a matter of time before death called at the door of the Gunnell family, as the final pauper letter in the series explains:

Gentlemen

About a fortnight ago my late husband wrote to you stating his and our Situation saying that he expected it would be his last time of troubling you which has been the case he Departed this transitory life on last Monday morning but one. We have not as yet received your answer I therefore write to inform you that I am left in a very precarious Situation as our Landlady has taken part of our Goods and I am in other Cases much embarrassed and in Respect to myself I am very Poorly in health and the worst is the loss of sight. I have long lost the sight of one Eye totally and have a Pearl far advancing upon the other. You will consider I am not able to provide for myself. Daughter

Elizabeth is still at home but it is not in her power to support me but is willing to do her utmost if you will assist her I therefore hope you will have the Goodness to take our Situation into consideration and favour me with an immediate Answer as I must have assistance from some Quarter and would much rather have it from you than to have the trouble of applying to this town which you are confident I must Do if you do not prevent it by doing it yourselves. I hope I have sufficiently explained my situation and shall patiently wait for a few days or a Week for your (I hope favourable) Answer.
And remain, Your Obliged servant,
Mary Gunnell[46]

Mary Gunnell had no money and implies she lacked funds for a pauper funeral. It is noteworthy that neither in her host or settlement communities does she beg for one. The Overseer in Manchester reveals why Mary was silent. He wrote another short letter to his counterpart in Great Oxendon explaining that the widow told him she wanted to bury her husband 'proper' but he was a cholera victim.[47] Most of the Manchester corpses were taken away for quick dissection before being placed in a lime pit to try to prevent contagion spreading.[48] There was evidently little point in asking for any further help in that direction. Mary requested instead that her rent be paid. She was desperate to repair her finances. The landlady of her lodging house had seized her domestic goods to sell in lieu of the arrears. Mary's daughter was still recovering from cholera and could not work full time. Mary was unable to get much work because she was blind in one eye and needed a cataract operation in the other, which was not available. This type of pauper agency was often a strategy to get more money, but in this case it spoke instead of how easy it was to fall into destitution in Manchester when fatal illness struck. The Gunnells were typical of those for whom dissection was an inevitable end to poverty. Recently this tale of misfortune came to light in another surprising way. In January 2010, builders found human remains when digging under Victoria railway station in central Manchester. Archaeologists, called in to investigate, located a pit of pauper bodies. They turned out to be the same cholera victims buried in the old workhouse graveyard.[49] Later this area became a major railway site. Most of the corpses were buried in parts, some whole, under platform one of the station. Daily, for over a century, commuters have been walking above dissection cases. These are the forgotten paupers who helped to advance medical education in Victorian times. John Gunnell's lost story is one part of a much larger economy of supply in Manchester.

Reconstructing the anatomy trade in Manchester is fraught with practical difficulties because there are a lot of gaps in the local records. Both the City and Poor Law authorities destroyed most of their files in 1929. Papers retained at the Local Government Board (hereafter LGB) indicate that they appear to have done so because the business of anatomy was considerable in the locality throughout the Victorian and early Edwardian eras. Recovering these details has, then, not been easy. Nevertheless alternative routes into the anatomy trade exist in the recordkeeping of the Victorian information state: a theme repeated throughout this book. It is fortunate that Fiona Hutton has researched the development of medical education until 1870 in the city. This has provided important context for this study. Nonetheless Hutton neither studied the New Poor Law in parallel with the Anatomy Act, nor did she examine the crusade against outdoor relief for which Manchester was renowned in the late-Victorian period. Research in this book indicates that this is an important omission since elsewhere the anatomy trade expanded considerably during that period. Retracing scattered burial records in the city, central government files, and emblematic case histories of paupers taken for dissection that did survive means that it is feasible for the first time to build upon and widen Hutton's work on medical education. New research reveals a strikingly Northern feature of the anatomy trade. In Manchester there was a very localised business in operation, more so than elsewhere. In the largest Northern cities the trade tended to be based at one and certainly no more than two carefully chosen large welfare institutions in close proximity to a nearby anatomy school. Manchester anatomists did not lack bodies after the passing of the Anatomy Act. The records show that anatomists developed a close business partnership. They came to rely on one main source of supply at Crumpsall workhouse in a Manchester suburb. Corpses from here were moved by local railway into and out of the city-centre, with some surplus sold in the summer months to Cambridge and Oxford.[50] To fully appreciate the scale of that trade it is important to consider briefly the complex nature of pauper burial provision in Manchester city-centre. Its inadequacy afforded anatomists ample opportunities to dissect corpses, as has already been suggested in the case of the Gunnell family.

Manchester pauper burial provision

"Bills, Bulls and Blunders"
The people make *game* of the Game Bill. The Coal Bill is all rubbish. The Bankruptcy Bill has broken down; and the projected Anatomy Bill is the *subject* of *cutting-up*. By-the-bye the last sample of law-mak-

ing contains two bulls, for it provides that *"not fewer than one!"* of the Inspectors to be appointed shall reside and transact the business of his office in London, and *"not fewer than one!"* in Edinburgh – We heartily wish that we had *"not fewer than one"* – [sic][51]

Contemporaries were sceptical about the Anatomy Act and its practical implementation. It seemed to lack a well-staffed inspectorate; its terms were ill-defined; and it targeted the unfortunate. These were common themes in penny journals printed in London and Manchester at the time. The satirical example quoted above is representative of a growing body of public opinion about just how unfair the Anatomy Act was on the poorest. Ruth Richardson has established that cheap journals were very critical of a lack of pauper funerals under the New Poor Law.[52] It was self-evident that this would penalise the bereaved living in destitution, anxious to prevent the dissection of a loved one. Manchester was soon the focus of considerable social commentary because the dead were everywhere within its city's environs.[53]

> Below the bridge you look upon the piles of *débris*, the refuse, filth and offal from the courts on the steep left bank; here each house is packed close behind its neighbour and a piece of each is visible, all black, smoky, crumbling, ancient, with broken panes and window frames... Here the background embraces the pauper burial-ground, the station of the Liverpool and Leeds railway, and, in the rear of this, the Workhouse, the "Poor Law Bastille" of Manchester, which, like a citadel looks threateningly down from behind its high walls and parapets, upon the working-people's quarter's below.[54]

Frederick Engels' famous remarks, published in his *Conditions of the Working Class of England* (1845), have been quoted frequently by social historians of the early Industrial Revolution in Manchester. He was a strong critic of pauper burial sites located beside Manchester workhouse on New Bridge Street. The graveyard was close to the station owned by the Lancashire and Yorkshire Railway Company in the city-centre. The stark contrast of a symbol of destitution and one of modernity beside one another was an irony not lost on Engels. His tone was forthright, fulsome, and up-front. It does not necessarily follow that his opinions were extreme. This was how another visitor from Paris described the area in 1844:

> Nothing is more curious than the industrial topography of Lancaster. Manchester, like a diligent spider, is placed in the centre of the web,

and sends forth roads and railways towards its auxiliaries, formerly villages, but now towns, which serve as outposts to the grand centre of industry... Amid the fogs which exhale from this marshy district, and the clouds of smoke vomited forth from the numberless chimneys, Labour presents a mysterious activity, somewhat akin to the subterranean action of a volcano.[55]

Despite the obvious Marxist overtones in these accounts, other Poor Law records retained by central government show that both men were correct to highlight the plight of the poorest when death came calling in the city-centre. The Statistical Society of Manchester sent in returns to the civil service which calculated 'in 1836, out of 169,000 inhabitants of Manchester and Salford, 12,500 lived in lodging-houses and more than 700 slept in cellars'.[56] In the whole of Lancashire in 1842 a survey showed that there were '454 charitable institutions worth £29,475 and 914 friendly societies worth £108,643'. Despite this level of charity, many died in destitution.[57] Although at the start of the early nineteenth century approximately one third of Manchester's residents were Church of England, another third were Roman Catholic, and the rest were labelled Non-Conformist faiths, few could afford a pauper funeral without some charitable assistance.[58] Many of those in regular work in Manchester tried to save with a friendly society or burial club. As Paul Johnson and, more recently, Timothy Alborn point out, 'working class spending patterns' were complex and a funeral payment was an important indication of not simply consumption but a person's place in a 'distinctive community' of the labouring poor.[59] This was why local public houses in poorer areas often ran subscription schemes. They were popular, good for business, and based on economic realities. It was critical to save for a pauper funeral when on the edge of absolute poverty. But it was also beyond many of the Irish (escaping the famine) who were refused poor relief on arrival in Manchester between 1847 and 1850.[60] An additional economic problem was that the presence of the Irish exacerbated unskilled labour supplies at a time when trade in raw cotton was in exceptional distress. Those unemployed were also often in poor health. Increasingly, burial clubs were overwhelmed with claims. The Jewish population redressed this social issue by establishing the Manchester Jewish Burial Board with its burial ground at Crumpsall. This ensured that the orthodox were buried intact within 24 hours of death. Other Non-Conformist churches ran small charitable funds to pay for parishioners' pauper funerals.[61] Meanwhile the Roman Catholic Church was charitable but unable to meet the demand of so many parishioners in poverty.[62] A lot of Catholic children were dying and they became the main focus of fundraising efforts. It was better to concentrate on the living rather than the dead. It meanwhile proved impossible to makeshift enough money to pay for a pauper funeral. These were the sorts of people who turned to the anatomy trade, selling a body for a small fee and the price of burial after dissection.

In simple numeric terms, then, Tristam Hunt explains what these religious and social trends meant: 'In 1841, life expectancy at birth was 26.6 years in Manchester, 28.1 years in Liverpool and 27 years in Glasgow – London was 37 – relatively healthy but whereas Camberwell was 34, Whitechapel was 26'.[63] This was the demography of the anatomy trade too and that fact was not lost on contemporaries. Concern was expressed that anatomy sales were being buried in the city. Commentators wanted to know if dissection cases were treated 'decently' and 'with dignity'. A number of leading radicals decided to find out. They were led by Abel Heywood (a councillor) who was determined to ask uncomfortable questions about anatomy burials in the local press. He also wrote complaints to central government and the anatomy school where Thomas Turner was in charge. Following extensive enquiries, Abel Heywood stated on 26 May 1845 in correspondence with the Poor Law Commissioners in London that, 'It has been the practice of the Medical School in the town to inter the bodies of paupers given to them for dissection after dark in the evening and without the burial service being performed on them'.[64] He consulted the Mayor, Alexander Kay, and both the Dean and Dr. Turner, the Demonstrator in Anatomy at the Medical School. He also undertook his own investigations on 'Sunday Evening May 9th 1845 at fifteen minutes past nine o'clock'. Abel Heywood claimed that 'a van stopped at an entrance to the burial ground, Walker's Croft'. He reported that 'from out of this van 4 coffins containing bodies, three of which were paupers, were taken and placed in the ground without the services of the Church being performed over them'. Abel Heywood mingled with the 'standers by' who told him that 'neither friends, relatives, mourners, or Ministers' were present at the covert burials. He approached 'the driver and the sexton' but they 'designed no reply'. At this,

> Indignant were exclamations by the crowd and expressions of disquiet and expressed from every lip – *"This is the way they treat the paupers said one"*; *"They bury them like dogs said another"*; *"Heaven help the poor"* – etc, etc. [sic]
>
> As an inhabitant and ratepayer of the borough I thought it my duty to lay the matter before the public in order that a stop may be

put to which I call disgraceful proceedings ... I respectfully appeal to you as Commissioners appointed by the government to supervise and watch over and direct the enforcement of the poor law to take such steps as will prevent the bodies of the poor misused and the sympathies of the people have been outraged.[65]

At first central government chose to downplay Abel Heywood's claims, but he kept on complaining. Civil servants then asked Thomas Turner to investigate. He promised to give the matter 'his personal attention' in a follow-up letter.[66] Charles Clements, Poor Law Inspector, confirmed in a lengthy report what happened next. Together, 'Mr Gardiner, clerk to the Board of Guardians' and 'Dr. Turner' looked into the matter. Gardiner explained that once a dissection sale had been made the body passed out of the jurisdiction of the Poor Law. The night burials had 'taken place beyond the control of the Board of Guardians'. This was confirmed by Thomas Turner:

> Dr. Turner stated in all the arrangements for providing anatomical subjects for the School over which he prescribed he had been strictly guided by the provisions of the Act of Parliament [Anatomy Act]. That as the bodies were sent from the School for interment it had for obvious reasons been considered advisable that the proceeding should take place at night, more particularly as the School [of Anatomy] is situated close to a very crowded thoroughfare; and that if the service had not been performed then he believed it had the following day; that the [funeral] fees and [burial] dues were invariably paid and the Certificate [of interment] duly forwarded [to the Anatomy Inspectorate] as the Act required -[67]

Local people were, however, unhappy about this outcome and so Thomas Turner agreed to make three new arrangements. He proposed that 'bodies should in future be deposited in the Chapel of the Burial ground at night, so that they might be committed to the ground the following day when the Clergyman should be ready to perform the service'.[68] If there was, however, 'any difficulty in depositing them in the Chapel he would arrange that they would be sent from the School of Medicine to the Dead House of the Infirmary at night, and that the interment should proceed with from thence in the usual manner during the day'. The third new procedure was that the Guardians hired 'William Gleaves to be employed to take care of the horses at the Workhouse; to remove all sick persons sent by the Relieving Officers, to attend all funerals of persons interred from the Workhouse and to make himself generally useful when he may be required with the horses at a weekly wage of twenty-one shillings'.[69] In a follow-up letter, the Workhouse Clerk stated that; 'The Guardians think it due to themselves to observe, that if blame attaches to the parties on the statement made, it does not rest with them'. They sold bodies and did not bury them. It was up to anatomists to avoid further bad publicity. Central government agreed in a file note that 'the responsibility for the interment of the corpses lay with the medical school'.[70]

It is rare to find a run of confidential correspondence on this controversial topic because civil servants were wary of writing down procedures, even in private. There evidently was a lot of concern about what was happening in Manchester because a 'draft letter was drawn up by senior civil servants telling Abel Heywood 'that no further interments will take place without a proper burial.'[71] Predictably there had been discrepancies. Private reports circulated among the same civil servants conceded that across the North of England it was common to skimp on funerals of dissection cases. In Sheffield, for instance, clergymen stood briefly over coffins at mass burials, refusing to have decomposing bodies stand by the altar to pollute and stink out the congregation,[72] a theme introduced in Chapter 1. The Victorian attitude was that the dissected were 'matter out of place', objects of dirt and disease.[73] They should be buried quickly, with little expense, and the minimum amount of publicity. Christian burial under the Anatomy Act was respected but those rites were often cursory. Pauper funerals, what these meant, and who should fund them were as controversial in the North of England as the Midlands and London. Meanwhile in Manchester public criticism of a lack of pauper funerals, which had implications for the anatomy trade, gathered pace in the press. On 20 March 1850, the *Manchester Times* announced:

> INTRAMURAL INTERMENTS – The Poor Law commissioners have recently ordered, under the 12th and 18th Vict., c. 3 sec. 9 and 11, returns from the different unions respecting intramural interments. From the returns made by the Manchester union, in accordance with this order, it appears that the total number of pauper interments during the last seven years has been as follows:
> Indoor paupers 3,881; outdoor paupers, removed from their own houses, 6,226, total 9,557, at a cost of £4,141 19s 10d; of which sum £2,337 8s 6d has been expended in coffins, at an average of 6s 7d for each adult, 3s 3d for each youth, and 1s 9d for each child. The average expense of burials in the union is £591 14s 3d.[74]

The editor was scathing about these calculations and he was right to be so. Manchester was spending just '6s 7d' despite the rising cost of living. Average burials in the 1830s were ten shillings under the Old Poor Law. What, the editor asked, was the real meaning of the expenditure figures? Did guardians pay these costs from local taxes or were any payments defrayed by the local medical fraternity? Where were the poorest being buried because the city-centre did not have an official pauper burial ground? Why were guardians so shy of explaining what a pauper funeral entailed? Did it mean just a parish coffin and no customary burial rites? Were bodies taken for dissection or not, and what happened to those human remains in the city-centre? Over the next 12 months, reporters pressed these questions on local clergy, city officials, and guardians. Letters to the editor appeared regularly expressing dismay about the lack of substantive detail. A typical example signed 'NECROPOLIS' remarked:

> You announced many weeks ago that the churchwardens of Manchester had arranged with the dean and canons to select a parish burial ground... I have been from week to week expecting to see an announcement... Can it be that the churchwardens are neglecting their important duty, or is it that the clergy of the parish church are still indifferent to the wants of the community.
>
> The mortality of the parish has, for the past few weeks been nearly 200 per week; and, I believe, extends to 10, 000 per annum. This is an enormous number of interments required in a parish which has for many years been deprived of its burial ground; a deprivation that costs the poorer inhabitants an increased sum for the interment of their deceased friends...[75]

This type of adverse publicity continued over the next decade. Often a burial scandal brought to light what was really happening to pauper corpses. One noteworthy account is representative of many at the time.

In June 1855, the location of pauper graves was exposed following a scandal at Harpurhey Cemetery. This burial site was a private venture owned by shareholders. It was located on the perimeter of the city limits. Manchester guardians made a private deal to send paupers there to be buried in mass graves. They were placed in deep pits at the edge of the new cemetery site. This was where dissection cases were interred too. A 'Visitor' to the cemetery explained what he found in a detailed letter to the press:

The want of one or more cemeteries in this city is becoming more and more felt, especially now that the government for the purpose of public health, has deemed it necessary to interfere with the churchyards and other public burial grounds; that at Harpurhey being the only one open to the public without restrictions...

I observed that the graves for the poorer classes of the community, called public graves, are of what I call the out-of-the-way holes of this prettily situated cemetery. So far is this saving of grass-land earthed, that an old water pit on its southerly side, adjoining Queen's Park, has been converted into a receptacle for such bodies as are sent by the guardians of the poor. At the latter pit I entered into a conversation with a young man, a sexton, who was making the place ready for interments, and who expressed himself thus:

Ah sir, this is only the grave we place them in first; for when funerals are very many, to save wages we take them out again, and we have placed a great many in the drain that was made yonder through that pit and now under the earth heaps that are continually being brought in by our governor: it saves ground and labour at the same time [sic].[76]

What, asked the anonymous author, did this say about the fate of paupers, being so 'freely and openly talked about'. It was surely 'repugnant to decency and the common feeling of our nature'. He pressed guardians to look into the matter, which they did reporting in the *Manchester Guardian* of 20 July 1855 that irregularities had occurred at Harpurhey Cemetery.[77] An investigation found that the sexton conducted a basic burial rite beside a mass grave but then bodies were removed to a larger pit because of inadequate space. There was no church service or marked grave, not even with a wooden cross. Guardians admitted that because '62 were received from the various unions in 5 days' there was not enough space to bury them and the graves that had been dug were too shallow. The sexton and gravediggers thought it would have been better 'to have notice of the required number of interments' so that the ground could be prepared properly for their arrival. The Chairman of the Manchester Board of Guardians expressed the view that 'the question of interment was one of great delicacy and upon which people's feeling were liable to be very much excited'.[78] He offered reassurances that procedures had been tightened but refused to be drawn further on the issue of what was happening to the corpses of the indoor poor between death and burial. It was also not made clear who was paying for the pauper burials – anatomists or guardians. To deflect attention and keep a tight grip on adverse publicity, it was reiterated that William

Gleaves now accompanied the sexton to the graves in Harpurhey Cemetery to ensure rites were supervised properly. The general public were not, however, satisfied. Over the next four years, pauper funeral rites featured in a lot of press stories. Consequently Manchester Union was forced to admit that burial space for the destitute was inadequate by 1859. In a tense open meeting in February that year, the Chairman of the Board conceded that there was a high death rate among the indoor poor: 'nearly one third of the persons who died in Manchester were buried by the guardians'.[79] He also admitted that there was sometimes a delay between the time of death and burial. He again preferred not to explain in detail what this meant. Evidence throughout this book has indicated that a delay between the day of death and each burial ceremony provided a time-gap for anatomists to dissect the destitute. A synthesis of press reporting at the time suggests that few local people were convinced that the bereaved were being treated appropriately. A follow-up letter in the next edition of the newspaper was for instance alarmed by the implication of the burial figures and high death rates: 'Surely sir, it cannot be the fact that one in three of the population in this centre of industry and wealth die in a state of pauperism?'[80] A third of the mortality pool was an important supply opportunity.

Meanwhile public health in the city deteriorated markedly and therefore pauper funerals continued to be contentious. Before leaving aside local pauper funeral controversies, it is necessary to briefly set out just how bad death and disease were in the city by the mid-Victorian period. In 1801, Manchester population was 84,000; by 1851 there were 391,000 recorded on the city census.[81] It is important to appreciate that this population explosion (a fivefold rise) would have been much worse had contagious diseases not been rife. Poor public health standards haunted the poorest neighbourhoods. Manchester's death rate was 37 per 1,000 residents in 1847 (when mortality rates peaked) compared with 22 deaths per 1,000 people nationally.[82] Predictably, death rates were disproportionate in the young and the elderly. Maternal mortality remained much higher among those in abject poverty.[83] Likewise, the big child killers were measles, tuberculosis, diphtheria, diarrhoea, scarlet fever, croup, typhoid, smallpox, typhus, and influenza. Even by 1910 the situation had not significantly improved in public health reports. A local coroner, for instance, calculated in 1910: '338 stillborn children had been buried in Phipps Park Cemetery [formerly Harpurhey where dissection cases where buried]'. He commented that, 'it would be the easiest thing in the world to murder a child in Manchester'.[84] Altogether in the city there were 3,224 child deaths and of these 26.4 per cent were under one year old. The coroner was very concerned that in the poorest areas 'midwives at present give certificates in case of stillbirth which are used as burial certificates'. He thought that the system was open to abuse. The midwife could hand over a certificate in infanticide cases. It was a worrying trend that child death remained omnipresent even in early twentieth-century Manchester.

The problem of where to bury those who died in misery never abated, especially among the Irish who, as we shall see in the next section, peopled the dissection table in Manchester. It is now well established in the historical literature that the Irish lived cheek by jowl in some of the worst overcrowded slums of the city.[85] One statistic speaks for many at the time. In the inner city area known as St. Michael's and Angel Meadow (where Abel Heywood witnessed dissected cases being buried in 1845), public health officials were alarmed by the scale of death along five streets where the Irish lived in temporary lodging houses. In 1851 the census showed 18,347 residents, and, of these, 8,048 or 44 per cent stated they were born in Ireland. Historical geographers have mapped the stratification of the poorest and shown that once population density increased there was a wave of out-migration of the merchant and middle classes (broadly defined).[86] Public officials knew about the scale of the social problems left behind but few blamed those that left to escape the deprivation and disease. The city authorities did point out that the Irish slums had one public health advantage. Their topography was hilly, sloping down to the river Irk in the Irwell valley. Manchester had high rainfall levels allowing gravity to wash the human waste down the city streets to the waterside. Newspapers pointed out, however, that disease was still rife. Everyday the smell of death was in the damp and stagnant air of central Manchester. The solution, it seemed, lay in Crumpsall, a suburb, where the city's public health, welfare, and medical fraternities turned their attention in desperation. The city's pauper burial controversy, secret anatomy trade, and public health crisis demanded a new solution. Crumpsall workhouse seemed to be the place to which the most pressing social problems could be removed. Here cadavers were sold to the local medical school, as well as Cambridge and Oxford. An analysis of Crumpsall's economy of supply is thus overdue. The third and final section outlines the development of the New Poor Law at Manchester and Crumpsall workhouse's construction after 1850. This will set in context the new data on the dead paupers sold for dissection from Crumpsall workhouse between 1860 and 1910.

'Better a third of a loaf than no bread':[87] the economy of supply

When Manchester Poor Law Union came into operation on 11 December 1840, 24 guardians represented 12 constituent townships in the immediate vicinity.[88] A decade later, the expansion of the city, flooded by the Irish, meant that the Township of Manchester had to become a Poor Law union in its own right. The other townships surrounding central Manchester were separated off to form the Prestwich Union.[89] Meanwhile, on New Bridge Street, a large workhouse accommodated the indoor poor. It was this scene that Engels reported in 1845. By the 1860s, the New Bridge Street site was accommodating up to '1644 inmates'. In reality the Poor Law in Manchester could only function by classifying and then farming out paupers to a number of ancillary workhouses. A large mill, for instance, on Tib Street functioned as a House of Industry for the able-bodied with '600–800 inmates'. Children were eventually sent to be boarded out at the Swinton Industrial School (for healthy children) and Swinton Home (for mentally defective or disabled children), housing around '800' by the 1881 census. Guardians meantime envisaged that to accommodate those in permanent poverty what was needed was a new workhouse devoted to the worst cases of destitution. A proposal was thus brought forward which was one part of a larger expansion plan to try to better manage urban pauperism by removing it physically to the suburbs. It also made it feasible to isolate and thereby expand a new economy of supply at the new site. The pauper population of Crumpsall workhouse thus became central to the anatomy trade at Manchester medical school.

A proposal to build a large workhouse at Crumpsall was brought forward by Manchester Poor Law Union over the course of 1854–1857. The idea was to get a leading firm of architects, Messrs Mills and Murgatrod, to design a purpose-built facility in which the indigent poor crowding the commercial districts could be moved out of the city-centre.[90] Crumpsall was situated in an area known locally as the 'crooked piece of land beside a river'.[91] It was a poor parish of 733 acres, located in a hilly district, rising to 280 feet above sea level, to the southwest of the river Irk in the Irwell valley. The hope was that public health would be improved here. A newspaper reporter explained that 'it was a healthier situation and one so free from smoke...it possesses all the advantages of the southerly and westerly breezes, and the east winds, so hurtful to delicate people, are tempered before they reach the workhouse by the rising ground in the direction of Harpurhey'.[92] When Manchester guardians announced the new scheme in July 1854, they also conceded that they were thinking of using Crumpsall as a burial site for all paupers. Crucially this could include bodies sold to the medical school. The *Manchester Guardian* reported that:

> It was proposed to take a portion of the land to be purchased, for the purposes of a Cemetery connected with the workhouse; for owing to the increased cost of interment at Harpurhey Cemetery the board was now, or soon would be, paying about £500 a year for the burial of paupers; in an economical view only, that was a great consideration.[93]

This was the first time that Manchester guardians admitted in print precisely how much the pauper burials in Harpurhey Cemetery were costing and that they planned to use Crumpsall as their new pauper cemetery. They refused, though, to elaborate on whether the expenditure on funerals was defrayed from dissection sales and if it would be in the future. Financial records indicate what an 'economical view' meant. Paperwork on dissection cases sent to Arthur Thomson at Oxford reveals that Manchester funded anatomy burials costing around '£500 a year' (average undertaking costs were '6s 7d for every adult pauper').[94] It is not clear how much each body cost but petty cash records seem to show that Manchester had an opportunity to buy around 100 bodies every year by the end of the 1860s. This was about 60 per cent more than the official returns to the Anatomy Inspectorate, which declared that: 'in 1869–70 there were 43 distributions [purchases], and in 1870–1 there were 37'.[95] There was, therefore, a discrepancy between Poor Law undertaking records, internal correspondence between provincial medical schools on supply levels, and central government's annual anatomy statistics. Manchester appears to have been acquiring more dissection cases than it was declaring, a new finding that will be discussed in more detail later.

Meantime, when Crumpsall workhouse was being built, controversy over pauper burials ignited. The *Manchester Guardian* reported a stand-off between local people, guardians, and city authorities about the need for a municipal cemetery to inter the destitute.[96] It was proposed that the only solution was to buy '50 acres of land' in one plot or several, to be located on both sides of the city.[97] This would ensure accessibility for the poorest travelling to funerals in each direction. The city corporation, however, wanted to defer the matter until a separate Burial Board could be created by an Act of Parliament. Many large ratepayers were critical of that move since the liability for paupers might come back to haunt local taxpayers in the suburbs. One leading spokesman repeatedly told the press that he calculated that if the burial scheme was adopted outside the city limits it 'would have relieved the corporation of all responsibility as to the 4000 or 5000 out of the 10,000 deaths in the city in the course of a year'. Other critics suggested that Manchester and Chorlton Poor Law unions should pay for a large pauper burial ground together, but this proposal was rejected too. Again ratepayers feared a rise in the poor rate, a form of indirect taxation, to pay for the new site. There was also the reaction of poorer people to consider. Another newspaper reporter pointed out, 'poor people took great pleasure in visiting the graves of their relatives; and apart from the question of distance, they would have a very strong and perfectly natural objection to going to the cemetery adjoining the workhouse'.[98] Predictably there was a lot of xenophobia. Nobody wanted a public cemetery for the destitute on their doorstep. It could devalue property. Pauper funeral traffic would bring the worst sort of people – undesirables – tramping across residential areas. These debates culminated in the *Manchester Guardian* explaining that the burial scheme:

> Providing for the interment of paupers in connection with the workhouses had been abandoned. The guardians, in fact, were now the most clamorous at the door of the corporation, in urging them to take steps for establishing a cemetery or cemeteries, for they said – and the corporation would be enabled to judge hereafter when the committee reported fully – that it was one of the greatest grievances of the guardians, that in consequence of no steps being taken, the expense of interring paupers was much greater than it ought to be and than it would be if a burial ground was established.[99]

Yet the poorest were not to be buried at Crumpsall or in Manchester. Instead they would be sold for dissection to Cambridge, Oxford, and local anatomists.[100] Some would be buried out of the area; others would continue be laid to rest in Harpurhey Cemetery. That arrangement, however, needed Crumpsall workhouse to be fully functioning. It was also essential to negotiate a railway contract to move the poorest in and out of the city-centre. Consequently, in 1854, Manchester Poor Law authorities purchased a large plot at the 'Bongs or Banks' in Crumpsall for their expansion plans (the site was just in front of the old Prestwich workhouse). The new T-shaped design housed 1,660 inmates.[101] By August 1858, on the eve of the Medical Act, the new site was functioning. It was an enormous enterprise and grew exponentially until it was absorbed back for official accounting purposes into the township of the city of Manchester in 1896. During World War I, in 1915, Manchester underwent another major re-organisation.[102] Now South Manchester Union (formerly Chorlton Union) and Prestwich Union (including Crumpsall workhouse) came back into the ambit of the Manchester Board of Guardians.

The three unity authorities were thus reunited. Anatomists now feared that their body-buying business at Crumpsall workhouse would be stopped. This would be a significant loss because further expansion plans at the Crumpsall site included the construction of a purpose-built infirmary that could accommodate nearly 2,000 inmates. They came by train on the Lancashire and Yorkshire railway line that travelled daily across Crumpsall. This reliable transportation link from Manchester to Bury also connected to an express going to the North and South of England. It continued to be used to move around the dead. Crumpsall remained pivotal to an expanding economy of supply.

Fiona Hutton has reconstructed from official returns sent into central government, the body count for British anatomical schools from 1834 until 1870.[103] Graph 7.1 takes those calculations and extracts the trends for Manchester. It shows that 53 corpses were declared to the Anatomy Inspectorate in the first year after the New Poor Law. These were bought from welfare agencies in the city-centre. The average then had peaks and troughs throughout the 1830s and 1840s. During the decade when the Medical Act became law, supply rates improved again to match student demand. There were thus 39 corpses acquired in 1860 from Crumpsall workhouse. The Cotton Crisis years produced on average 27 bodies per annum because more charity was available in death. This meant that by the time that the crusade against outdoor relief was being promoted there were on average 40 bodies sold per year by 1871. What these figures tell is a success story compared to other provincial medical schools but not to their competitors in the capital. At Cambridge, anatomists bought smaller numbers in the 1850s, around ten per year on average. Oxford had very low levels of supply until the 1880s, a maximum of five. Only St. Bartholomew's outstripped Manchester with, on average, 80 per year from 1832 to 1842, emphasising the attraction of a metropolitan education. Given the high mortality rates in Manchester it is striking that official figures were not higher. There was some manipulation of returns, outlined above, but even allowing for the fact that an extra 60 per cent, or 60 bodies, seems to have been undeclared. As a total proportion of the deaths in the city,

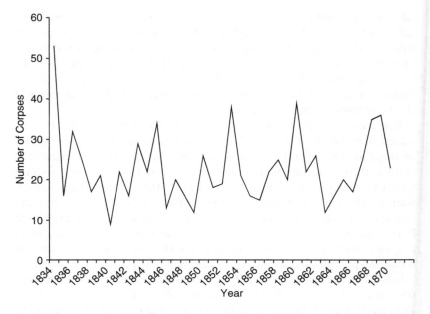

Figure 7.1 Anatomy Inspectorate Returns from Manchester Anatomists on Poor Law Purchases, 1834–1871.

Source: F. Hutton, 'Medicine and Mutilation: Oxford and Manchester and the Impact of the Anatomy Act of 1832' (Oxford Brookes University; unpublished PhD, 2007), reconstructed from figures extracted from Anatomy Inspectorate returns, appendix 2, pp. 296–325.

the economy of supply was low. Manchester anatomists took what they needed and not a lot more. Storage issues made it sensible to buy fresh, and at source, from Crumpsall workhouse Dead House. Besides, their supply chain broadly matched their rising student numbers, so there was no need to be greedy. Occasionally the medical press hinted about the disparity between official returns and the real situation, but there was no scandal since an expert eye would know that supply matched demand.[104] It is possible to expand on this picture despite the patchy nature of local burial records.

At first St. Michael's Angel Meadow and its extension into Walker's Croft was where dissection cases were buried and then later Harpurhey Cemetery was used for anatomy burials. This makes it feasible to reconstruct from fragile records the sorts of typical cases sold for dissection from three key workhouses sources in the Manchester area. Salford workhouse was a supplier up to 1854; others were sold from Cheetham workhouse after 1858; whereas those from Crumpsall workhouse were the predominant source of supply by 1870. These have been double-checked against LGB files to ensure that the calculations, though conservative, are accurate. What is striking about those burials that can be accurately identified as dissection cases is that some 80 per cent were interred in the Roman Catholic parts of the various graveyards used for anatomical burials. They were young, generally single, and the majority were Irish, buried in 'section F'. More painstaking record linkage work adds detail to this view. Until 1854, pauper remains were interred together in St. Michael's Angel Meadow.[105] July 1849 was a busy month before the start of the student vacation. At this time, corpses came to the Anatomy School from Salford workhouse. They were later buried at a cost of 'two shillings' per infant and 'four shillings six pence' for an adult. This was meagre indeed even by the standards of the poorest part of Manchester city-centre. The fee covered the sexton who said the last rites over the burial pit. There were a lot of infants in the supply chain, despite religious considerations.

July 1849

1st	Thomas William Hutton	11 days	2 shillings
7th	Mary Ann Beswick	1 month	2 shillings
8th	Edward Edwards	1 month	2 shillings
9th	Lawrence Rannigan	No age [unknown]	4s 6d pence
14th	Elizabeth McMaster	6 months	2 shillings
15th	Elizabeth Clarke	3 months	2 shillings
22th	William Worrall	7 months	2 shillings
24th	Robert Taylor	31 years	4s 6d pence
24th	George Slater	4 months	2 shillings
27th	Winifred Gilligan	5 months	2 shillings
27th	Mary Holden	1 week	2 shillings
29th	Robert Knott	5 months	2 shillings[106]

Moving ahead to the start of the crusade against outdoor relief in 1873–1874 the dissection cases now came from two main sources, 'Crumpsall and Cheetham Poorhouses'. All were buried in batches of three; adults and children were placed in the same pauper pit. Again representative cases are detailed here. They symbolise many more, who will feature later in the entire economy of supply data (see Figure 7.3). Thus new research shows that 'Anne Gillow 50 years old a widow' was buried alongside 'Eliza Clancy aged 31 years a married woman' and on top of both of them was placed 'Mary Hopkins aged 3 weeks an infant' on '27th June 1871'. Likewise 'Peter Bradley an infant of 3 years' was buried with 'James Crawford a labourer aged 42 years, single', and 'Michael Hayes, aged 30 years, single, dyer' on '1st July 1871'. During the summer months medical students were often on vacation and so there was less need for bodies. Ironically, this was the best time to acquire infants because summer fevers were rife and the death toll rose. In August 1871, for instance, 'Patrick McCartney 5 months', 'Mary McDonald 1 year', Thomas Henraty 1 month', Robert Johnson 1 year', 'John Shandley 5 months', Patrick Caudle 6 years', Patrick Burns 70 years a labourer', 'John Crawford 60 years a porter', and John Rud 5 months' were buried together. The number of Irish names is noteworthy, as is the frequency of infants. In order to examine individual life histories in more detail it is necessary to focus on the 1881–1888 period. In this period Crumpsall workhouse communicated with London Poor Law officials and that evidence sets lost lives in context.

Centrally collected Poor Law records make it feasible to piece together the population of Crumpsall workhouse-infirmary that constituted the anatomy trade. Taking the 1881–1888 accounting period as representative, there were on average 4,007 persons claiming poor relief each year.[107] Of those, 3,283 were residents of the workhouse-infirmary and 724 lived on outdoor relief at home. This meant that the number of available beds on the workhouse-infirmary wards often fell short of inmates. There were 1,550 beds for men, 1,450 for women, a total of 2,575 paupers. So there was a shortfall of 708 beds in an annual cycle. Of course there were seasonal peaks and troughs but staff lacked about 59 beds a month. In segregated wards the same sexes often slept two to a bed. An Inspector added that 'it is rather hard to maintain proper classification as infirm adults are under the same roof as schoolchildren' despite a policy of boarding out.[108] The sick wards were so busy that there was often an overflow onto the other wards. It needed '68 staff – 53 day nurses and 15 night nurses' to care for the range of medical cases. To save money, the infirmary had a policy of employing trainees. Of the 68 nurses, '38 were paid nurses including probationers and 30 (paying) pupil nurses'. They were supervised by the 'Lady Superintendent of the Infirmary' because of the high turnover of maternity cases. Good communication was needed between the central Manchester and Crumpsall workhouse sites to redistribute cases according to bed capacity. The Inspector noted that, 'Miss Hanmen the Lady Superintendent of the Infirmary has not found the telephone a success and prefers to communicate by messenger'.[109] This was also the way that the medical school was contacted about 'unclaimed' bodies on a daily basis and it should be repeated that the majority of those inmates were Roman Catholics. In Crumpsall workhouse there were on average '1160' who attended Mass per year compared to '120 Dissenters' recorded during the 1880s.[110] The design of the entrance to Crumpsall facilitated their regular anatomy sales. In an 1888 report, the LGB Inspector pointed out that everyone who passed through the main gates of the workhouse had to sign in and out of a porter's book kept in the main lodge. Once inside the site, however, it was possible to walk between the infirmary and workhouse without being seen or recorded. The inspector noted that this meant that everyone could enter the 'Post Mortem Room' too.[111] Access to body parts and dead bodies was unregulated. Crumpsall workhouse was always open to the business of anatomy. A representative sample in 1883–1884 analyses life histories and general trends that could be recovered from local sources in the city.

As Figure 7.2 shows, '195 bodies' were traded in 1883–1884. That figure matches what Poor Law records, rather than official Anatomy Inspectorate returns, revealed. Usually at each anatomy burial there was a minimum of one infant or stillbirth among the Irish populace sold for dissection. In January 1884, for instance, 'William Hall aged

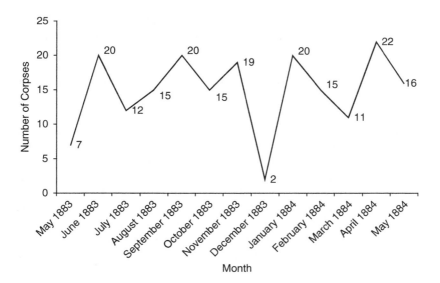

Figure 7.2 Number of Pauper Bodies from Crumpsall Workhouse, May 1883–May 1884, where n = 194.

Source: Manchester Local Studies Room, MFRR/704, MFPR/706, Phipps Cemetery Records, 1871–6, 1883–91.

5 minutes infant' was buried with 'Elizabeth Steward's child stillborn' and 'Harriet Hodgkinson's child, stillborn' in the cold winter. Likewise in March 'Eliza Joyce a widow aged 49' was interred with 'Ada Wren 8 months, an infant' and 'Celia David's child, stillborn'. Following on from these general observations it is then important to note that April seemed to be the time when the elderly reached a pauper grave. Most died in the coldest part of the year and three months later their dismembered remains were put in a pauper pit in the cemetery. Medical students could therefore take strategic decisions about when to get hold of the sort of human material they needed to qualify. The aged were generally used in the spring season. In 1883–1884 this was the fate of 'Bridget Wheelan, 66 widow', 'Helma Twine, 79 widow', John Burke 47 widower', Jane Hawks, 75 widow', 'William Bowker, 42, painter, single', and 'Mary Gannon, 78 widow'. In these cases, financial crisis and general debility meant that few left behind loved ones who could makeshift a pauper funeral. Dissected and dismembered human remains were interred together at one brief ceremony for committal on 24 April 1884.

What, though, were pauper funerals and anatomy burials like? At the Royal Commission of the Aged Poor, the Clerk who oversaw all of Manchester's workhouses including Crumpsall gave evidence about pauper burials. There had been 'alleged deficiencies in the coffin or minor details' and in a subsequent LGB report it was felt that 'guardians should see for themselves that the arrangement for funerals are carried out with care and attention'.[112] The Clerk stated that at Crumpsall workhouse:

> We have an official who attends every interment of a pauper inmate, and there are four inmates who wear a distinctive black dress which is furnished to them, to act as bearers or carriers. The coffin is provided with a pall, and there is a coffin-plate painted with the name and age and date of death of every person interred, and the name is also sewn on the shroud inside. There is a proper shroud for every inmate interred, and the name is shown on it. That is seen by the nurse and particulars are given as to the ward in which the deceased died, the description of the disease, and all other necessary particulars are placed inside the coffin.[113]

At Oxford, Arthur Thomson's confidential files contain evidence that contradicts this rosy picture. His secret report, written in 1921, stated that:

Manchester – Manchester Union
  This, which should be the chief source of supply for Manchester University, is unfortunately a difficult problem to deal with. As now constituted it includes three [Poor Law] Unions which have recently been amalgamated. One of these Unions had for long been the main, and a very large source of supply, of the University of Manchester & indeed in vacation time at Manchester of Cambridge as well. The Clerk of that section, whom I interviewed on Nov. 3rd, is now Clerk of the amalgamated Board, & he advises very strongly against the Minister's letter to the Board, as he feels certain the other two sections would be opposed to co-operation. Meanwhile, though that source is very much diminished in comparison to what it used to be, he is carrying on supply from his own original area of the city, on his own responsibility. He promises too, to feel his way gradually to the attitude likely to be adopted by the present Board as a whole. His advice for the present is, 'Better a third of a loaf than no bread'. It is only from the *Crumpsall Institution* [sic], therefore, that subjects are being sent to the Medical School.[114]

This was why civil servants praised Manchester in public. They did not want to explain the inside of the anatomy trade that made the level of checking and re-checking, labelling and cross-matching, identifying clothing, and putting names in coffins so necessary at Crumpsall workhouse. The Clerk was running an anatomy business personally from the premises. Evidently the lessons of court cases like *Rex versus Feist* (1858) in Chapter 1 had been learned. Manchester Poor Law officials would not countenance a single scandal coming to light. They moved the dissection business to the suburbs. Organisation was the key to avoiding any bad publicity at Crumpsall workhouse. Nobody at the Royal Commission on the Aged Poor thought to ask why the burial system was so rigorous. It was assumed that guardians were generous, not profit-making. Evidence again which reiterates that the anatomy trade should never be taken at face value. What was furtive often lies forgotten in the financial recordkeeping of the Poor Law system and surviving medical school archives.

Returning to the representative sample of dissection cases in 1884, their number and variety is noteworthy. In Figure 7.2, 14 paupers were for instance put in a pit between 20 and 22 May 1884. The eldest was '44', the youngest '1 ¼' years old. They symbolise the spectrum of people that made up the economy of supply at Crumpsall on a regular basis:[115]

20 May	Patrick Cain	76 years	Dyer	Single
20 May	Thomas Hanaghan	1¼ years		Infant
20 May	Alfred Rooney	6 years		Infant
20 May	Emily McGuire	7 years		Infant
20 May	James Oaks (found drowned)	28 years	Joiner	Single
21 May	Margaret Handley	70 years		Widow
22 May	John Newcombe	3 years		Infant
22 May	John Higgins	9 months		Infant
22 May	Lucy Dunn	44 years	Milkmaid	Single
22 May	Mary Gettings	44 years	Rag Sorter	Single
22 May	Joseph Brennan	38 years	Painter	Single
22 May	Mary Furlong	26 years	Servant	Single
22 May	Henry McMahon	35 years	Umbrella Maker	Single
22 May	Agnes Murray	35 years	Charwoman	Single

Retrieving this sort of burial information after a dissection sale means that historians can get at features of the trade lost in the Crumpsall records. Although only broad trends can be recovered they nevertheless provide an important glimpse of pauper life *and* death stories. A lot of cases in the sample had worked in hazardous trades before their untimely deaths. A dyer, painter, and rag sorter had died in this selection. Being poisoned by chemicals or contracting diphtheria from dirty clothes were everyday occurrences. Patrick Cain had thus lived much longer than Mary Gettings, although she was six years older than Joseph Brennan. Being in service did not protect either Mary Furlong from destitution or Agnes Murray, a charwoman. St. Bartholomew's study in Chapter 4 showed that there were many emotional reasons why a young female in service entered the general economy of supply. Likewise the faceless corpse in Chapter 2 symbolised how a man like James Oaks could end up dead, floating in the deep water of the River Irwell in Manchester. Murders and suicides were all too common on the dissection table. At Manchester medical school there were also a lot of children aged below five as well as infants. Evidence on the economy of supply might be patchy but it does confirm lower life-expectancy trends for children and young people. It is striking that public health reports for the city-centre stressed improvements in mortality statistics. The economy of supply indicates another outcome. It was debatable whether the benefits of clean drinking water, fresher air, and better housing had penetrated the poorest neighbourhoods around Manchester. In many Northern cities, that fact of life explains why so many paupers supplied the dissection room once the Medical Act was extended in 1885. It has been difficult to find paupers hiding in the shadows of Manchester's lodging houses and welfare institutions, and yet piecing together fragile records puts a human face onto an anonymous anatomy trade. Those bodies that travelled out of Crumpsall by railway can be seen in more detail because their records survived in other places where burial records are complete, like Cambridge and Oxford.

A profile of the corpses sold from Crumpsall to Cambridge University provides new evidence of a typical cross-section of the economy of supply transported out of the area. Each corpse was sent from central Manchester by train on the orders of the Clerk of Crumpsall workhouse. Taking 1906 as a representative sample we can again glimpse the paupers sold:

16th February 1906 – dissection cases from Crumpsall workhouse[116]
John Rowland aged 51
James Beddows aged 61
Richard Johnson aged 62
Thomas Connor aged 50
Thomas Toole aged 68
Mary Healey aged 78

All were sent on the fast train, dissected into parts, and then buried together in a single burial plot in Mill Road cemetery, central Cambridge. Most were middle-aged men who were born on the threshold of the Medical Act, providing evidence once more of its tangible impact on pauper death rites: a finding that is matched elsewhere throughout this book. One female, Mary Healey, was Irish and aged, with no family left to fend for her. The second batch of bodies in 1906 was similar. It again confirms demographic and ethnicity trends that we have seen in other case studies.

20th April 1906 – dissection cases from Crumpsall workhouse[117]
John Slavin aged 68
Isabella Quinn aged 73
William Hall aged 43
James Gavin aged 58
William Burgess aged 65
Jane McDonald aged 48

Aging Irish women were sold on a regular basis. Most men were spent forces by 50 years of age and permanent residents in the workhouse at

Crumpsall because they drained family economies. They simply disappeared in death.[118]

Bringing all of this record-linkage work together, Figure 7.3 reconstructs the body sales from Crumpsall workhouse between 1860 and 1910.[119] In total there were not less than 7,801 bodies sold to Manchester medical school. This calculation is conservative for reasons of accuracy. In the future, if further research material comes to light, there is every expectation that its cautious assumptions would be changed. The sales made in the later period between suppliers again seem to contradict official returns to the Anatomy Inspectorate in London. These are patchy, but an extra 25 per cent was obtained than seems to have

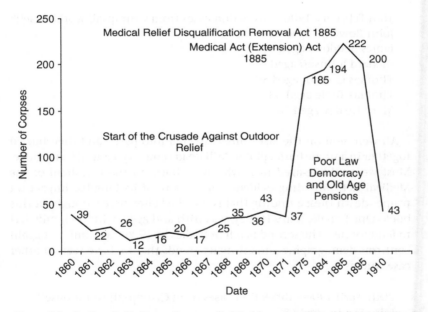

Figure 7.3 Bodies Sold from Crumpsall Workhouse to Manchester Medical School, 1860–1890.

Source(s): The economy of supply has been painstakingly reconstituted in three ways to ensure accuracy. Firstly, mortality statistics in the MH12/6042 series from Manchester to the Poor Law Board and its predecessor the LGB have been utilised, 1860–1910. These have been cross-matched to burial records where it has been possible to identify dissection cases at Manchester Local Studies Room, MFRR/704, MFPR/706, Phipps Cemetery Records, 1871–6, 1883–91. Then a third check has used pamphlets published by leading guardians about mortality rates, see, A. McDougall (1885) *Further inquiries into the causes of pauperism in the township of Manchester: a paper read before guardians* (Manchester: Charles Sever Printer), pp. 1–24, Crumpsall mortality statistics at pp. 13–18.

been declared in this later period. And that figure does not allow for body parts, which, as we have seen in Chapter 6, matched sales for whole corpses. This seems to be one key reason why so many records for that period were destroyed in central government. It was sensible to eradicate evidence that the crusade against outdoor relief had a pivotal role in the expansion of the anatomy trade in Manchester and elsewhere. Likewise the Medical Relief (Disqualification Removal) Act was a catalyst for the anatomy trade in 1885. Once more these policy changes coincided with the Medical Act (Extension) benefiting anatomists, findings repeated in chapters 4 to 6. It did not matter in what area of the country a pauper lived. Geographically a harsher New Poor Law underpinned the Anatomy Act in the late-Victorian era. The Clerk at Crumpsall workhouse told the LGB that he calculated that the cost per corpse for an 'anatomy burial' was small-scale – the 'Burial rate was 1¼d in the £'.[120] A Manchester pauper sold for dissection was worth little more than a single penny in death.

Conclusion

Alexander McDougall (Chair of the Board Guardians and an exponent of the *Manchester rules*) carried out an investigation into death rates in Crumpsall workhouse in 1884–1885. He left a detailed account of how destitution underpinned the anatomy trade. McDougall calculated that there were '2297 deaths' and of these '393' were 'paupers' in the first six months of 1884–1885.[121] This meant that 'one in every 5.84 deaths which occur in the township is that of a pauper'. He then broke these figures down by investigating how long they had been in the infirmary before death.

> Of the 359 deaths, 77 males and 60 females, or 137 persons, died within 1 month
>
> Of the 359 deaths, 24 males and 27 females, or 51 persons, died in 1–2 months
>
> Of the 359 deaths, 17 males and 7 females, or 24 persons, died in 2–3 months
>
> Of the 359 deaths, 18 males and 19 females, or 37 persons, died in 3–6 months
>
> Of the 359 deaths, 11 males and 5 females, or 16 persons, died in 6–9 months
>
> Of the 359 deaths, 9 males and 5 females, or 14 persons, died in 9–12 months[122]

The rest survived up to two years residency before dying. McDougall observed that: 'It will be noticed that the death-rate is higher amongst males than females, which bears out what was found when investigating causes of pauperism – that men bring themselves into conditions of destitution and disease more frequently than women'.[123] Therefore men, rather than women, tended to end up on the dissection table (repeated in chapters 4 to 6). He explained that most people came into the infirmary too late to be treated. This meant that altogether there were '393 deaths up to 31st December 1884'. Of these, some '171 were buried by friends and not at the cost of the Guardians – In many cases the deceased had been a member of a Burial Society'.[124] This left another '222' unaccounted for who entered the supply chain of the anatomy trade – a stable average after 1872–1874 – all sold from Crumpsall workhouse until 1910.

Altogether McDougall calculated that 9.78 per cent or 14,552 of a total population of 148,799 were paupers who died in destitution each year in Manchester during the late-Victorian era.[125] Of these, over half(7,960), or 54.7 per cent, were potential anatomy sales, such was the high rate of annual mortality in the city. Yet just 200 bodies were traded on average in a 12-month teaching cycle. Anatomists took what they needed and guardians buried the rest to avoid bad publicity, despite the latter's penny-pinching Poor Law schemes. Along the chain of supply, 95 per cent of cases were those defined as 'single' (children, young people, and those widowed without any personal ties). They lived lonely lives lost in abject poverty. Historians often debate the meanings of terms like pauperism, being poor, and so on. In this study, destitution meant the disappearance of 8,701 dissection cases between 1860 and 1910. Yet more sales continued until the New Poor Law ended in 1929, and beyond.

It has been a more painstaking research process trying to piece together the economy of supply at Manchester than at other regional centres. The fact that a glimpse of the general trends has been reconstituted reiterates how important it is to think laterally about the nature of the Victorian information state. Despite alternative types of record-keeping there was a concerted attempt to destroy evidence of dissection in 1929. There are gaps in the run of records and information about individual pauper life cycles is more fragile. Nevertheless a range of sources used in this chapter facilitates a novel perspective. The trade in Manchester was sufficient to sustain an expanding school of medicine, which was to train up to 250 medical students by the late-Victorian era. Bad publicity was minimised by making the decision to use one main workhouse source at Crumpsall. There soon developed an efficient system of supply. The workhouse Clerk was an important link in the chain. His unflinching support meant that anatomists could rely on a steady flow of corpses. 'Pauper burials' were processed at Manchester; 'anatomy burials' from Crumpsall. The crusade against outdoor relief made this system, and the expansion of human anatomy teaching, viable. Moreover, it must be remembered that research could only retrace *whole* corpses in the supply chain. There was a body-part business but little record survives of its supply networks or financial dealings. Discrepancies in the archives have made a longitudinal study unfeasible. Primary research does indicate, however, that it was active. Again those financial records disappeared from central government sources in 1929. It was not just very sensitive material but potentially embarrassing for all concerned. This final case study thus ends where we began, with forgotten human faces.

In the summer term of 1908, 'John Butters 45, Edward Jandy 70, Thomas Johnson 70, Hugh Chapman 62, William Findley 82, Emos Bibby 64' were packed into death boxes.[126] The Clerk sent them on the fast train from Crumpsall workhouse to Oxford anatomy school. They were buried in three lots, two bodies in each grave space. There was a brief graveside ceremony at Botley cemetery on the outskirts of the city. There they lie today in an unmarked plot. Did they fail in death or live lives of meaning? Only by listing their names is it possible to take the first research step into blank entries in the historical record. It turns out that there are historical parallels that still need to be appreciated about those lost medical lives. In 2009, a 'Health Profile for Manchester' was produced by the Department of Health. It highlighted that:

> The health of people in Manchester is generally worse than the England average...There are inequalities by deprivation, gender and ethnicity. For example, men in the least deprived areas can expect to live six years longer than men in the most deprived areas, and for women the difference is four years. Over the last ten years there have been decreases in death rates from all causes and in early death rates for heart disease and stroke, and cancer. However, these rates remain well above the national averages. Manchester has the worst early death rate from heart disease and stroke in England...Some 66.6 % of Manchester's residents, 299,950 people, still live in the third most deprived inner city area of the UK. Of these 39,911 children, 48.2% are living in poverty.[127]

Many of today's health inequalities were similar in the past (a drinking culture, heart disease, teenage pregnancy, and higher infant mortality rates). It is important to not neglect the relevance of historical material in the medical humanities. In the conclusion this book reflects on the broader lessons that have been learned from the hidden history of human faces in Victorian times to the present.

Conclusion

> Wherever the art of medicine is loved, there also is love for humanity.
>
> Hippocrates, *Precepts*, vi (*Littré*, ix, 258)

This book is 125,000 words long. By taking the spine in your hand and thumbing the pages the scale of the human stories can be visualised in the flickering black and white print. Each word represents an equivalent corpse sold in Victorian times. This was the scale of the anatomy trade found in many forgotten entries in burial books or dissection registers. A lot more still need to be added to the 17,500 lives recovered for posterity in the four chosen case studies. This human face of an economy of supply, wrote one social commentator in the *British Medical Journal* in 1870, was memorable:

> Coherent in statistical despairs
> With such a total of distracted life...
> To see it put down in figures on a page –
> Plain, silent, clear, as God sees through the earth
> The scene of all the graves; that's terrible
> For one who is not God, and cannot right
> The wrong he looks on.[1]

Only a recurring dialogue between the medical humanities and the scientific community can ensure that this hidden history continues to inform consensual medical ethics in the twenty-first century. For this past is not that of another country. It is what is happening on everyone's doorstep in a global marketplace.[2] The poorest remain vulnerable in all societies. Despite tight regulations, governments are running to catch up with one of the internet's fastest growing business sectors. A black-market network buys and sells organs, limbs, brains, bones, hips, hair, teeth, and torsos to the highest bidder. In America 'each body can be sold for anywhere between $10,000 and $100,000 or more, depending on the skill of the broker and the need of the moment'.[3] The *Daily Mail* reported in May 2010 that in some parts of Europe a human head 'will set the buyer back £1400 while a cross-section of the human body costs £14, 000'.[4] Anatomy acquisitions are also the latest commodity trend for a new type of consumer. Medical tourism connects rich patient to poor person.[5] This business report might seem far-fetched, the product of economic determinism, but in reality it is simply a stark financial fact of life for those who have to sell part of their body in abject poverty.

There is a crisis in modern medicine. We are living longer in the Western world. Life expectancy has been improved by new drugs and better technology. Nobody, though, can hold back the aging process. Cosmetic surgery cannot change the biological fact that everyone's body clock runs out of time. Some have the hard cash to defy human nature. The wealthier spend money in a world with a population explosion. In the hands of a skilled surgeon, the heart, lungs, kidneys, and liver are transplanted by modern health care. Yet what the dedicated healer studies to treat humanity, by contrast the body trader seeks to exploit for profit. To the unscrupulous, the operating theatre is the body-repair shop of the twenty-first century.[6] The past and present often run in parallel. Despite legislation, governments still need to stop 'a macabre industry that operates in the shadows, harvesting corpses for body parts which are then sold on as a commodity'.[7] There are six findings in this book that provide new insights into how this was once organised, indicating what more now needs to be done.[8]

First and foremost, the pauper histories at the heart of this book are a major achievement and the method for finding them is a new template. They are not just individuals. The regularities of their life and death stories make them powerful exemplars. Out of the shadows, from crowded lodging houses, busy red-light districts, slums filled with migrants, have emerged the faces and names of hidden lives. These were the sorts of people who read the Anatomy Act poster on the workhouse wall or crowded round to hear the performance of a new form of injustice. Their human material was the 'matter out of place' that became the anatomy trade in Victorian times. Ruth Richardson's rediscovery of the Anatomy Act was crucial in the 1980s. Building upon that perspective has deepened historical appreciation of dissection as a lived, rather than simply a death, experience. The economy of supply was more extensive and built

Conclusion 305

up over a longer duration than previous studies have documented. The New Poor Law lasted from 1834 until 1929. It was in the late-Victorian period that changes in medical education converged with a drive to limit welfare expenditure. Tax savings advantaged anatomists anxious to buy more bodies. In each region, the tightening of Poor Law funding converged with the timing of the expansion of the medical profession. Many infirmaries and workhouses denied their involvement when they were secretly selling bodies. It is not accurate to say that the Anatomy Act faded out after a mid-Victorian heyday. The opposite happened. Legislation really took hold around 1870. The anatomy trade turned into a profitable business in both corpses *and* body parts. These were bought in equal numbers, a material fact filed by central government. The voice of the poor protesting on behalf of the deceased was written down in pauper letters, but the survival of those precious accounts is sporadic. So it has been important to find alternative historical routes into the experience of being poor. Looking down onto the dissection table has been one traditional approach in standard histories of medical education. In this book the aim has been to look around what was happening. The dead cannot speak, but the records of their deaths can be used more creatively. Lateral thinking has produced new research recovered from the Victorian information state. It tells the history of dissection in a different way. By working backwards into the archives from the point of death it has been possible to restore a name, an age, a sex, an identity. The poorest are finally not just props in the dissection room drama but walk-on actors with speaking parts in the history of medicine.

The scale of the anatomy trade is the second new discovery in this book. In London alone, not less than 60,000 corpses bought whole changed hands among body dealers. On every street corner in a slum-clearance area, like central Holborn or the East End, familiar faces paid for information and access to available dead bodies. Robert Hogg, the so-called undertaker, but really a body dealer, turned state's witness at the Old Bailey in 1858. He symbolises the furtive nature of the business of anatomy across the capital. Historians of medicine and welfare need to interrogate more undertakers' business records. These will reveal the diverse ways that the funeral trade earned their profits in the poorest sectors of Victorian society. Many profited from the pauperism of their clientele along the body supply chain. Large wicker laundry baskets were wheeled along streets, parked at the back of public houses, and returned to the back entrance of major teaching hospitals or workhouses at nightfall. Inside were corpses for dissection and dismemberment. All were buried at cost in large unmarked graves away from public attention. In the provinces, a large medical school like Cambridge traded almost 3,000 compared to Oxford's much smaller supply of some 500 corpses. In larger urban centres like Manchester, over 8,000 were traded from a central workhouse location like Crumpsall. Most were purchased by local anatomists, with surplus distributed around the country. This was more than was declared officially to the Anatomy Inspectorate. This book's calculations are cautious about the figures for those bodies broken up for a more profitable sale. The Oxford case study, however, confirms that body parts matched the sales of complete corpses. Extrapolating from the economy of supply that has been compiled confirms that the anatomy trade in its entirety was not less than double the official estimates by central government. Compared to the very high mortality statistics of the time this was a small proportion of all the human material that could have been made available. Nevertheless every body part or corpse lost to the anatomy trade meant a lot to a poor family. Once future regional case studies have been completed the precise figures will emphasise that the business of anatomy was more extensive and far-reaching than previous historical research could have contemplated. It certainly created a complicated medical market in human research material, coroners competing with anatomists for corpses to enhance their professional reputations in ways that have been misunderstood.

A third new finding has been the reliance of the anatomy trade on the railway system in Victorian England. It is impossible to understand the geography of supply lines without appreciating the transportation networks that got a fresh corpse from source to the dissection table. On a daily basis, where there was a fast train, there was a corpse for anatomy being transported in a rear carriage in a dead box. Smaller towns in East Anglia or surrounding Oxford supplied a disproportionate number of dead bodies because they were ideally located on the railway line to do so. This was why Macalister and Thomson became travelling anatomists in their public engagement work. They worked out the best routes, negotiated the carriage rates, and outbid their competitors to get supply chains underway. It was often onerous and time-consuming work but also necessary if the anatomy trade was to expand to meet increasing medical student demand. The Anatomy Inspectorate took little interest in the organisation of the railway network system. Their primary concern was to avoid adverse publicity. The anatomy trade was a truly Victorian enterprise. It was inventive, always seeking practical solutions to overcome its operational problems in the most pragmatic way. The implementation of the Anatomy Act came to personify a very British character trait. Individual medical schools made up body-finding schemes as they went along. There was no blueprint or grand plan. This is why regional studies are pivotal when assessing what really happened inside the business of anatomy on a national basis. Taking a micro-approach does not mean losing historical perspective. Only by synthesising quantitative and qualitative primary research is it possible to bring together the tiny jigsaw pieces that activated the Anatomy Act. The poorest are a crucial part of that mosaic of medical progress.

The importance of bringing together a national and local perspective is the fourth new finding. There needs to be an English local history of the anatomy trade. Historians of medicine know about the impact of the Anatomy Act following Ruth Richardson's groundbreaking work which established that the poorest individuals were used by the state to advance medical education.[9] Nevertheless there was a neglected complex regional picture. Throughout this book new research has shown that what really mattered was what was happening on people's doorsteps. At St. Bartholomew's hospital a wide cross-section of the capital's population were dissected. A servant class, street prostitutes, and suicides were common cases. Ethnicity trends – Irish and Italian corpses – confirm historic patterns of poverty density. It was a consistent fact of Victorian life that some areas of Britain provided a regular supply of anatomical material compared to others. No matter where the immigrant population settled, from Italy or Ireland, whether in Manchester or London, they had to take a pragmatic view of dissection sales, and this despite their cultural and religious inclinations to act the opposite. This finding raises the intriguing proposition that a certain brand of Irish and Italian Catholicism has always provided most of the human material for human anatomy teaching. Traditionally Northern Europe is sensitive about the cutting of the body, whereas Ireland and Southern Europe has taken a more pragmatic view. That historical pattern may have implications for policy-makers tackling sensitive areas of medical ethics like future organ donation today.

The fifth new finding is that more men than women ended up in the welfare institutions of the New Poor Law. That demography reflects the economic problems of family life in the poorest households of Victorian England. In the provinces, a ratio of three men to one woman typifies the anatomy trade. In the capital, one man to one woman was sold for most of the nineteenth century and then three men to one woman after 1900. Most were traded from overcrowded workhouses and infirmaries where men aged over 50 resided, spent forces from hard field work or manufacturing labour. Women could engage in calculative reciprocity (childcare, washing, mending, and household duties) to avoid the New Poor Law until they could no longer contribute to family economies in old age. The shortage of female bodies and rateable changes persuaded asylums to sell dead women by the turn of the century. In places like Oxford, however, it was common for male medical students to train on male corpses before entering general medical practice. Future research is needed to investigate what this may have meant for the development of well-woman medicine outside the large London teaching hospitals. In the meantime, many families were reluctant to sell children but had to face grim economic facts. A deceased infant could earn a poor relative up to a year's wage and save the cost of an expensive funeral. In abject poverty there were few welcome choices. For the grief-ridden it was best never to speak about what had to be done. Other accounts tell of some paupers overcoming their private feelings to make a necessary sale. Records do survive of stillbirths, notably in the St. Bartholomew's and Manchester studies. The coming of Poor Law democracy ironically increased their value in ways that have been overlooked, even today. It is this observation that brings this book full circle in its research journey. A sixth and final finding is thought-provoking.

Alexander McPhial, Anatomy Inspector, wrote from the Ministry of Health to Professor Arthur Thomson at Oxford on 2 March 1928. He clarified the law with regard to the use of stillborn children for anatomy and their burial. Although the letter is lengthy it is worth quoting in full:

Dear Sir,

Utilisation and disposal of the still-born

I have received a number of requests for information as to how far the practice of using stillborn bodies for teaching and research may be affected by recent legislation; and it has thus appeared probable that a general statement upon the subject might be found of convenience.

I desire in the first place to make it clear that the Anatomy Act does not apply in any way to the reception and disposal of these bodies.

Since the provisions of the Births & Deaths Registration Act of 1926, requiring the registration of stillbirths appear to have give rise to some misapprehension in the minds of hospitals and other authorities in charge of the bodies of the stillborn regarding their power to continue to place them, as heretofore, at the disposal of medical schools, I have consulted the Registrar-General on the matter & have the authority for stating that there is nothing new in the Act or in the rules made there under which in any way hinders or restricts the utilisation of these bodies as hitherto practised.

I am informed, however, that for the purposes of the ultimate disposal of the remains in a burial ground or by cremation, it is necessary to produce to the burial or cremation authority the registrar's certificate of registration of the stillbirth (Certificate for the Disposal – Stillbirth).

When therefore, for the purpose of burial, stillborn remains are included in the same coffin as those of the deceased person, all that is required is the production to the burial authority of the registrar's certificate of registration with regard to the stillbirth (in addition of course, to the registrar's Certificate of Disposal in respect of the deceased person [i.e. adult multiple burial]) & the endorsement of that certificate in writing by the undertaker to the effect that the stillborn remains to which the certificate relates have been enclosed in the coffin in question...

No notification of disposal is required by the registrar in the case of the stillborn.[10]

This is a bureaucrat's letter, written in formulaic language. It has a medical mentality and paternalistic attitude about the use of stillbirths for anatomy. A standard legal definition set out that stillbirths were not independent human beings of their birth mother. All had not survived outside the womb for one continuous circulation of the blood. A certificate was needed to bury the remains but not for the purposes of the Anatomy Act. Each could be disposed in an adult coffin or buried in a multiple grave. Today the same system survives.

An investigation by the *Evening Standard* in March 2010 alleged that doctors were placing stillborn children in pauper graves organised by London hospitals.[11] Under the Human Tissue Act it is illegal to dissect any human remains without consent and this is respected. Nevertheless it remains the case that most stillbirths are buried in a communal grave to save NHS costs. A journalist named David Cohen calculated that in the capital there were '1,010 child deaths' between 2007 and 2010. Of these, he found that '177 were stillborn' and that the parents had requested that the NHS hospital in question organise the burial. In Kensington and Chelsea a 'maximum of 20 bodies per common grave' were interred in one of the richest London districts. In Southwark that figure was much higher, not less than '30 stillbirths' together. There is still very little historical distance from the scene of *Rex versus Feist* at the Old Bailey in 1858. According to recent press reports, the boroughs of 'Hackney, Wandsworth, Brent, Southwark, Lewisham and Barnet...account for 85 per cent of communal burials' in London during 2009–2010. This was ironically where the body trade was active in the past. It is easy to lose sight of the fact that new legislation is just the first step in closing loopholes in the law. There is always more work to be done to close the gap between rhetoric and reality in the body trade. Victorian pauper graves, for people like the lady in the café encountered in this book's preface, are still just a coffee cup apart for the bereaved who request an NHS burial.

Afterward

Professors Alexander Macalister and Arthur Thomson regretted the subterfuge that surrounded their anatomical work. Victorians were not told about how some of the major advances in medicine were developed at Cambridge where the dead were being dissected on a daily basis. Yet this is now standard in hospitals around the world. Three of the most groundbreaking happened in the vicinity of a vibrant economy of supply. The evidence suggests that the first X-ray in Britain relied on early skiagraph work and some images were of dead pauper children like that published in the *British Medical Journal* on 29 February 1896. Cambridge scientists likewise pioneered tissue culture work in a location where human pauper material was in plentiful supply. Again, how much human material was passed on is unclear, but pilot-project research for the next phase of this study suggests that there was overlap. It is perhaps a fitting and curious fact of medical history that the famous model of DNA created by Crick and Watson was constructed in a laboratory setting where nineteenth-century anatomists cut up the poor to understand the fundamental biology of life. Those findings reiterate the contribution that the medical humanities can make, given the opportunity, to future debates on bioethics. Meantime the New Poor Law was coming to a close in May 1929. It was just six months before the Wall Street crash. Lots would be sold in the severe economic downturn of the 1930s. Edith Sitwell, the poet, wrote to a guardian of the poor, Mr. Frederick Wills of Hendon, North London, that:

> It is indeed terrible to be friendless and poor. I have been aghast, lately, at the cases of unhappy creatures who have fallen from starvation and cold, and who have died from these on reaching hospital. One cause of this is, that the feeling has been encouraged deliber-

ately that it is "a disgrace to go into the Workhouse". It is no disgrace whatsoever, in my opinion. It is a fearful disgrace to be lazy and useless, but it is no disgrace whatsoever to fail...[12]

This book has been about that economic failure and its scientific success. Absolute poverty has been the basis of so much medical progress, and it goes on being so. This is why this history is not just in our keeping, but in our making too. The dissection room door is now ajar for future research opportunities. Other contributions are needed. At the close it seems therefore fitting to keep in mind advice Alexander Macalister gave in a speech to new medical students at Cambridge in 1908. He said that were three 'golden rules for medical writers':

(1) Don't write unless you have something to say.
(2) Say it.
(3) Stop when you have done...[13]

In *Dying for Victorian Medicine*, a history of silence is now at an end. For here begins a new chapter in our health-care history, one that looks into the human face of poverty and sees finally *all* of pauper kind in the scientific progress of modern medicine.

Notes

1 Chalk on the Coffin: Re-Reading the Anatomy Act of 1832

1. 'The Weather in London', *Times*, 23 February 1858.
2. 'Mean Temperatures at Camden Square London at 9 am, 1858–1881' in Anon. (1882), *London Winter Temperatures from 1858 to 1881* (London: John Murray), pp. 1–2.
3. The first spring flowers were reported to have bloomed at Helston, Cornwall in *The Times*, 14 January 1858. By February London parks had spring colour from 'more than 100 blossoms'.
4. 'John Welsh, 1824–59, meteorologist' in S. Lee, ed. (1860), *The Dictionary of National Biography*, Vol. 60 (London: John Murray), p. 239.
5. S. Wise (2008), *The Blackest Street: The Life and Death of a Victorian Slum* (London: Random House) recounts typical social conditions.
6. M. Brodie (2004), *The Politics of the Poor: The East End of London, 1885–1914*, (Oxford: Oxford University Press), for essential context.
7. R. Richardson (2008), *The Making of Mr. Gray's Anatomy: Bodies, Books, Fortune and Fame* (Oxford: Oxford University Press), pp. 117–39.
8. There is an extensive historiography on the Medical Act (1858). An excellent appraisal of the literature can be found in K. Waddington (2003), *Medical Education at St. Bartholomew's Hospital 1123–1995*, (Woodbridge: Boydell and Brewer), pp. 13–44.
9. 'The New Medical Act', *Times*, 23 October 1858.
10. Ibid., for a useful summary of its stipulations.
11. M. W. Weatherall (1996), 'Making Medicine Scientific: Empiricism, Rationality and Quackery in Mid-Victorian Britain', *Social History of Medicine*, IX, pp. 175–94, explains the new education standards. When the Medical Act was extended in 1885 all doctors then had to qualify in both medicine and surgery; this was applicable to midwives too.
12. See, S. V. F. Butler (1981), 'Science and the Education of Doctors during the Nineteenth Century: A Study of British Medical Schools with Particular Reference to the Development and Uses of Physiology' (unpublished Ph.D., UMIST).
13. D. Burgh (2007), *Digging up the Dead: Uncovering the Life and Times of an Extraordinary Surgeon* (London: Random House), p. 3.
14. See, E. T. Hurren and S. A. King (2005), 'Begging for a Burial: Form, Function and Conflict in Nineteenth Century Burial', *Social History*, XXX, pp. 321–41.
15. Contemporary attitudes are elaborated in N. Jones and P. Jones (1976), *The Rise of the Medical Profession: A Study of Collective Mobility* (London: Croom Helm); M. J. Peterson (1978), *The Medical Profession in Mid-Victorian London* (Berkeley: University of California Press); A. Digby (1994), *Making a Medical Living: Doctors and Patients in the English Market of Medicine, 1720–1911* (Cambridge: Cambridge University Press); T. Bonner (1995), *Becoming a*

Physician: Medical Education in Britain, France, Germany and the United States, 1750–1945 (Oxford: Oxford University Press).
16. M. J. Durey (1976), 'Body Snatchers and Benthamites: The Implications of the Dead Body Bill for the London Schools of Anatomy, 1820–42', *London Journal*, II: pp. 200–25.
17. Summarised in, R. Richardson (1991), ' "Trading Assassins" and the Licensing of Anatomy', in R. French and A. Wear, eds, *British Medicine in the Age of Reform*, (Cambridge: Cambridge University Press), pp. 74–91.
18. See, notably, J-M. Strange (2005), *Death, Grief and Poverty in Britain, 1870–1914* (Cambridge: Cambridge University Press).
19. All of the quotations in *Rex versus Feist* [1858] have been reconstructed from record linkage work on www.oldbaileyonline.org, *The Proceedings of the Old Bailey, London's Central Criminal Court, 1674–1913*. All quotations hereafter are taken from the original OB case number: t18580222-354. Reports in *Times*, 21 January, 25 February, 5 March, and write-up of the quashed judgement 24 April 1858, are also used. To test for bias in the court and newspaper record, poor law accounts have also been consulted at the London Metropolitan Archives (hereafter LMA), St. Mary's Newington, P92/MRY.
20. For a brief summary, see D. W. Meyers (2006 edn), *The Human Body and the Law: A Medico-Legal Study* (Chicago: Aldine Transactions), p. 107; it is also cited, but not analysed, in H. MacDonald (2009), 'Procuring Corpses: The English Anatomy Inspectorate, 1842–1858', *Medical History*, 53, III, pp. 379–96. It is regrettable that in recent years the popular end of the publishing market has tended to exploit pauper stories like this by not doing anywhere like enough record linkage work. This book argues that such strategies compound the problem of the voices of the poor not being heard in their proper historical context. Chapter 4 returns to *Rex versus Feist* with new scholarship.
21. LMA, St. Mary's Newington, P92/MRY, 1858; Southwark Local History Library, Parish of St Mary Newington, 'Legal Papers about the Indictment of Alfred Feist, Workhouse Master, for the Removal of Paupers' Bodies to Guy's Hospital School of Anatomy, 1858'; Lambeth Police Court, Monday, 20 January 1858 and 'Copy of evidence taken by Mr. Farnall, Poor Law Inspector, on 22 January 1857'.
22. S. Lewis, ed. (1848), *A Topographical Dictionary of England* (London: Institute of Historical Research), pp. 389–93, recounts poor law arrangements, parish boundaries and population size.
23. 'Sick Poor', *British Medical Journal*, 26 September 1868, II, p. 348, summarises medical care standards across London for the decade 1858–1868, including St. Mary Newington.
24. For the most recent summary of Poor Law rules and regulations, see E. T. Hurren (2007), *Protesting about Pauperism: Poverty, Politics, and Poor Relief in Late-Victorian England* (Woodbridge: Boydell & Brewer), pp. 17–58.
25. All figures taken from 'The Poor Law Statistics in the Metropolis', *Times*, 22 April 1858, issue 229874, p. 12, column d.
26. See 'London Destitution and Its Remedy – Letter to the Editor from An East End-Incumbent', *Times*, 2 February 1858, issue 22906, p. 9, column f – states that pauperism returns commonly show 'that certain portions of London are almost exempt from poor and poor rates, and that other portions – the

destitute ones alluded to – are overwhelmed with the burdens which they cannot possibly support'.
27. LMA, St. Mary's Newington, P92/MRY, 1858.
28. OB, case number: t18580222-354, 22 February 1858.
29. S. Tarlow (2007), *The Archaeology of Improvement in Britain, 1750–1850*, (Cambridge: Cambridge University Press), pp. 142–8, for basic architectural layout of Dead Houses.
30. J. Litton (2002), *The English Way of Death: The Common Funeral since 1450* (London: Robert Hale Ltd).
31. See another common example in *The Examiner*, 17 October 1848, p. 686, which contains a report of the suicide of a young woman at Twyford in Winchester called 'Rebbecca Mabbett' who died 'from taking poison' aged just 'seventeen'; she was accompanied to the grave by a crowd who checked that her name was chalked on the coffin and that her corpse went into the earth: 'The coffin consisted of a few thin boards, hastily nailed together, but a white cloth for a pall, was thrown over it, and the mourners had provided themselves with crape hat-bands ... a heavy clod [of earth] was thrown in [to the grave] which seemed as if it would break the fragile coffin and a shudder of horror ran through the spectators'. The mourners waited until the grave was 'covered with turf'. Chapter 2 revisits this theme.
32. OB, case number: t18580222-354, 22 February 1858; verified in LMA, St. Mary's Newington, P92/MRY, 1858.
33. Ibid.
34. His full address was given in an early report of the allegations, 'The Dead Bodies of Paupers, Extraordinary Revelations', *Liverpool Mercury*, 8 January 1858, issue 3087, p. 1, column a; evidence cross-checked to LMA, St. Mary's Newington, P92/MRY, 1858 and Census Returns, 1861.
35. OB, case number: t18580222-354, 22 February 1858. All the quotations now come from the same source and are hereafter not footnoted individually.
36. See 'Police report', *Times*, 21 January 1858, issue 22896, p. 11, column f.
37. It was opened in 1853 and closed in 1876, being located in Hackney, Bullards Place, in East London, not far from Bow, just opposite the western end of Old Ford Road. Its burial records are held at The National Archives (hereafter TNA) RG8/42–51, where registers are arranged alphabetically.
38. The original allegations as reported proved remarkably accurate in court. See early editions all printed on 9 January 1858 in regional newspapers like *Manchester Times*, issue 5, *Leeds Mercury*, issue 6740, *Bristol Mercury*, issue 3538.
39. The *Times* reported the case extensively on 21 January, 25 February, 5 March, 1858.
40. Bodleian Library (hereafter Bodl. Lib.), University Archives (UA), Anatomy Department Records (ADR), HA89, Thomson to Home Office, report, 17 February 1920, reported Arthur Thomson, head of anatomy at Oxford University, that since the Anatomy Act had been passed in 1832 the body certification system was 'unworkable'. Chapter 6 elaborates this context.
41. R. Richardson (2001 edn), *Death, Dissection and the Destitute* (London: Phoenix Press), first identified the 'bureaucrat's bad dream', pp. 239–61.
42. OB, case number: t18580222-354, 22 February 1858.
43. For example, in the TNA, MH12 series of poor law records, they were pinned by civil servants to pauper complaints. Likewise in the Macalister

Notes 315

Papers, Human Anatomy Department, Cambridge University, discussed in Chapter 5.
44. E. Higgs (2004), *The Information State in England: The Central Collection of Information on Citizens, 1500–2000* (Basingstoke and New York: Palgrave Macmillan) provides important context on how bureaucracy can betray what it wants to hide.
45. This new methodology providing empirical evidence of the anatomy trade was first established in E. T. Hurren (2004), 'A Pauper Dead-House: The Expansion of Cambridge Anatomical Teaching School under the late Victorian Poor Law, 1870–1914', *Medical History*, 48, I, 69–94.
46. OB, case number: t18580222–354, 22 February 1858.
47. See *Daily News*, 25 February 1858 (third most popular daily newspaper in the nineteenth century); *Liverpool Mercury*, 27 February 1858; *Baner Cymru*, 3 March 1858; *Trewman's Exeter Flying Post*, 4 March 1858; *Birmingham Daily Post*, 15 March 1858.
48. *The Bury and Norwich Post*, 2 March 1858, p. 4. This can also be consulted online: www.foxearth.org.uk/1858BuryNorwichPost.html – weblink by The Foxearth and District Local History Society.
49. *Rex versus Feist* [1858], in H. R. Dearsly and T. Bell (1858), *Crown Cases Reserved for Consideration, and Decided by the Judges of England* (London: Stevens & Norton, H. Sweet & W. Maxwell), p. 1135.
50. See, notably, M. Sappol (2002), *A Traffic of Dead Bodies: Anatomy and Embodied Social Identity in Nineteenth-Century America* (Princeton: Princeton University Press); MacDonald, *Human Remains*; T. Buklijas (2008), 'Cultures of Death and Politics of Corpse Supply: Anatomy and Vienna, 1848–1914', *Bulletin of the History of Medicine*, 82, III, 570–607; E. T. Hurren and I. Scherder (2011), 'Dignity in Death? The Dead Body as an Anatomical Object in England and Ireland c. 1832 to 1929', in A. Gestrich and S. A. King (eds), *Dignity of the Poor* (Oxford: Oxford University Press), pp. 1–30.
51. Richardson, *Death, Dissection and the Destitute*.
52. See, www.publications.parliament.uk/pa/.../g/cmhuman.htm, quotations from transcript of Standing Committee G can be consulted, seventh sitting, 5 February 2004, column 221.
53. Ibid.
54. See seventh sitting, 5 February 2004, column 221.
55. Ibid.
56. It is not the purpose of this chapter to repeat the important work that Richardson did on the various select committees and pressure groups that pushed for the Anatomy Act during the 1820s and 1830s. Context can be read in Richardson, *Death, Dissection and the Destitute*, Chapter 7, pp. 159–92.
57. Refer Karel Williams (1981), *From Pauperism to Poverty* (Manchester: Manchester University Press), pp. 96–102.
58. W. C. Maude (1903), *The Poor Law Handbook* (London: Poor Law Officers Journal), p. 25.
59. Historians of poverty have begun to appreciate the meaning of official poor law posters from the perspective of the poorest; see, for instance, K. D. M. Snell (2006), *Parish and Belonging: Community, Identity and Welfare in England and Wales, c. 1700–1950*, (Cambridge: Cambridge University Press), pp. 339–65 and Hurren, *Protesting about Pauperism*, pp. 56–8, 163–5, 259–61.

60. Richardson, *Death, Dissection and the Destitute*, p. 207.
61. Ibid.
62. For a summary of the literature, see Hurren, *Protesting about Pauperism*, pp. 159–60.
63. Refer also Snell, *Parish and Belonging*, pp. 339–65 and Hurren, *Protesting about Pauperism*, pp. 56–8, 163–5, 259–61.
64. See www.parliament.uk/parliamentary.../parliamentary.../archives_electronic. cfm for all quotations taken from House of Commons parliamentary papers online (2006), Proquest Information and Learning Company.
65. See, recently, T. Borgstedt (2009), *Topology of the Sonnet: Theory and History of its Genre* (Verlag Germany: Max Niemeyer), English translation.
66. www.parliament.uk/parliamentary.../parliamentary.../archives_electronic. cfm – 1831–32 (419), 2 William IV (Session, 1831–2) – A Bill [as amended on the second re-commitment] For Regulating the Schools of Anatomy, ordered by the House of Commons to be printed 8 May 1832, ref: 419, pp. 1–6, quote at p. 1.
67. On Victorian ballads and their culture, see D. E. Gregory (2006), *Victorian Songhunters: The Recovery and Editing of English Vernacular Ballads and Folk Lyrics, 1820–1883* (Lanham, MD: Scarecrow Publications).
68. Refer T. Watt (1991), *Cheap Print and Popular Piety, 1550–1640* (Cambridge: Cambridge University Press); M. Brodie (2003), 'Free Trade and Cheap Theatre: Sources of Politics for the 19th-Century London Poor', *Social History*, 28, III, pp. 346–60.
69. See P. Fumerton (2006), *Unsettled: The Culture of Mobility and the Working Poor in Early Modern England* (Chicago: Chicago University Press).
70. Refer J. Childs (2006), *Henry VIII's Last Victim: The Life and Times of Henry Howard, Earl of Surrey* (London: Vintage), Chapter 10, 'Poet Without Peer', pp. 164–218.
71. Refer M. Schoenfeldt, ed. (2007), *A Companion to Shakespeare's Sonnets* (Oxford: Blackwell).
72. See S. H. L. Walter (2006), *John Milton, Radical Politics, and Biblical Republicanism* (Newark: University of Delaware Press); T. P. Anderson (2006), *Performing Early Modern Trauma from Shakespeare to Milton* (Aldershot: Ashgate); S. Eliot (2006), 'What Price Poetry? Selling Wordsworth, Tennyson, and Longfellow in Nineteenth- and Early Twentieth-Century Britain', *Papers of the Bibliographical Society of America*, 100, IV, pp. 425–45.
73. 2 William IV (Session, 1831–2), 8 May 1832, ref: 419, p. 1.
74. See Northampton Central Library Local Studies Collection, popular and theatre posters collection; Cambridge Central Library Local Studies Collection, poster collection for the nineteenth century. I am grateful to Shropshire History Society for alerting me to a rare poster in their local study collection.
75. J. Lawrence (1998), *Speaking for the People: Party, Language and Popular Politics in England, 1867–1914* (Cambridge: Cambridge University Press), p. 1.
76. Brodie, 'Free Trade and Cheap Theatre', p. 360.
77. 2 William IV (Session, 1831–2), 8 May 1832, ref: 419, p. 2.
78. Ibid. pp. 2–3.
79. 2 William IV (Session, 1831–2), 8 May 1832, ref: 419, pp. 1–6.

80. Ibid., p 3. It is noteworthy that p. 5, clauses 12–15, states 'That nothing in this Act contained shall be construed to extend to Post-Mortem Examination of any human body'. In other words, many thought Anatomical Examination and Post-Mortem Examination were two different things. In practice they were the same.
81. Full text: http://www.publications.parliament.uk/pa/cm200304/cmbills/049/04049.32–35.html#j033g.
82. Anon. (1858 edn) *Oxford English Dictionary* (Oxford: Oxford University Press).
83. Anon. (1858 edn) *Webster's Medical Dictionary* (London: John Murray).
84. I. Burney (2000), *Bodies of Evidence: Medicine and the Politics of the English Inquest, 1830–1926* (Baltimore: Johns Hopkins Press), pp. 2–5.
85. 2 William IV (Session, 1831–2), 8 May 1832, ref: 419, p 3.
86. Ibid. pp. 4–5.
87. See Hurren and King, 'Begging for a Burial', pp. 321–41.
88. Cambridgeshire Record Officer (hereafter CRO), P25/1/21–23.
89. Oxfordshire Record Office (hereafter ORO), ENG/1/A1/2.
90. Corpses from Birmingham Medical School were buried in Witton cemetery (one of the largest common cemeteries in Britain); copies of burial entries are held on location and at Birmingham Central Library.
91. OB, case number: t18580222–354, 22 February 1858.
92. St. Bartholomew's dissection records, MS/81–6.
93. An issue discussed in, for instance, G. D. Jones & M. I. Whitaker (2009), *Speaking for the Dead: Cadavers in Biology and Medicine* (Aldershot: Ashgate).
94. Hurren & Scherder, 'Dignity in Death?' pp. 1–30, elaborates this theme.
95. In *Times*, 14 June 1862, for example, Rev. Livesey, who buried corpses for Sheffield medical school, told a shocked jury: 'How could he take into his church a box of putrid matter, emitting a most offensive smell, and read a service over it?'
96. This term was coined by M. Douglas (1966), *Purity and Danger: An Analysis of the Concepts of Pollution and Taboo* (London: Ark Paperbacks).
97. 2 William IV (Session, 1831–2), 8 May 1832, ref: 419, p 6.
98. Refer *An Act to Amend the Regulating Schools of Anatomy* [34 & 35 Vict. C. 16, 1871] and Bodleian Library (hereafter Bodl. Lib.), University Archives (UA), Anatomy Department Records (ADR), HA89, Oxford medical school to H. M. Inspector, London re: timeframes for dissecting corpses.
99. See a Sheffield scandal, reported in *Times*, 14 and 24 June, 1862.
100. 2 William IV (Session, 1831–2), 8 May 1832, ref: 419, p 6.
101. Richardson, *Death, Dissection and the Destitute*.
102. What so few anticipated was the longer-term relevance of her pathbreaking research. Interest in her scholarship revived after the NHS organ scandals; see R. Richardson (2006), 'Human Dissection and Organ Donation: A Historical and Social Background'. *Mortality*, 11, II, 151–65.
103. Richardson, *Death, Dissection and the Destitute*, p. xiv.
104. Ibid. Richardson wrote at the time, 'inevitably there are gaps and it is really only now that I am able to let you read it [her book] that I can myself see its shortcomings, and very much feel the necessity of a further decade's work to do the subject justice'. This book's aim is to build on that observation with a new methodology.

105. Richardson, *Death, Dissection and the Destitute*, p. 239.
106. This viewpoint has recently been revisited in MacDonald, 'Procuring Corpses', p. 379.
107. Richardson, *Death, Dissection and the Destitute*, p. 245.
108. Ibid.
109. Ibid., p. 271.
110. Richardson's estimates are based on TNA, MH74/16, Anatomy Inspectorate, London returns for supply rates – see *Death, Dissection*, fn. 112, table, pp. 368–9.
111. Ibid., p. 266.
112. Strange, *Death, Grief*, p. 8, in reviewing Richardson takes issue with her claim that the poorest went on resenting the Anatomy Act, concluding that fears were overstated because 'by the latter decades of the 19th century, poor law guardians increasingly refused to co-operate with the demands of medical schools'. This book tests Richardson's and Strange's remarks for the first time with empirical evidence.
113. For example, S. Wise (2005), *The Italian Boy: Murder and Grave-Robbery in 1830s London* (London: Jonathan Cape), p. 287, states that: 'Though the Anatomy Act had killed off the trade [digging up bodies] by 1844, resurrection it seems remained a potent folk memory.'
114. Richardson, *Death, Dissection and the Destitute*, pp. 263–4.
115. It is generally assumed in standard medical historiography that paupers were passive or silent historical actors who left little record behind of their true feelings about the anatomy trade. Hurren, *Protesting about Pauperism*, pp. 191–213, refutes this view and throughout this book additional evidence will be presented.
116. Higgs, *The Information State in England*.
117. In TNA, MH12 series, there is a general neglect of pauper correspondence about the anatomy trade, evidence elaborated in Chapters 4–7.
118. See Hurren, *Protesting about Pauperism*, pp. 248–63.
119. E. T. Hurren (2008), 'Whose Body Is It Anyway? Trading the Dead Poor, Coroner's Disputes and the Business of Anatomy at Oxford University, 1885–1929', *Bulletin of the History of Medicine*, 82, IV, 775–819, first made this point and substantiated it with empirical research. It has started to become the subject of wider commentary in MacDonald, 'Procuring Corpses', pp. 379–96.
120. In 2009, the United Nations Development Fund for Women (UNIFEM) highlighted the importance of stopping the body trade for the living and the dead. Likewise *The Sunday Times* front page on 27 September 2009 led with the headline, 'Crunch Victims Sell their Kidneys to Pay off Debt'.
121. See D. Dickenson (2009), *Body Shopping: The Economy Fuelled by Flesh and Blood* (New York: One World Publications).
122. Williams, *From Pauperism to Poverty*, p. 102.
123. Hurren, *Protesting about Pauperism*, p. 250.
124. C. Dickens, 'Use and Abuse of the Dead', *Household Words*, volume 17, issue 419, pp. 361–5 on *Rex versus Feist* [1858].
125. Ibid., p. 364.
126. Ibid.
127. Ibid., p. 361.
128. Ibid., p. 364.

129. Ibid.
130. Ibid.

2 Restoring the Face of the Corpse: Victorian Death and Dying

1. Estate agent conversation with author, Cambridge 2003.
2. S. M. Gilbert (2006), *Death's Door: Modern Dying and the Ways We Grieve* (New York: Norton and Co.), p. xvii.
3. A point first made by R. Richardson (2001 edn), *Death, Dissection and the Destitute* (London: Phoenix Press).
4. See, for example, J. S. Stephenson (1985), *Death, Grief and Mourning: Individual and Social Realities* (New York: Free Press).
5. See J. Whaley (ed.) (1981), *Mirrors of Mortality: Studies in the Social History of Death* (New York: St. Martin's Press).
6. 'Lonelier lives lead to rise in pauper funerals', first printed in *The Observer*, 15 February 2009, and *Guardian* online: www.guardian.co.uk/uk/2009/feb/15/state-funded-funerals.
7. Ibid.
8. For an overview, see J. Benson (1989), *The Working Class in Britain, 1850–1939*, (New York and London: I. B. Tauris & Co. Ltd).
9. Context, in J. Curl (2000 edn), *The Victorian Celebration of Death* (London: History Press Ltd).
10. Notably H. George (1879), *Poverty and Progress: An Inquiry into the Cause of Industrial Depressions and of Increase of Want with Increase of Wealth – The Remedy* (London: Doubleday).
11. A. Tennyson (1859 8th edn), *In Memoriam* (London: Edward Moxon & Co.), stanza vi, p. 6.
12. See, for example, D. S. Wilson (2003 edn), *Darwin's Cathedral: Evolution, Religion and the Nature of Society* (Chicago: Chicago University Press).
13. J-M. Strange (2000), 'Review Article – Death and Dying: Old Themes and New Directions', *Journal of Contemporary History*, 35, III: pp. 491–9, q. at p. 494 and for context, J. Dollimore (2000), *Death, Desire and Loss in Western Culture* (New York and London: Routledge).
14. E. Hallam, J. Hockney, and G. Howarth (eds) (1999), *Beyond the Body: Death and Social Identity* (New York and London: Routledge).
15. Notably, T. Laqueur (1983), 'Bodies, Death, and Pauper Funerals', *Representations*, I: pp. 109–31; T. Laqueur (1994), 'Cemeteries, Religion and the Culture of Capitalism' in J. A. James and M. Thomas (eds), *Capitalism in Context: Essays on Economic Development and Cultural Change in Honour of R. M. Hartwell* (Chicago: Chicago University Press), pp. 138–55; Richardson, *Death, Dissection*.
16. J. Litton (1998), 'The English Funeral, 1700–1850', in M. Cox (ed.), *Grave Concerns: Death and Burial in England, 1700–1850* (York: Council for British Archaeology), pp. 3–16, q. at pp. 15–16; J. Litton (1991), *The English Way of Death: The Common Funeral since 1450* (London: Robert Hale).
17. For example, J-M. Strange (2003), 'Only a Pauper Whom Nobody Owns: Reassessing the Pauper Grave c. 1880–1914', *Past & Present*, CLXXVIII: pp. 148–75; E. T. Hurren and S. A. King, 'Begging for a Burial: Form, Function and Conflict in 19th Century Pauper Burial', *Social History*, XXX: pp. 321–41.

18. Elaborated in E. T. Hurren (2008), 'Whose Body Is It Anyway: Trading the Dead Poor, Coroner's Disputes and the Business of Anatomy at Oxford University, 1885–1929', *Bulletin of the History of Medicine*, 82: pp. 775–819.
19. 'Restoring the Face of a Corpse', *London Review* (May 16, 1863), vol. 6, issue 150: pp. 530–1.
20. Ibid., p. 530.
21. S. Lee (ed.) (1901), *Dictionary of National Biography, Supplement 3* (London: Smith, Elder & Co.) provides a summary of Sir Benjamin Richardson's career and important work on chemical analysis in forensic medicine (1828–1896).
22. He was physician to the Blenheim Street Dispensary (1854), the Royal Infirmary for Diseases of the Chest, City Road (1856), the Metropolitan Dispensary (1856), and the Marylebone and Margaret Street Dispensaries (1856).
23. 'Faceless Corpse', p. 530.
24. Ibid., p. 531.
25. Ibid.
26. Ibid.
27. Ibid.
28. See P. J. R. King (2006), *Crime and Law in England, 1750–1840: Remaking Justice from the Margins* (Cambridge: Cambridge University Press).
29. See H. Taylor (1998), 'Rationing Crime: The Political Economy of Criminal Statistics since the 1850s', *English History Review*, 3rd series, IX: pp. 569–90.
30. See, notably, V. A. C. Gattrell, B. Lenman, and G. Parker (eds) (1980), *Crime and the Law: The Social History of Crime in Western Europe since 1500* (London: Europa) pp. 238–339.
31. Further research complements J. Archer (2008), 'Mysterious and Suspicious Deaths: Missing Homicides in North-West England, 1850–1900', *Crime, Historie et Societies/Crime, History and Societies*, XII: pp. 45–63.
32. Refer I. Burney (2000), *Bodies of Evidence: Medicine and the Politics of the English Inquest, 1830–1926* (Baltimore: Johns Hopkins University Press).
33. See Hurren, 'Whose Body', pp. 775–819.
34. Key trends are in C. Gittings and P. C. Jupp (eds.) (1999), *Death in England: An Illustrated History*, (Manchester: Manchester University Press).
35. M. E. Hotz (2008), *Literary Remains: Representations of Death and Burial in Victorian England* (New York: State University of New York Press).
36. Selectively, see G. Rowell (1974), *Hell and the Victorians* (Oxford: Oxford University Press); R. Houlbrooke (ed.) (1989), *Death, Ritual and Bereavement* (New York and London: Routledge), pp. 136–50 (children), pp. 118–35 (cremation), pp. 105–117 (death industry); Litton, *Common Funeral*; P. Jalland (1996), *Death and the Victorian Family* (Oxford: Oxford University Press); P. Jupp and G. Howarth (eds) (1997), *The Changing Face of Death: Historical Accounts of Death and Disposal* (Basingstoke and New York: Palgrave Macmillan); G. Avery and K. Reynolds (eds) (2000), *Representations of Childhood Death* (Basingstoke and New York: Palgrave Macmillan); J. Wolffe (2000), *Great Deaths: Grieving, Religion and Nationhood in Victorian and Edwardian Britain* (Oxford: Oxford University Press); J-M. Strange (2005), *Death, Grief and Poverty in Britain, 1870–1914* (Cambridge: Cambridge University Press).

37. See, notably, D. Vincent (1980), 'Love and Death in the 19th Century Working Class', *Social History*, V: pp. 223–47; P. Johnson (1985), *Saving and Spending: The Working Class Economy in Britain, 1800–1939* (Oxford: Clarendon Press); S. V. Bailey (2000), *This Rash Act: Suicide Across the Life Cycle of the Victorian City* (Stanford: Stanford University Press).
38. Johnson, *Saving and Spending*, p. 59.
39. Refer S. V. Barnard (1990), *To Prove I'm Not Forgot: Living and Dying in the Victorian City* (Manchester: Manchester University Press).
40. Johnson, *Saving and Spending*, p. 59.
41. A point first made in A. Howkins (1991), *Reshaping Rural England: A Social History 1850–1929* (New York and London: Routledge).
42. See, notably, S. A. King and A. Tomkins (eds) (2003), *The Poor in England 1700–1850: An Economy of Makeshifts* (Manchester: Manchester University Press).
43. A. Sen (1992), *Inequality Re-examined* (Oxford: Oxford University Press), pp. 109–10.
44. See J. Flanders (2003), *The Victorian House: Domestic Life from Childbirth to Deathbed* (London: Harper Collins).
45. See, for example, C. Waters (2003), ' "Trading Death": Contested Commodities in *Household Words*', *Victorian Periodicals Review*, 36, IV: pp. 313–30.
46. Jalland, *Death and the Victorian Family*, p. 12.
47. Richardson, *Death, Dissection*, provides important context, pp. 3–30.
48. E. T. Hurren (2007), *Protesting about Pauperism: Poverty, Politics and Poor Relief in Late-Victorian England, 1870–1900* (Woodbridge: Boydell and Brewer), pp. 71–3.
49. A history of silence is understudied, except in J. Bellamy (1988), 'Barriers of Silence: Women in Victorian Fiction', in E. M. Sigworth (ed.), *In Search of Victorian Values: Aspects of Nineteenth Century Thought and Society* (Manchester: Manchester University Press), 131–46.
50. T. Crook (2008), 'Accommodating the Outcast: Common Lodging Houses and the Limits of Urban Governance in Victorian and Edwardian London', *Urban History*, 35, III, pp. 414–36.
51. Strange, *Death, Grief*, p. 12.
52. Ibid., p. 37.
53. Refer Strange, 'Only a Pauper', pp. 148–75; King and Hurren, 'Begging for a Burial', pp. 321–41; S. A. King (2007), 'Pauper Letters as a Source', *Family & Community History*, 10, II, pp. 167–70.
54. D. Cannadine (1981), 'War and Death, Grief and Mourning in Modern Britain', in J. Whalley (ed.), *Mirrors of Mortality: Studies in the Social History of Death* (London: Europa), pp. 187–241.
55. Hurren and King, 'Begging for a Burial', pp. 321–41.
56. Ibid.
57. See, for instance, G. W. Cunningham (1881, 4th edn), *The Law Relating to Burial of the Dead: Including the Burial Acts, 1852 to 1871* (London: Shaw and Sons).
58. Extract from Anon. (1851), *The Overseers Handbook* (London: John Murray), p. 453.
59. Berkshire Record Office, Bradfield Union, Correspondence, 1845–1841, Tilehurst parish, D/P 132/19/7, letter, 10 January 1839.

60. R. Porter (2001), *Bodies Politic: Disease, Death and Doctors in Britain, 1650–1900* (London: Reaktion), p. 246.
61. Northamptonshire Record Office (hereafter NRO), Brackley Union, letter book, PL/1/47, 9 June 1842.
62. A Poor Law Union was a large geographic unit of local government that combined up to forty Old Poor Law parishes. Generally it was centred around a market town, where a workhouse was built to accommodate the destitute.
63. *Overseers Handbook*, p. 453
64. Note: author's emphasis in bold.
65. W. C. Maude (1903), *The Poor Law Handbook* (London: Poor Law Officers Journal publication), p. 24 nn. 2, 7 and 8, Vict. c. 101. Again, author's emphasis in bold.
66. See S. Burrell and G. V. Gill (2005), 'The Liverpool Cholera Epidemic of 1832 and Anatomical Dissection – Medical Mistrust and Civil Unrest', *Journal of the History of Medicine and Allied Sciences*, 60, IV, pp. 478–98.
67. Refer Hurren and King, 'Begging for a Burial', pp. 321–41.
68. NRO, Brackley Union, letter book, 1835–1843, PL/1/47, 3 August 1840.
69. D. L. Roter and J. A. Hall (2006), *Doctors Talking with Patients/Patients Talking with Doctors: Improving Communication in Medical Visits* (Westport, USA: Praeger Publishers, Westport), p. 18, sets out calculative reciprocity as 'things people can do for, or give each other, in the spirit of exchange' notably child care by an elderly relative for a young family, or a young child looking after a sick person in the household economy.
70. 'Police' reports, *Times*, 20 August 1868, issue 26208, column e, p. 11.
71. Ibid.
72. See King and Hurren, 'Begging for a Burial', pp. 321–41.
73. Strange, *Death, Grief*, p. 21.
74. NRO, Brackley Union, PL/1/37, pasted scrap of undated paper put into a book kept about nameless and friendless medical and asylum cases. Northamptonshire corpses were generally sold to Cambridge medical school, as Chapter 5 suggests.
75. Strange, *Death, Grief*, p. 14.
76. 'Sunshine and Shadow: A Tale of the 19th Century by Thomas Martin Wheeler late Secretary to the National Charter Association and the National Land company', *Northern Star and National Trades Journal of Leeds*, 10 November 1849, Issue 629, Column 1, pp. 1–2.
77. Ibid., p. 1
78. Ibid.
79. Ibid.
80. 'Sunshine and Shadow', p. 1, col. 2.
81. Ibid.
82. Ibid.
83. Refer Hurren, *Protesting about Pauperism*, pp. 17–59.
84. 'Sunshine and Shadow', p. 1, col. 2.
85. See, for example, S. T. Anning and W. J. K. Walls (1982), *A History of the Leeds School of Medicine: One and a Half Centuries, 1831–1981* (Leeds: Leeds University Press); M. Parsons (2002), *Yorkshire and the History of Medicine* (York: Sessions of York) recounts similar circumstances that led to Anatomy Act riots in 1832.

86. Richardson, *Death, Dissection*, p. 281.
87. 'More Mutilation', *Figaro in London*, 29 April 1837, issue 282, p. 1, col. 1.
88. G. A. Walker (1839), *Gatherings from Graveyards: particularly those of London. With a concise history of the modes of interment among different nations and a detail of the results produced by the custom of exhuming the dead amongst the living* (London: Ayer Co. Publications [1977 edition]).
89. 'Review of Walker's *Gathering from Graveyards*', *Lancet*, 1839, p. 542.
90. This was the same area in which the Newington workhouse scandal was exposed twenty years later in 1858.
91. Walker, 'Gatherings', pp. 201–2.
92. Ibid., pp. 139–40.
93. Ibid., p. 152.
94. Ibid.
95. Ibid.
96. Further context in J. Rugg (1998), 'A New Burial and Its Meanings: Cemetery Establishment in the First Half of the 19th Century' in M. Cox (ed.), *Grave Concerns: Death and Burial in England 1700–1850* (London: Council for British Archaeology), pp. 44–54.
97. Litton, 'The English Funeral' in Cox (ed.), *Grave Concerns*, pp. 3–16, q. at pp. 9–10.
98. Ibid., pp. 9–10.
99. L. Picard (2005), *Victorian London: The Life of a City, 1840–1870* (London: Orion), p. 378.
100. Litton, 'The English Funeral', pp. 9–10.
101. Ibid.
102. St. Bartholomew's Hospital Archive (hereafter St. BHA), Medical School records, MS 81/1, miscellaneous funeral cards in dissection records.
103. Exceptionally, see S. A. King (2008), 'Friendship, Kinship and Belonging in the Letters of Urban Paupers, 1800–1840', *Historical Social Research*, 33, III, pp. 249–77, q. at p. 125.
104. See, notably, King, 'Pauper Letters', pp. 167–70.
105. Refer Hurren and King, 'Begging for a Burial', pp. 321–41.
106. See, notably, T. Sokoll (2006), 'Writing for Relief: Rhetoric in English Pauper Letters, 1800–1834' in A. Gestrich, S. A. King, and L. Raphael (eds), *Being Poor in Modern Europe: Historical Perspectives, c. 1840–1900* (Oxford: Peter Lang), pp. 91–112.
107. NRO, bills of supply 4 March 1848 from 'Mary March haberdasher and hatter and funerals neatly furnished Market Place Kettering'
108. NRO, 42p/17/3, Letter re John Wiggins to the Relieving Officer of Brackley Union, 1 May 1844.
109. D. Englander (2000), 'From the Abyss: Pauper Petitions and Correspondence in Victorian London', *London Journal*, 25, I: pp. 71–83, noted the importance of appreciating pauper perspectives in their rhetorical context.
110. See, for instance, D. R. Green (1995), *From Artisans to Paupers: Economic Change and Poverty in London, 1790–1870* (London: Ashgate), and D. R. Green (2006), 'Pauper Protests: Power and Resistance in Early Nineteenth-Century London Workhouses', *Social History*, 31, II: pp. 137–59.
111. Refer Hurren, *Protesting about Pauperism*, pp. 191–214.
112. Ibid., pp. 17–59.

113. 'West Bromwich Guardians and Unclaimed Pauper Bodies', *Birmingham Daily Post*, 13 March 1877, issue 5826, pp. 1–2, col. a.
114. Ibid.
115. TNA, MH74/36, letter to Anatomy Inspectorate, 3 February 1890, Professor Bertram C. A. Windle, Professor of Anatomy, Birmingham Medical School re: James Clarke's corpse.
116. Ibid. Mrs. Margaret Longmine's copy letters & legal opinion, February to March 1890.
117. TNA, MH74/36, 3 February 1890.
118. Ibid.
119. TNA, MH74/36, case file of James Clarke and Birmingham Medical School.
120. NRO, 281p/102, Roade Burial Board Minute Books, recent re-cataloguing, 281p/517.
121. St. BHA, dissection records, MS1/8, 1834 corpses.

3 A Dissection Room Drama: English Medical Education

1. St. Bartholomew's Hospital Archive (hereafter St. BHA), MS81/1, Dissection register, 1834, body number 181.
2. Building on R. Porter (1987), 'The Patient's View: Doing Medical History from Below', *Theory and Society*, XIV, pp. 167–74
3. There is a large literature on the history of medical education; see an excellent summary in K. Waddington (2003), *Medical Education at St. Bartholomew's Hospital, 1123–1995*, (Woodbridge: Boydell and Brewer), pp. 1–13, 76–113.
4. M. Sappol (2002), *A Traffic in Dead Bodies: Anatomy and Embodied Social Identity in Nineteenth-Century America* (Princeton: Princeton University Press), p. 8.
5. R. Richardson (2001 edn), *Death, Dissection and the Destitute* (London: Phoenix Press), pp. 3–52; Sappol, *Traffic in Dead Bodies*, pp. 168–211, q. at p. 173, provides an excellent summary of the 'avid audience' fascinated by 'popular anatomy' in nineteenth-century America; H. MacDonald (2006) *Human Remains: Dissection and its Histories* (London: Yale University Press), pp. 1–11, summarises 'performing anatomy' in Victorian times.
6. R. Richardson (2008), *The Making of Mr. Gray's Anatomy: Bodies, Books, Fortune and Fame* (Oxford: Oxford University Press).
7. Sappol, *Traffic in Dead Bodies*, p. 314.
8. Refer J. Reinarz (2005), 'The Age of Museum Medicine: The Rise and Fall of the Medical Museum at Birmingham's School of Medicine', *Social History of Medicine*, 18, III, pp. 419–37.
9. Richardson, *Death, Dissection*, pp. 3–52.
10. I am grateful for an advance copy of S. A. King (2011), *The Sick Poor in England, 1750–1850* (Basingstoke and New York: Palgrave Macmillan).
11. Sappol, *Traffic in Dead Bodies*, p. 316.
12. Refer S. A. King (2000), *Poverty and Welfare in England, 1700–1850: A Regional Perspective* (Manchester: Manchester University Press); E. T. Hurren (2007), *Protesting About Pauperism: Poverty, Politics and Poor Relief in Late-Victorian England*, (Woodbridge: Boydell and Brewer).

13. D. Livingstone (2007), 'Science, Site and Speech: Scientific Knowledge and the Spaces of Rhetoric', *History of Human Sciences*, XX, I, pp. 71–98, first identified the need to explore how 'scientific "claim" is talked about, how responses to it' can be both 'enabled and constrained', q. at p. 72.
14. Sappol, *Traffic in Dead Bodies*, p. 210.
15. Ibid., pp. 98–135, shows that the poorest Irish and Negro bodies supplied a vibrant 'buyer's market' in dissection subjects.
16. Goldsmith's quote was often used in Victorian times to deride medical students; see 'The Medical Session', *Times*, 2 October 1885, issue 31567, p. 13, col. e.
17. *Contemporary Review*, volumes 34 and 35 (June-July 1879), pp. 582–690, contained plenty of articles on doctors and medical students.
18. Summarised in R. Porter (1997), *The Greatest Benefit to Mankind: A Medical History of Humanity from Antiquity to the Present* (London: Harper Collins), pp. 304–428.
19. See Waddington, *Medical Education at St. Bartholomew's*, pp. 2–9.
20. Ibid., p. 6.
21. Ibid., pp. 6, 49.
22. For context, refer I. Loudon (1986), *Medical Care and the General Practitioner, 1750–1850* (Oxford: Oxford University Press).
23. See S. A. King and G. Timmins (2001), *Making Sense of the Industrial Revolution: English Economy and Society 1700–1850* (Manchester: Manchester University Press).
24. E. A. Wrigley (2004), 'British Population during the "Long" Eighteenth Century, 1680–1840' in R. Floud and P. Johnson (eds), *The Cambridge Economic History of Modern Britain: Volume 1, Industrialisation* (Cambridge: Cambridge University Press), chapter 3, pp. 57–95, q. at p. 57.
25. King, *The Sick Poor in England*.
26. See A. Digby (1994), *Making a Medical Living: Doctors and English Patients in the English Market for Medicine, 1720–1911* (Cambridge: Cambridge University Press).
27. R. Porter (1989), *Health for Sale: Quackery in England, 1660–1850* (Manchester: Manchester University Press).
28. T. N. Bonner (2000 edn), *Becoming a Physician: Medical Education in Britain, France, Germany and the United States, 1750–1945* (Maryland: Johns Hopkins Press).
29. M. Pelling and F. White (2003), *Medical Conflicts in Early Modern London: Patronage, Physicians and Irregular Practitioners, 1550–1640* (Oxford: Oxford University Press).
30. The Barbers and Surgeons had been compelled to unite by King Henry VIII in 1540.
31. M. Pelling (2006), 'Corporatism or Individualism: Parliament, the Navy, and the Splitting of the London Barber-Surgeons' Company in 1745', in I. A. Gadd and P. Wallis (eds), *Guilds and Association in Europe, 900–1900* (London: Centre for Metropolitan History Publication), pp. 57–82.
32. Refer M. Pelling (1998), *The Common Lot: Sickness, Medical Occupations and the Urban Poor in Early Modern England* (London and New York: Longman).
33. Refer J. G. L. Burnby (1983), *A Study of the English Apothecary from 1660–1760* (London: Medical History, Supplement III, Wellcome Trust).

34. See, notably, S. Lawrence (1991), 'Private Enterprise and Public Interest: Medical Education and the Apothecaries' Act, 1780–1825', in R. K. French and A. Wear (eds), *British Medicine in the Age of Reform*, (London: Routledge), pp. 45–73.
35. Digby, *Making a Medical Living*, p. 139.
36. S. Lawrence (1993), 'Educating the Senses: Students, Teachers and Medical Rhetoric in Eighteenth Century London', in W. F. Bynum and R. Porter (eds), *Medicine and the Five Senses* (Cambridge: Cambridge University Press), pp. 154–78, quote at p. 155.
37. For context, see I. Loudon (1992), 'Medical Practitioners and Medical Reform in Britain 1750–1850' in A. Wear (ed.), *Medicine in Society: Historical Essays* (Cambridge: Cambridge University Press), pp. 219–37.
38. See overview in C. Lawrence (1994), *Medicine in the Making of Modern Britain, 1700–1920*, (London: Routledge).
39. Waddington, *Medical Education at St. Bartholomew's*, p. 30.
40. Context outlined in V. Nutton and R. Porter (eds) (1995), *The History of Medical Education in Britain* (Amsterdam: Rodopi).
41. Porter, *Greatest Benefit*, pp. 316–17.
42. R. Cooter (1983), *Surgery and Society in Peace and War: Orthopaedics and the Organisation of Modern Medicine, 1880–1948* (Basingstoke and New York: Palgrave Macmillan), p. 2.
43. On French dissection, see R. Maulitz (1987), *Morbid Appearances: The Anatomy of Pathology in the Early Nineteenth Century* (Cambridge: Cambridge University Press).
44. J. L. Grainer, 'A Medical Student in Search of a Supper', *Penny Satirist*, 6 June 1840, issue 164, pp. 3–4.
45. Ibid., p. 4.
46. See, for instance, 'Police Intelligence', *The Illustrated Police News*, 7 March 1885, issue 1099, p. 1, which reported at length a large fight between medical students at Piccadilly Circus.
47. There was general concern that unruly behaviour in London would pollute the minds of medical students elsewhere; see 'London Gossip', *Birmingham Daily Post*, 13 November 1885, issue 8541, p. 1, col. 1, student fights at the Lord Mayor's Show.
48. Ibid., p. 4. A Taglioni Coat was a comfortable greatcoat or overcoat said to have been designed by a celebrated family of Italian professional dancers. It symbolises being decadent enough to take anatomy training abroad in Padua, Italy. Count Alfred Guillaume Gabriel D'Orsay was a renowned Parisian dandy, an obvious reference to the fashion for French anatomy. Biccory is a wood often used to make silver-topped canes that dandies carried in the nineteenth century.
49. Grainer, 'A Medical Student in Search of a Supper', p. 4.
50. Ibid., p. 4.
51. 'A Dissecting Room', *Penny Satirist*, 7 November 1840, issue 186, p. 1.
52. Ibid., p. 1.
53. 'The Morgue', *Once a Week*, December 1864, issue 286, volume 11, pp. 714–19.
54. 'A Dissecting Room', p. 1.
55. Ibid., p. 1.
56. St. BHA, Dissection register, 1834, body number 181.

57. Case file reconstructed from record linkage work on St. BHA, MS81/1, Dissection register, 1834, body number 181; 'Death by Poison', *Times*, 26 November 1834, issue 15635, p. 3, col. B; and London Metropolitan Archives (hereafter LMA), undertaking records ACC/1416/47 & BRA898/B/186. All original quotations cited in *Times* source.
58. St. BHA, MS81/1, Dissection register, 1834, body number 181: an explanation was attached to the body-entry in the duty-anatomist's handwriting.
59. *Times*, 26 November 1834.
60. Waddington, *Medical Education at St. Bartholomew's*, p. 66.
61. Ibid., p. 55.
62. See S. Paget (ed.) (1901), *Memoirs and Letters of Sir James Paget* (London: Longman, Green & Co.), pp. 41–2, also cited in Porter, *Greatest Benefit*, p. 316.
63. Ibid., p. 54.
64. Paget, *Memoirs and Letters*, p. 48.
65. St. BHA, MS81/1, Dissection register, 1834, body number 181; again an explanation was attached to the body-entry in the anatomist on duty's handwriting.
66. Ibid.
67. The false names are cited in coroner's report published in *Times*, 26 November 1834.
68. St. BHA, MS81/1, Dissection register, 1834; at the back in the anatomist's handwriting is a full list of the hospital's undertakers and their standard fees. Some random bills are also to be found in the medical school archives. These have been checked against leaseholds and freehold conveyances held in the LMA to make sure they are accurate.
69. 'The Difficulties of the Anatomist', *Chamber's Journal*, October 1858, volume 248, p. 218.
70. Ibid.
71. Quoted in "Aesculapius" (1818, 2nd edn), *The Hospital Pupil's Guide* (London: E. Cox and Sons), p. 38
72. Ibid.
73. Waddington, *Medical Education at St. Bartholomew's*, p. 49.
74. See, recently, A. W. Bates (2008), '"Indecent and Demoralising Representations": Public Anatomy Museums in Mid-Victorian England', *Medical History*, 52, I, 1–22.
75. See, notably, strong feeling, in S. Burrell and G. Gill (2005), 'The Liverpool Cholera Epidemic of 1832 and Anatomical Dissection – Medical Mistrust and Civil Unrest', *Journal of the History of Medicine and Allied Sciences*, 60, IV, pp. 478–98.
76. Paget, *Memoirs and Letters*, p. 78.
77. Refer Charles Darwin [1758–1778] (1780), *Experiments establishing a criterion between mucaginous and purulent matter. And an account of the retrograde motion from the absorbent vessels of animal bodies in some diseases* (Edinburgh: Green).
78. 'The Life of Charles Darwin (1758–1778)' (1778), *Medical and Philosophical Commentaries*, V, pp. 329–36.
79. R. H. Syfret (1950), 'Some early reactions to the Royal Society', *Notes Rec. Royal Society of London*, VII, 207–58, quote at p. 233. Also discussed in Bates, '"Indecent and Demoralising Representations"', pp. 5–6.

80. Selectively, see J. Kahn (1855), *The Evangel of Human Nature; being fourteen lectures, on the various organs of the human frame, in health and disease* (London: James Gilbert); S. J. M. M. Alberti (2008), 'Wax Bodies: Art and Anatomy in Victorian Medical Museums', *Museum History Journal*, 2, I, pp. 7–36.
81. Bates, '"Indecent and Demoralising Representations"', p. 6.
82. P. J. Bowler (1996), *Charles Darwin* (Cambridge: Cambridge University Press), pp. 40–1, for example, points out that when the younger Charles Darwin was a medical student at Edinburgh, aged just 16 in 1825, he described the 'unpleasant experiences in the operating theatre' and expressed his 'disgust at how bloody dissections were' at the time. He left after two years to eventually pursue the Natural Sciences at Cambridge.
83. C. Dickens, 'A Great Day for the Doctors', *Household Words* (9 November 1850), 32, II, pp. 137–9, quote at p. 137.
84. Ibid. p. 139.
85. Case reconstructed from successive articles all entitled 'Anatomy in the City', *Daily News*, 16 January, 22 January, and 23 January 1877, issues 9590, 9595, & 9596.
86. Ibid., 16 January 1877, pp. 1–3, q. at p. 1.
87. 'Anatomy in the City', 16 January 1877, pp. 1–3, q. at p. 2.
88. Ibid., 16 January 1877, pp. 1–3.
89. 'Anatomy in the City', 16 January 1877, pp. 1–3, q. at p. 1.
90. Ibid., 22 January 1877, pp. 1–2.
91. 'Anatomy in the City', 22 January 1877, pp. 1–2.
92. Ibid., 23 January 1877, letter, p. 1.
93. S. Paget, *Memoirs and Letters*, pp. 41–2.
94. Bodleian Library (hereafter Bodl. Lib), University Archives (UA), Anatomy Department Records (ADR), HA107, Thomson's notes on staff, buildings and fittings.
95. See E. L. Hussey (1894), *Miscellanea Medico-Chirurgica: 3rd part, Occasional Papers and Remarks* (Oxford: Horace Hart), p. 274.
96. See The National Archives (hereafter TNA), MH12/9721, Hussey to Local Government Board report styled "Inquests at Oxford, 1877–1888".
97. Refer Downing College, Macalister Papers, box retained in the Human Anatomy Department containing assorted details of dissection work.
98. 'The Subject for Dissection or the Student's Joke', *Reynolds Miscellany*, 17 November 1864, volume 33, issue 857, p. 333.
99. See, for example, MacDonald, *Human Remains*.
100. A. Bashford (1998), *Purity and Pollution: Gender, Embodiment and Victorian Medicine* (Basingstoke, Hampshire: Macmillan Press Ltd), p. 115.
101. Ibid., p. 114.
102. Bashford, *Purity and Pollution*, p. 115.
103. L. Jordanova (1989), *Sexual Visions: Images of Gender in Science and Medicine Between the Eighteenth and Twentieth Centuries* (Madison, WI: University of Wisconsin Press).
104. Ibid., pp. 87–100, discusses this sexualised treatment.
105. An observation also made by Bashford, *Purity and Pollution*, p. 115
106. Porter, *Greatest Benefit*, p. 356.

107. On the Victorian discourse on dirt, see M. Douglas (1966), *Purity and Danger: An Analysis of the Concepts of Pollution and Taboo* (London and New York: Routledge).
108. Wellcome Collection, Slide Number, L0013321, 'The dissection of a young beautiful woman directed by J. CH. G. Lucas (1814–1885) in order to determine the ideal female proportions', chalk drawing by J. H. Hasselhorst (1864). The image was popular in English medical circles and is one of a number used to recruit students to the capital's hospitals in the late Victorian era.
109. Bashford, *Purity and Pollution*, p. 114.
110. B. Dijkstra (1986), *Idols of Perversity: Fantasies of Feminine Evil in Fin-de-Siècle Culture* (Oxford: Oxford University Press).
111. Jordanova, *Sexual Visions*, pp. 29–30, points out that there is a long history of focusing on the image of the breast in medical imagery, emphasizing natural nurturing of children and the science of procreation.
112. 'Charles Darwin Obituary: Collected Essays II' (1888), *Notes of the Proceedings of the Royal Society*, quote at p. 263.
113. Wellcome Collection, Slide L0002687, 'Photograph of the Interior of the Department of Anatomy at Cambridge University, 1888'.
114. Refer A. Macalister (1891), *The History of the Study of Anatomy in Cambridge* (Cambridge: Cambridge University Press).
115. Downing College, Macalister Papers, private notes on taking up his appointment.
116. Wellcome Collection, Slide L0013441, 'Photograph of Edinburgh University Dissection Room, 1889'.
117. Ibid., Slide L0014980, 'Photograph of The Dissection Room, Medical School, Newcastle Upon Tyne, 1897'.
118. Wellcome Collection, slide L0039195, 'Photograph of The Interior of a Dissecting Room: Five Students and/or Teachers Dissect a Corpse at University College, London'.
119. M. Morrell, 'Is Medicine a Progressive Science?' *Fortnightly Review*, 18 June 1860, volume 39, issue 234, p. 845.
120. Waddington, *Medical Education at St. Bartholomew's*, p. 110.
121. Ibid., p. 100 n. 154.

4 Dealing in the Dispossessed Poor: St. Bartholomew's Hospital

1. 'Terrific Hail and Thunderstorm', *Observer*, 10 July 1836, p. 3.
2. Ibid.
3. St. Bartholomew's Hospital Archive (hereafter St. BHA), dissection registers, MS81/1–6.
4. 'Dreadful Hurricane in the Metropolis and in the Country', *Observer*, 5 December 1836, p. 4.
5. K. Waddington (2003), *Medical Education at St. Bartholomew's Hospital, 1123–1995* (Woodbridge: Boydell and Brewer), p. 110.
6. S. Wise (2005), *The Italian Boy: Murder and Grave-Robbery in 1830s London*, (London: Pimlico Publishers), p. 20.
7. Ibid., p. 10.

8. Waddington, *Medical Education at St. Bartholomew's*, p. 18.
9. See summary in L. Young (2002), *The Book of the Heart* (London: Harper Collins), pp. 32–5.
10. Waddington, *Medical Education at St. Bartholomew's*, p. 25.
11. Ibid., p. 29.
12. Ibid., p. 39.
13. Ibid., p. 40.
14. See a summary of the East End in S. Wise (2008), *The Blackest Streets: The Life and Death of a Victorian Slum* (London: Bodley Head).
15. Smithfield had a Metropolitan Cattle Market that often caused traffic chaos and was closed in 1855, relocating to Islington. It was replaced by a purpose-built wholesale meat market in 1868. Refer J. White (2008), *London in the 19th Century: A Human Awful Wonder of God* (London: Vintage Books Ltd), pp. 42, 46.
16. Ibid., pp. 209–10.
17. J. White, *London*, p. 209. St. Bartholomew's Fair ran until 1849.
18. Ibid.
19. Refer I. Burney (2000), *Bodies of Evidence: Medicine and the Politics of the English Inquest, 1830–1926* (Baltimore, MD: Johns Hopkins University Press).
20. See E. T. Hurren (2010), 'Remaking the Medico-Legal Scene: A Social History of the late-Victorian Coroner in Oxford', *Journal of the History of Medicine and Allied Sciences*, 65, IV, pp. 207–52.
21. See I. Burney (1996), 'Making Room at the Public Bar: Coroners' Inquests, Medical Knowledge and the Politics of the Constitution in Early Nineteenth-Century England', in James Vernon (ed.), *Re-reading the Constitution: New Narratives in the Political History of England's Long Nineteenth Century* (Cambridge: Cambridge University Press), pp. 123–53.
22. This is why the political economy of coroners' statistics has been questioned by, notably, H. Taylor (1998), 'Rationing Crime: The Political Economy of Criminal Statistics since the 1850s', *English Historical Review*, 3rd series, IX, pp. 569–90.
23. Hurren, 'Remaking the Medico-Legal Scene', pp. 207–252.
24. Quoted in Anon. (1886), *St. Bartholomew's Hospital Reports*, Vol. XXII, pp. 34–35.
25. Quoted in White, *London*, p. 169.
26. Ibid., p. 168.
27. See S. King and A. Tomkins (eds) (2003), *The Poor in England 1700–1850: An Economy of Makeshifts* (Manchester: Manchester University Press).
28. White, *London*, p. 169 points also to 'a small epidemic of suicides noted in the press' in 1862.
29. See, for example, 'The Northern and Central Bank Crisis', *Observer*, 5 December 1836, p. 5.
30. 'Suicide from Seduction', *Observer*, 8 March 1835, p. 2.
31. Ibid.
32. Ibid.
33. Ibid.
34. Refer two key examples. One is an anonymous account, written by a man called Walter (a pseudonym), who paid for Victorian street –prostitutes: *My Secret Life, 11 vols., 1888–1894* (New York: Grove Press edn, 2 vols, 1966); the

other is J. R. Walkowitz (1992), *City of Dreadful Delight: Narratives of Sexual Danger in Late-Victorian London* (Chicago: Chicago University Press).
35. London Metropolitan Archives (hereafter LMA), HO107/670, Holborn Union, Guardian Minute Books, 1838–1930 and financial records, 1848–1930.
36. Selective examples are elaborated later where record linkage work is detailed.
37. Selectively, see M. Mason (1994), *The Making of Victorian Sexuality* (Oxford: Oxford University Press), R. Porter and M. Teich (eds) (1994), *Sexual Knowledge, Sexual Science* (Cambridge: Cambridge University Press); L. Nead (1988), *Myths of Sexuality: Representations of Women in Victorian Britain* (Oxford: Blackwell); O. Moscucci (1999), *The Science of Women: Gynaecology and Gender in England, 1800–1929* (Cambridge: Cambridge University Press).
38. White, *London*, p. 297.
39. St. BHA, dissection registers, MS81/1–6, body entry 18/02/1833.
40. St. BHA, order in Council, HAL/20, p. 504, states that the Hospital Burial Ground was first opened in 1744 and closed in 1853. A 74-year lease was then signed with St. Luke's vestry at an annual rent of 1s to turn the area into green space in which 'the ground [is] not [to] be disturbed or broken up'. See also St. BHA, Hospital Reports (1886), pp. 35–6.
41. St. BHA, dissection registers, MS81/1–6, body entry 10/11/1836.
42. Refer, notably, S. Burrell and G. Gill (2005), 'The Liverpool Cholera Epidemic of 1832 and Anatomical Dissection – Medical Mistrust and Civil Unrest', *Journal of the History of Medicine and Allied Sciences*, 60, IV, pp. 478–498,
43. St. BHA, dissection registers, MS81/1–6, body entry 19/01/1837.
44. Ibid. MS81/1–6, body entry 10/10/1847 and LMA, St. Giles parish records. Likewise White, *London*, p. 299, explains that each St. Giles brothel had about '200 beds' and that of those, '500–1000 prostitutes a night' were for sale on the nearby Strand.
45. The National Archives (hereafter TNA), Anatomy Inspectorate, MH74/15, report dated 4 November 1843, John Dix discussed on p. 54.
46. TNA, Anatomy Inspectorate, MH74/12, Reports of James Somerville, show that in 1834–5, he was aware of, and indeed supported, the fact that St. Bartholomew's was paying '£5 per body' to dealers on a regular basis.
47. St. BHA, dissection registers, MS81/1–6, body entry 23/09/1850.
48. St. BHA, dissection registers, MS81/1–6, body entry 02/11/1851.
49. St. BHA, dissection registers, MS81/1–6, body entries 17/01/1856 and 21/01/1856, both stillbirths.
50. Note that under the Anatomy Act certification was still not required: an important loop-hole in the law revisited in this book's conclusion.
51. Stillbirths were born dead and therefore officialdom held that they were not a complete human being for the purposes of parish records.
52. Ibid.
53. St. BHA, Treasurer and Almoner's Book, 30 July 1874–27 December 1877 volume, request styled '*Dead Bodies – Removal from the Wards*', dated 2 September 1875, from Joseph J. Miles Esq. to Robert Philipson Barrows Esq., both almoners.
54. St. BHA, Hospital Accounts, HA1/23, 1866–72 volume, 24 February 1870, p. 371.

55. All data in this section is compiled from St. BHA, dissection registers, MS81/1–6.
56. R. Richardson (2001 edn), *Death, Dissection and the Destitute* (Phoenix Press: London), p. 266.
57. See E. T. Hurren (2007), *Protesting about Pauperism: Poverty, Politics and Poor Relief in late-Victorian England c. 1870–1900* (Woodbridge: Boydell and Brewer), pp. 17–58.
58. See E. T. Hurren and S. A. King (2005), 'Begging for a Burial: Form, Function and Conflict in Nineteenth-Century Pauper Burial', *Social History*, XXX, pp. 321–41.
59. Richardson, *Death, Dissection*, p. 263.
60. Ibid., p. 248.
61. Hurren, *Protesting about Pauperism*, p. 23.
62. Ibid., 248–58, summarises its cost-saving initiatives.
63. Hurren, *Protesting about Pauperism*, p. 63.
64. M. W. Weatherall (1996), 'Making Medicine Scientific: Empiricism, Rationality and Quackery in mid-Victorian Britain', *Social History of Medicine*, IX, 175–94, outlines both the 1858 and 1885 medical standards.
65. For context see A. Digby (1994), *Making a Medical Living: Doctors and Patients in the English Medical Market, 1720–1911* (Cambridge: Cambridge University Press).
66. Hurren, *Protesting about Pauperism*, pp. 214–47.
67. Ibid.
68. See J. Storey (1895), *Historical Sketches of some Principal Works and Undertakings of the Council of the Borough of Leicester by the Town Clerk, September 1874 to October 1894* (Leicester: Leicester Publishers), p. 52. In Leicester the increase in annual rates was notable, from £44 to £540. Leicester thus began to sell its asylum cases for dissection to Oxford (see Chapter 6).
69. Their burial location was in the East End but the burial records are fragile.
70. 'The General Medical Council – News', *Times*, 15 May 1885, issue 31447, column c, p. 3.
71. Again building on Richardson, *Death, Dissection*, p. 271.
72. Waddington, *Medical Education at St. Bartholomew's*, p. 110
73. R. Porter (1997), *The Greatest Benefit to Mankind: A Medical History of Humanity from Antiquity to the Present* (London: Harper Collins), p. 356.
74. B. Hayes (2009), *The Anatomist: A True Story of Gray's Anatomy* (New York: Bellevue Literary Press), pp. 36–7.
75. TNA, HO45/10062/B2694, Document added to the Appendix of a Home Office (Anatomy Inspectorate) Report dated '1905'.
76. Downing College, Cambridge, Macalister Papers, A. Keith to A. Macalister, letter dated 6 May 1903.
77. St. BHA, MS13/19, 'Letter from Holborn Union 7 King's Road, Bedford Row dated 21 February 1871 to Mr. Paget, St. Bartholomew's Hospital Medical School'.
78. St. BHA, Medical School Cash Payments, MS 42/1, cash book, 1879–1891, p. 238, school expenses and receipts for the session of 1889–1890.
79. Ibid., p. 250.
80. See, for example, BBC News, 22 February 2010, 'How Much is a Human Body Worth?' – www. news.bbc.co.uk/2/hi/business/8519611.stm

81. Recent newspaper coverage about the credit crunch shows that the body trade is vibrant even in wealthy America – *Times*, 21 November 2009, 'Unburied bodies tell tale of city in despair – Detroit and its Dead Poor – World News', p. 61.
82. See, notably, J-M. Strange (2005), *Death, Grief and Poverty in Britain, c. 1870–1914* (Cambridge: Cambridge University Press).
83. St. BHA, dissection registers, MS81/1–6, entry for George Chapman's cadaver and body dealing details dated 31 March 1857.
84. The case of George Chapman has undergone record linkage work at the LMA, St. Mary's Newington, P92/MRY.
85. Listed in St. BHA, dissection registers, MS81/1–6.
86. All of the quotations in *Rex versus Feist* [1858] have been reconstructed from record linkage work on www.oldbaileyonline.org, *The Proceedings of the Old Bailey, London's Central Criminal Court, 1674–1913* (hereafter OB), case number: t18580222-354, 22 February 1858, and reports in *Times* of the case proceedings, 21 January, 25th February, 5 March, and report of the quashed judgement 24 April 1858. To test for bias in the court and newspaper record, poor law accounts have also been consulted at the LMA, St. Mary's Newington, P92/MRY.
87. Ibid.
88. St. BHA, dissection registers, MS81/1–6, entry for George Chapman's cadaver and body dealing details dated 31 March 1857.
89. LMA, St. Mary's Newington, P92/MRY, 1858.
90. St. BHA, dissection registers, MS81/1–6, Annual list of Robert Hogg's fees paid.
91. See summaries in D. W. Meyers (2006 edn), *The Human Body and the Law: A Medico-Legal Study* (Chicago: Aldine Transactions), p. 107; H. MacDonald (2009), 'Procuring Corpses: The English Anatomy Inspectorate, 1842–1858', *Medical History*, 53, III, pp. 379–396.
92. Ibid.
93. All data compiled from St. BHA, dissection registers, MS81/1–6.
94. 'Holborn Hill Improvements', *Observer*, 31 August 1835, p. 1.
95. Waddington, *Medical Education at St. Bartholomew's*, p. 70, identifies the 'body-trafficking' controversy but not the actual supply contracts or body sources which were beyond the scope of his excellent study on medical education.
96. White, *London*, p. 31.
97. Ibid.
98. Ibid.
99. L. Picard (2005), *Victorian London: The Life of a City, 1840–1870* (London: Phoenix Press), p. 30.
100. 'Pauper Funerals: Holborn', *Times*, 6 September 1883, issue 30918, p. 3, col. g.
101. Ibid.
102. White, *London*, p. 313.
103. A. Hardy (1993), *The Epidemic Streets: Infectious Disease and the Rise of Preventative Medicine, 1856–1900* (Oxford: Clarendon Press), p. 202.
104. After 1870, St. George's Hanover Square was absorbed by St. George's Union.

334 Notes

105. "The *Lancet* Sanitary Commission for Investigating the State of the Infirmaries of Workhouses", reported in *British and Foreign Medico-Surgical Review*, Vol. 40 (October 1887), I, pp. 171–6.
106. White, *London*, p. 14.
107. See, selectively, A. S. Wohl (1977), *The Eternal Slum: Housing and Social Policy in Victorian London* (London: Transaction Publishers); L. Hollen Lees (1979), *Exiles of Erin: Irish Migrants in Victorian London* (Manchester: Manchester University Press); L. Sponza (1988), *Italian Immigrants in Nineteenth Century Britain: Realities and Images* (Leicester: Leicester University Press); White, *London*, pp. 131–9.
108. Prior to the passing of the Anatomy Act, the murder of an Italian boy sold on for anatomy gripped the London press; see Wise, *The Italian Boy*.
109. All data compiled from St. BHA, dissection registers, MS81/1–6, 1835–6.
110. St. BHA, MS81/1–6, 1854.
111. St. BHA, MS81/1–6, 1839.
112. St. BHA, MS81/1–6, 1843. Erysipelas was sometimes called *St. Anthony's Fire* because it was a serious skin condition caused by Streptococcus pyogenes. It was common and often fatal in the young or elderly who lacked proper nourishment.
113. S. Lawrence (1996), *Charitable Knowledge: Hospital Pupils and Practitioners in Eighteenth Century London* (Cambridge: Cambridge University Press), p. 196.
114. Waddington, *Medical Education at St. Bartholomew's*, p. 54.
115. Refer S. A. King (2000), *Poverty and Welfare in England, 1700–1850: A Regional Perspective* (Manchester: Manchester University Press); T. Sokoll (2001 edn), *Essex Pauper Letters, 1731–1837* (Oxford: Oxford University Press).
116. St. BHA, Treasurer's Letter Book, volume 1867–74, HA10/1, 5 July 1872, p. 716.
117. TNA MH12/74 series is a rich and yet under-researched source for pauper protests of this nature. See Hurren and King, 'Begging for a Burial', for more details.
118. This quote from Honoré de Balzac, *The Physiology of Marriage* (1829), is often cited in anatomy books; see the most recent – and an excellent study – in Hayes, *The Anatomist*, p. 176.
119. Summarised in S. Szreter (1994), 'Mortality in England in the 18th and 19th Centuries: A Reply to Sumit Guha', *Social History of Medicine*, 7, II, pp. 269–82. For recent debates, see F. Condrau and M. Worboys (2009), 'Second Opinion: Final Response – Epidemics and Infections in 19th Century Britain', *Social History of Medicine*, 22, I, pp. 165–171.
120. Hurren, *Protesting about Pauperism*, pp. 128–155.
121. See, notably, R. A. Emmons and M. E. McCullogh (eds) (2004), *The Psychology of Gratitude* (Oxford: Oxford University Press).
122. Hurren, *Protesting about Pauperism*, pp. 79–90
123. For an appraisal of female wax models, see A. McGregor (2007), *Curiosity and Enlightenment: Collectors and Collections from the Sixteenth to the Nineteenth Century* (New Haven: Yale University Press), pp. 159–176.
124. All case histories compiled from St. BHA, dissection registers, MS81/1–6, 1833 and record linkage work on Poor Law and Prison sources held at LMA.
125. Wise, *The Italian Boy*, p. 30.

126. See, for example, J-M. Strange (2000), 'Menstrual Fictions: Languages of Medicine and Menstruation, 1850–1930', *Women's History Review*, 9, III, pp. 607–28.
127. See G. Mooney (2007), 'Infectious Diseases and Epidemiologic Transition in Victorian Britain? Definitely', *Social History of Medicine*, XX, pp. 595–606 and response by F. Condrau and M. Worboys, 'Second Opinions', pp. 165–171.
128. Jack-the-Ripper historiography is vast. For one of the best recent scholarly books, see P. Begg (2009), *Jack the Ripper: The Facts* (London: Portico Publishers).
129. Ibid., pp. 21–32.
130. Refer St. BHA, dissection registers MS1/1–6, 1834 register had 9 cases of 'body had been opened' but not coroner's cases. Likewise, ' "Sarah West" a 63 year old female was admitted on 9th November 1837 and died 5 days later on the 14th'. Her medical notes state 'the abdomen had been opened'. She died alongside a 44-year-old male pauper called 'Henry Any' described as 'much mutilated' before death on '25th November 1837'. Similarly, 'Jane Sheriiff [sic] aged 60 from St. Giles Workhouse' died on the 2 December 1857 from chronic bronchitis'. Again her medical notes state that she had an 'opened body' when admitted. She had been cut up by someone with a knife before admission.
131. Begg, *Jack the Ripper*, pp. 101–135.
132. Ibid., p. 79.
133. I am grateful to several anatomists who were kind enough to share their expertise with me in the course of writing up the research for this book in its early stages, notably Logi Barrow at Cambridge University and Zoltan Molnar at Oxford University.
134. Begg, *Jack the Ripper*, p. 135.
135. Ibid., p. 134.
136. See, St. BHA, dissections registers MS1/1–6, marginalia notes for 1840.
137. Begg, *Jack the Ripper*, p. 52.
138. Ibid., p. 78.
139. Robert Hogg appears on a regular basis in the 1888–1889 anatomy register and was busy trading in the vicinity of the hospital and the murder locations. Other regular body dealers were too and these have been reconstituted by the author. On forthcoming publications in production on Jack the Ripper, refer: www.ah.brookes.ac.uk/staff/details/hurren/
140. Hayes, *The Anatomist*, p. 15.

5 Pauper Corpses: Cambridge and Its Provincial Trade

1. A. Macalister (1883), 'Introductory Lecture on the Province of Anatomy by Fellow of St. John's College and Professor of Anatomy in the University of Cambridge', *British Medical Journal* (27 October), issue 1191, II, pp. 808–11, q. at p. 811.
2. His famed London–Cambridge walks were recorded in many of his obituaries; see, for example, W. L. H. Duckworth (1919), 'Obituary: Alexander Macalister', *Man (published by the Royal Anthropological Institute of Great Britain)*, XIX, pp. 164–8.

3. Ibid.
4. J. Barclay-Smith (1919), 'Macalister Obituary', *Journal of Anatomy*, CIV, I, p. 97.
5. His activities and press coverage are summarised in 'Obituary: Professor Macalister', *Times* (3 September 1919) issue 42195, p. 14, col. f.
6. See lifelong commitment in A. Macalister (1891), *A History of the Study of Anatomy at Cambridge: A Lecture* (Cambridge: Cambridge University Press), pp. 1–28.
7. Macalister, 'Introductory Lecture', p. 810.
8. Ibid.
9. A. Macalister (1908), 'An Address on Fifty Years of Medical Education Delivered at the Opening of the Winter Session at King's College, London', *British Medical Journal*, Volume 2 (3 October), issue 2492, pp. 957–60, q. at p. 960.
10. Ibid., p. 960.
11. Northamptonshire Record Office (hereafter NRO), Brackley Union, Letter Book, 1835–43, p. 129, series of resignation letters, 12 March 1839, from Rev. Pryse Jones, onwards.
12. NRO, Brackley Union, Letter Book, 1835–43, p. 141, copy letter from eight petitioners to the Poor Law Board by 'James Fairbrother, George Shouldson, William Bayliss, William Farmer, William Panister, John Henry Parkins, Thomas South and James Pool'.
13. Ibid., p.130, discusses 'Alice Rubra the nurse'.
14. NRO, ZA2037/I, 'Evidence of Mr Rosebrook Morris, surgeon to the Brixworth Union before the *Select Committee on Medical Poor Relief* (1844), questions 7071–7121.'
15. Ibid., question 7101–2.
16. Ibid., question 7104.
17. Ibid., question 7107.
18. E. T. Hurren (2007), *Protesting about Pauperism: Poverty, Politics and Poor Relief in late-Victorian England, c. 1870–1900* (Woodbridge: Boydell & Brewer), p. 83, Table 1.
19. I am grateful to S. A. King (2011), *The Sick Poor in England, 1750–1850* (Basingstoke and New York: Palgrave Macmillan), for sharing with me an advance copy.
20. 'A Curse on Bozeat Bodysnatchers', *Northamptonshire Herald*, 30 March 1840, p. 1, col. 4.
21. See, notably, E. T. Hurren (2010), 'Remaking the Medico-Legal Scene: A Social History of the Victorian Coroner in Oxford, c. 1877 to 1894', *Journal of the History of Medicine and Allied Sciences*, 66, IV, pp. 207–52.
22. NRO, Inquest Returns, Dean and Chapter of Peterborough, Box X.88, 1813–1842, The Case of Lucy Ladds, 9 March 1841.
23. Ibid.
24. Macalister Papers, Downing College, Cambridge, letter from W. W. Driffield of Chelmsford Union to Macalister, 19 May 1885.
25. Ibid., Circular 9 October 1884, written by Macalister for distribution.
26. Macalister Papers, Downing College, Cambridge, notes on old and new sources of bodies, dated 1912.
27. Ibid., Private notes on body sales dated 1912.

28. Refer, context, in S. French (1978), *The History of Downing College, Cambridge, Volume Two, 1888–1894* (Cambridge: Cambridge University Press). The anatomy school was located within the Downing College site on what was the old botanical gardens area. It remains there to this day.
29. Macalister Papers, Downing College, Cambridge, Private notes on taking up his appointment, dated 1883.
30. E. T. Hurren (2004), 'A Pauper Dead-House: The Expansion of the Cambridge Anatomical School, 1870–1914', *Medical History*, 48, I, pp. 69–94. It should be noted that this chapter contains new archive material and therefore builds on rather than repeats previously published research. A great deal of time and effort has been taken to ensure that there is minimal overlap in terms of work already in print. Data is shared but it is introduced in a new write-up.
31. Macalister Papers, Downing College, Cambridge, Private notes on taking up his appointment, dated 1883.
32. Ibid.
33. M. Weatherall (2000), *Gentlemen, Scientists and Doctors: Medicine at Cambridge 1800–1940* (Woodbridge: Boydell & Brewer), pp. 216, 219.
34. E. Barclay-Smith (1919), 'In Memoriam: Professor Alexander Macalister M.D. F. R.S, etc., 1844–1919', *Journal of Anatomy*, 54, I, pp. 96–9, quote at p. 97.
35. Macalister Papers, Downing College, Cambridge, private notes and circulars contain numerous paternalistic phrases.
36. Ibid., H. E. Jenkins (Hull) to Macalister, letter dated 24 June 1895.
37. Ibid.
38. Reconstructed from railway invoices and receipts in Macalister Papers and D. I. Gordon (1968), *A Regional History of the Railways of Great Britain, Volume 5: The Eastern Counties* (Newton Abbott: David and Charles), chapters 5 & 6.
39. The London Necropolis Company ran a train service every Sunday from Waterloo station in London to their private burial ground at Brookwood Cemetery in Surrey. Bodies were left outside the station and travelled on the London and South Western Railway trains. There was also a funeral station at King's Cross run by the great Northern London Cemetery Company to New Southgate. Most were middle class interments, not paupers, since the latter could not afford the subscriptions; see J. M. Clarke (1995), *The Brookwood Necropolis Railway, Locomotion Papers No. 143* (London: Oakwood Press).
40. A. Macalister (1900), 'Anatomical Teaching in 1800', *British Medical Journal*, Volume 2987, II, pp. 1839–41, q. at p. 1839.
41. Weatherall, *Gentlemen, Scientists and Doctors*, p. 216.
42. Today, plasticization has revolutionised views inside the dissected human body.
43. All cases are reconstructed from record linkage work on Cambridge Record Office (hereafter CRO), P25/1/21–23, St. Benedict's parish, burial records.
44. *Cambridge Review* (1888–9), vol. xi, pp. 115–6 and quoted in Weatherall, *Gentlemen, Scientists and Doctors*, p. 216.
45. Weatherall, *Gentlemen, Scientists and Doctors*, p. 211.
46. Ibid., p. 212.
47. Weatherall, *Gentlemen, Scientists and Doctors*, p. 212.
48. Ibid., p. 217.

49. R. Richardson (2001 edn), *Death, Dissection and the Destitute* (London: Phoenix Press), p. 245.
50. TNA, MH74/10, Anatomy Inspectorate Returns, for example, letter dated 25 February 1848, G. Cursham to G. M. Humphrys confirms Cambridge's situation.
51. A viewpoint in H. D. Rolleston (1932), *The Cambridge Medical School: A Biographical History* (Cambridge: Heffer), p. 68, and subsequently in Weatherall, *Gentlemen, Scientists and Doctors*, p. 100 n. 122.
52. Ibid., p. 100.
53. All cases cited can be found in CRO, PR25/21–23, St. Benedict's burial records. Refer PL25/1/21, for individual entries identified by date, month and year, 1855–1894.
54. TNA, MH74/10, letter from Mr. Ewbank to G. M. Humphrys, 9 May 1851, confirms the timing in respect of Addenbrookes hospital; MH74/10, letter from G. Cursham to G. M. Humphrys, 6 July 1855, sets out Cambridge Union's agreement.
55. For example, R. E. Hamrighaus (2001), 'Wolves in Women's Clothing: Baby Farming and the British Medical Journal, 1860–72', *Journal of Family History*, XXVI, pp. 350–72, discusses the sale of middle-class babies in East Anglia. This poster, though, is concerned with infants of the poorest being sold in Cambridge.
56. Richardson, *Death, Dissection*, p. 89, on Cambridge riot after 1832.
57. Cambridge Library Local Studies Room, poster collection.
58. Context in Hurren, *Protesting about Pauperism*.
59. CRO, St. Benedict's parish records, burial books, PL25/1/22, 1877 register.
60. Context, in Hurren, *Protesting about Pauperism*, pp. 136–43.
61. Macalister Papers, Downing College, Cambridge, private notes on appointment and teaching reforms.
62. 'The Dissection of Unclaimed Dead Bodies: Report of Nottingham Board of Guardians', speech by A. Macalister, *Nottingham Evening Post*, 12 October 1897 and draft copy confirming new standards in Macalister Papers, Downing College, Cambridge.
63. Outlined in D. Englander (1998), *Poverty and Poor Law Reform in 19th century Britain, 1834–1914: From Chadwick to Booth* (New York: Longman), p. 25.
64. A. Macalister (1908), 'An Address On Fifty Years of Medical Education', *British Medical Journal*, 2492, II, pp. 957–60, q. at p. 958.
65. CRO, St. Benedict's parish records, burial books, PL25/1/22, 1887 register.
66. 'Local Gossip', *The Hull Packet and East Riding Times*, 6 February 1885, issue 5235, pp. 1–2, quotes at p. 2, col. 1.
67. 'The Dissection of Pauper Bodies: Luton', *The Essex Standard, West Suffolk Gazette and Eastern Counties' Advertiser*, 7 March 1885, Issue 2830, p. 9, col. 1.
68. CRO, St. Benedict's parish records, burial books, PL25/1/22, 1887 register.
69. Ibid.
70. On Lang's career, see P. Bury (1952), *The Colleges of Corpus Christi and the Blessed Virgin Mary: A History, 1922–1952* (Cambridge: Cambridge University Press), pp. 278–89.
71. I am grateful to the archivist at Cambridge Record Office who in the summer of 2003 helped me to retrace the extent of the anatomy school burials in local parish records.

72. CRO, St. Benedict's parish records, burial books, PL25/1/23, 1900 register.
73. Ibid.
74. See context in Hurren, *Protesting about Pauperism*, pp. 214–41.
75. Refer Countesthorpe inoculation controversy in, for example, *Leicester Chronicle and Mercury*, 3 January 1891, issue 4169, p. 8, col. 2.
76. Macalister Papers, Downing College, Cambridge, memo from Claude Douglas, New Walk, Leicester, Medical Office of Health and Guardian, 7 October, 1987.
77. Refer, notably, S. A. King (2005, reissued in paperback 2010), *'We Might Be Trusted': Women, Welfare and Local Politics c. 1880–1920* (Brighton: Sussex Academic Press).
78. It was built on a 63-acre site, had 512 hospital beds, and opened in 1905. From its inception it traded dead bodies with Oxford anatomy school, see Chapter 6.
79. 'The Dissection of Unclaimed Dead Bodies: Report of Nottingham Board of Guardians', *Nottingham Evening Post*, 12 October 1897, copies kept in Macalister Papers, Downing College, Cambridge.
80. Macalister Papers, Downing College, Cambridge, Letter from Eugene J. O'Mulhane (Nottingham Union) to Dr. Ranson & Alexander Macalister (Cambridge Anatomy School), 17 August 1897.
81. Ibid., Macalister's handwritten note on the voting patterns of Nottingham guardians.
82. Macalister Papers, Downing College, Cambridge, notes on Nottingham controversy.
83. Ibid.
84. Macalister Papers, Downing College, Cambridge; Dr. James Bolton of Nottingham Union sent a number of undated, quick notes to Macalister.
85. Ibid., letter from Dr. D. G. Thomson, Medical Superintendent, Norfolk County Asylum, to Dr. James Barclay-Smith, 16 September 1912. Thomson confirmed that his asylum committee would not help because they were 'sentimental', did not want any bad publicity, disliked the policy, and with the coming of democracy feared a backlash about 'cases dying here...sent away to *be cut up* [sic] in the dissecting room'.
86. See, for example, 'District News', *Birmingham Daily Post*, 17 November 1891, issue 10422, p. 1, col. b, which makes reference to the possibility of an 'Anatomical League' in response to an article that had appeared in the *Pall Mall Gazette*. This was ten years before Macalister unified anatomists in 1912.
87. Bodleian Library (hereafter Bodl. Lib), University Archives (UA), Anatomy Department Records (ADR), HA89, 'Copy of memo handed to the Ministry, 26 November 1919 prepared by Professor Arthur Thomson on the Need for an Anatomical Supply Committee', pp. 1–11, q. at p. 5.
88. Macalister Papers, Downing College, Cambridge; A. Keith (later Sir A. Keith) a former Cambridge student based at UCL London, wrote to Macalister 6 May 1903 that the London Anatomical Committee would not permit bodies to go outside the city limits.
89. 'The Demand for Paupers', *Funny Folks*, 27 November 1880, Issue 313, p. 381, col. 1. This satirical weekly magazine often mocked the medical profession.

90. J. Lawrence (1998), *Speaking for the People: Party, Language and Popular Politics in England, 1867–1914* (Cambridge: Cambridge University Press), termed the notion 'politics of place', p. 1, to describe how much local issues shaped Victorian lives.
91. Richardson, *Death, Dissection* (2001 edn).
92. J-M Strange (2005), *Death, Grief and Poverty in Britain, c. 1870–1914* (Cambridge: Cambridge University Press).
93. Quoted in '"Sleeping with a Corpse, Shoreditch" London', *Cheshire Observer*, 11 January 1890, issue 1953, p. 7. The case was also reported in other London and provincial newspapers but this extract gives the fullest description.
94. See burial strategies, E. T. Hurren and S. A. King (2005), 'Begging for a Burial: Form, Function and Conflict in 19th-Century Pauper Burial', *Social History*, XXX, pp. 321–41.
95. Quoted in 'Summary of News: Spalding Union', *The Sheffield and Rotherham Independent*, 7 September 1886, issue no 9988, p. 6, col. 1; 'General News: Spalding Union', *Birmingham Daily Post*, 8 September 1886, issue 8797, p. 2, col. 1.
96. Quoted in, 'The Provinces: Another Croydon Scandal', *Lloyds Weekly Newspaper*, 25 May 1890, issue 2479, p. 1, col. 1.
97. Quotations, 'A Glimpse of Colchester Workhouse', *The Essex Standard, West Suffolk Gazette & Eastern Counties' Advertiser*, 7 March 1885, issue 2830, p. 8, col. 1.
98. Ibid.
99. 'Chelmsford, Thursday, from our Special Commissioner, August 15th 1891 edition', in L. Bellamy and T. Williamson (eds) (1999), *Life in the Victorian Village: The Daily News Survey of 1891*, Vol. 1 (London: Caliban Books), p. 16.
100. Ibid., p. 175.
101. 'Chelmsford, Thursday, August 15th 1891 edition', p. 112.
102. Macalister Papers, Downing College, Cambridge, petty cash receipts.
103. All cases cited reconstructed from record linkage work on CRO, 1881 census, Cambridge workhouse inmates, cross-checked to CRO, St. Benedict's parish records, burial books, PL25/1/22, 1881 register.
104. The 1881 census said that Sarah Perkins was 68, 'age heaping' being common.
105. The 1881 census said that Mary Trayler was 50.
106. I am grateful to the staff at CRO who assisted me in this mapping exercise when I did the original research in the summer of 2003, sharing their street geography expertise.
107. See, notably, E. Garrett, A. Reid, K. Schurer and S. Szreter (eds) (2001), *Changing Family Size in England and Wales: Place, Class and Demography, 1891–1911* (Cambridge: Cambridge University Press).
108. See, for example, J. M. Eyler (1997), *Sir Arthur Newsholme and State Medicine, 1885–1935* (Cambridge: Cambridge University Press).
109. Weatherall, *Gentlemen, Scientists and Doctors*, p. 216.
110. *Medical Society Magazine* (1931), August edition, p. 50, quoted also in Weatherall, *Gentlemen, Scientists and Doctors*, p. 216 n. 34.
111. Macalister Papers, Downing College, Cambridge, letter from A. Keith (later Sir. A. Keith) UCL to A. Macalister, 6 May 1903.

112. Ibid., letter from W. H. Elkins, superintendent, Three Counties Asylum, to Dr. Barclay-Smith, Cambridge Anatomical School, copied to Macalister, 2 October 1912.
113. *Lancet* (1889), vol. 2, p. 1881.
114. Record linkage work on the Macalister Papers and surviving burial registers show that this happened a lot at Cambridge. It will be the subject of more research after this book.
115. I am grateful to a former member of the department in which early skiagraphs were taken at Cambridge for alerting me by confidential letter, following my *Medical History* article on this topic in 2004, about the use of dead children in medical photography.
116. See, notably, S. Rowland, E. Gray, J. Poland and R. L. Bowles (1896), 'Report on the Application of the New Photography to Medicine and Surgery', *British Medical Journal*, 1836, I, pp. 620–2.
117. See, for instance, S. Rowland and A. C. C. Swinton (1896), 'Report on the Application of the New Photograph to Medicine and Surgery', XII, *British Medical Journal*, 1846, I, pp. 1225–6; this mentions work on necropsies but is short on personal detail.
118. Rowland, Gray, Poland and Bowles, 'Report on the New Photography', p. 621.
119. 'Medical Society of London', *British Medical Journal* (November 14th 1896), 1872, II, pp. 1446–51, mentions at p. 1447 work on necropsies and doing skiagraphs of sick children but no personal details are given.
120. M. Sappol (2002), *A Traffic in Dead Bodies: Anatomy and Embodied Social Identity in Nineteenth-Century America* (Princeton: Princeton University Press), pp. 39–42.
121. See http://www.webhistoryofengland.com/?cat=161, 16 October 2009, 'Dead Photos – Victorian Post-Mortem Photographs'.
122. A. Macalister, 'An Address on Fifty Years of Medical Education', p. 960.

6 Balancing the Books: The Business of Anatomy at Oxford University

1. 'Oxford – News – Inquest before Mr. Cecil Coroner', *Jackson's Oxford Journal*, 5 April 1834, issue s4223, p. 2, col. 2.
2. Ibid.
3. This claim was incorrect. If there were enough parts to make up an entire corpse then officially they had to be buried together; if not, there was no legal protection.
4. See, for example, 'Assize Intelligence – Lincoln', *Jackson's Oxford Journal*, 19 March 1836, issue 1323, p. 2, cols. 1–2. Andrews Roberts, a surgeon's apprentice, was one of the first prosecutions under the Anatomy Act.
5. See E. T. Hurren and S. A. King (2005), 'Begging for a Burial: Form, Function and Conflict in 19th Century Pauper Burial', *Social History*, XXX, pp. 321–41.
6. Refer also E. T. Hurren (2008), 'Whose Body Is It Anyway? Trading the Dead Poor, Coroner's Disputes and the Business of Anatomy at Oxford University c. 1885–1929', *Bulletin of the History of Medicine*, 82 (2008), pp. 775–819. Please take note that although this chapter uses the same data it introduces a lot of new primary research material. A lot of time and effort has been taken to ensure minimal overlap with previous publications in this new write-up.

7. 'City Police Court, Tuesday', *Jackson's Oxford Journal*, 1 April 1876, issue 6418, p. 1.
8. 'Oxford City Court, Tuesday, Daniel Hookham, plasterer', *Jackson's Oxford Journal*, 26 August 1871, issue 6178, p. 3, col. 1, details his personal history and unemployment.
9. Ibid., p. 3.
10. Violent nature recounted in 'Oxford City Court report – Daniel Hookham an old offender', *Jackson's Oxford Journal*, 30 August 1873, issue 6283, p. 1.
11. He was previously charged for 'absconding from the workhouse' and 'stealing clothes' given to him on entry but 'dismissed with a caution'; see 'Daniel Hookham', *Jackson's Oxford Journal*, 19 July 1873, issue 6277, p. 1, col. 2.
12. Refer, notably, S. A. King (2000), *Poverty and Welfare in England, 1700–1850: A Regional Perspective* (Manchester: Manchester University Press).
13. 'City Police Court, Tuesday', 1 April, 1876, p. 1.
14. J. Bellars (1714), *Essay Towards the Improvement of Physick in Twelve Proposals by which the lives of Many Thousands of the Rich as well as the Poor may be SAVED Yearly* (London: J. Sowle), p. 25.
15. Quoted in T. Hearne (1884), *Remarks and Collections, 1705–1735*, 11 vols. (Oxford: Oxford Historical Society Publication), vol. 1, introduction.
16. On links between hospital history and anatomy teaching, see K. J. Franklin (1936), 'A Short Sketch of the History of the Oxford Medical School', *Annals of Science* I, pp. 431–46.
17. Refer M. G. Brock and M. C. Curthoys (eds) (1997), *The History of the University of Oxford, Volume VI, 19th Century, Part One* (Oxford: Clarendon Press), pp. 563–70.
18. 'The President's Address to the Provincial and Medical Association', *Provincial Medical and Surgical Journal*, 26 August 1848, 34, X, pp. 393–412, quote at p. 395.
19. See, notably, R. Humphreys (1995), *Sin, Organized Charity and the Poor Law in Victorian England* (Basingstoke and New York: Palgrave Macmillan).
20. Rules summarised in 'Out-Door Relief', *Jackson's Oxford Journal*, 6 January 1872, issue 6197, p. 1.
21. 'Oxford Board and Guardian: Unfounded Charges against the Chairman of the Board and the Relieving Officer', *Jackson's Oxford Journal*, 10 February 1872, issue 6202, p. 1.
22. 'The Cemetery Question', *Jackson's Oxford Journal*, 23 October 1876, p. 2.
23. Ibid.
24. See Editorial, 'Medical Education at Oxford', *Oxford Magazine*, 1885, III, p. 309.
25. H. M. Sinclair and A. H. T. Robb-Smith (1950), *A History of Anatomy in Oxford* (Oxford: Clarendon Press), p. 74.
26. Refer R. Richardson, *Death, Dissection and the Destitute* (London, 2001 edn).
27. Recently revisited in H. MacDonald (2005), *Human Remains, Dissection and its Histories* (London and Melbourne: Yale University Press), pp. 11–41.
28. Sinclair and Robb-Smith, *History of Anatomy*, p. 74.
29. J. Litton (1991), *The English Way of Death: The Common Funeral since 1450* (London: Robert Hale Ltd), pp. 26–31.
30. See, notably, S. Burrell and G. Gill (2005), 'The Liverpool Cholera Epidemic of 1832 and Anatomical Dissection – Medical Mistrust and Civil Unrest', *Journal of the History of Medicine and Allied Sciences*, 60, IV, pp. 478–98.

31. Studies of major figures, rather than less well-known anatomists, can be found in J. Morrell and A. Thackray (1981), *Gentlemen of Science: The Early Years of the British Association for the Advancement of Science* (Oxford: Oxford University Press); A. Desmond (1989), *The Politics of Evolution: Morphology, Medicine and Reform in Radical London* (Chicago: Chicago University Press); A. Desmond and J. Moore (1992), *Darwin* (Chicago: Chicago University Press); J. Browne (2003 edn), *Charles Darwin: The Power of Place* (New York and London: Pimlico); R. Fortley (2008), *Dry Store Room No. 1: The Secret Life of the Natural History Museum* (London: Harper Perennial).
32. See, notably, J. Lawrence (1998), *Speaking for the People: Party, Language and Popular Politics in England, 1867–1914* (Cambridge: Cambridge University Press).
33. *DNB*, 'Arthur Thomson (1858–1935)', entry 101036498.
34. A. Thomson (1899), *Handbook of Anatomy for Arts Students* (Oxford: Clarendon Press), was dedicated to his father.
35. *DNB*, p. 76.
36. Hurren, 'Whose Body Is It Anyway?', pp. 784–6, expands on his career trajectory.
37. J. P. D. Dunbabin (1978), 'Oxford College Finances: A Reply', *Economic History Review*, 2nd series, XXII, I, p. 439.
38. Bodleian Library (hereafter Bodl. Lib), University Archives (UA), Anatomy Department Records (ARD), HA50, 'Arthur Thomson's files notes'.
39. Bodl. Lib. UA, ADR, Box HA107.
40. Ibid., Box HA50, papers relating to Anatomy Department extension plans.
41. 'The University Museum: Opening of the New Anatomy Department', *Jackson's Oxford Journal*, 21 October 1893, issue 7335, pp. 1–3, quote at p. 1.
42. See praise in *Oxford University Gazette* (26 May 1891), xxi, p. 499.
43. Brock and Curthoys, *History of Oxford*, pp. 570–80, provides a useful summary.
44. See *British Medical Journal* (5 January 1878), pp. 34–5 and (12 January 1878), p. 66.
45. Ibid., p. 66.
46. I am grateful to Arthur Thomson's niece for sharing her family history with me in 2007; A. Thomson (1912), *The Anatomy of the Human Eye, as Illustrated by Enlarged Stereoscopic Photographs* (Oxford: Clarendon Press), pp. 1–61, and 67 plates, that pioneered the use of medical photography in the anatomy of the human eye.
47. See, for example, a review in J. H. Vaidya (1921), 'A Point in Favour of Professor Arthur Thomson's Theory of the Production of Glaucoma', *The British Journal of Opthalmology*, 5, IV, 172–5.
48. See, notably, E. Nettleship (FRCS) (1875), 'The State of the Eyelids in the Orphan and Deserted Children as compared to those who have Parents', *4th Annual Report of the Local Government Board Report, 1874–5* (London: George Edward Eyre and William Spottiswoode), pp. 113–31.
49. Refer P. Rivière (ed.) (2007), *A History of Oxford Anthropology* (Oxford: Oxford University Press), pp. 21–62.
50. A. Thomson (1899), *Anatomy for Art Students* (Oxford: Clarendon Press), pp. 1–405 – the first to be published for art teaching by OUP.
51. There is still a famous 'Caricature of Professor Thomson, 7 June 1922' drawn by Francis Derwent Wood in the Royal Academy of Arts Collection.

52. Bodl. Lib. UA, ADR, 'Thomson to Fraser, letter dated 29 Jan. 1908', Box 105.
53. Hurren, *Protesting about Pauperism*, pp. 136–43, for a rural recession synopsis.
54. Refer debate with A. J. Engle, in Dunbabin, 'Oxford College Finances', p. 439. Both agree that 'annual university spending on science at Oxford between 1892 and 1900 ran somewhat under 75% of spending at Cambridge.'
55. Hurren, 'Whose Body Is It Anyway?', pp. 786–802, recounts related supply problems.
56. E. Barclay-Smith (1937), *The First Fifty Years of the Anatomical Society of Great Britain and Ireland: A Retrospect* (London: John Murray); L. Gray (2006), 'The early days of the Anatomical Society of Great Britain and Ireland', *Anastomosis*, I, pp. 1–8.
57. Ibid.
58. Refer E. T. Hurren and I. Scherder (2011), 'Dignity in Death? The Dead Body as an Anatomical Object in England and Ireland, c. 1832 to 1929' in A. Gestrich and S. A. King (eds), *The Dignity of the Poor* (Oxford: Oxford University Press), pp. 1–31.
59. Bodl. Lib. UA, ADR, HA89, 'Thomson's private papers and notes containing a copy of a paper handed to the Ministry of Health dated 26 Nov. 1919'.
60. Hurren, 'Whose Body Is It Anyway?', p. 787, table 1, p. 794, illustration 1, and p. 795, table 2, provide more details.
61. Bodl. Lib. UA, ADR, HA89, 'Thomson's notes kept on file'.
62. Ibid., HA64, reconstructed from 'Inventory of the Department of Human Anatomy'.
63. Bodl. Lib. UA, ADR, HA64.
64. Hurren, 'Whose Body Is It Anyway?' p. 799, demography of his economy of supply.
65. Notably, N. Hopwood (1999), '"Giving Body" to Embryos: Modeling, Mechanism and Microtone in Late-Nineteenth Century Anatomy', *Isis*, CXXXX, I, pp. 462–96.
66. All the data quoted in this section has been reconstructed from record linkage work at the Bodl. Lib, UA, ADR, HA/1/1-3, 1885–1929 and Oxfordshire Record Office hereafter OXR), C/ENG/1/A1/2, Botley Cemetery Records, 1894–1938, cadavers buried by Oxford.
67. Ibid.
68. Paupers' cases taken for dissection reconstructed from record linkage work on Oxfordshire Health Archives (hereafter OHA), RI/I/50, Admission Registers of Medical Inpatients, 1890–1917; RI/I/58, Admission Registers of Surgical In-Patients, 1898–1924; RI/9/B1/3–6, Registers of Death, 1878–May 1908; RI9/b1/7–9, Registers of Death, March 1916–April 1920; RI/I/23, RI Letter books, 1893–1901 and 1897–1909. Mary Coolling entry 14 July 1886.
69. Ibid., Joseph Jeffrey entry 22 December 1894.
70. Bodl. Lib. UA, ADR, HA1/1, regular fees in Thomson's petty cash books.
71. Ibid., HA1/1-3, reconstructed from daily financial records and petty cash expenses.
72. Bodl. Lib. UA, ADR, HA1/1HA64, Thomson's letter book, miscellaneous entry, 18 February 1899, recalling changes to teaching methods.
73. Ibid., HA1/1, Thomson petty cash book entry.
74. Bodl. Lib. UA, ADR, HA1/1, petty cash.
75. Ibid.

76. Bodl. Lib. UA, ADR, HA1/1, petty cash.
77. Ibid., HA/1/1, petty cash book entries confirm that the dissections of Matthew May and James Francis were completed on 10 June 1887 & Anne Hargreaves on 1 July 1887.
78. Bodl. Lib. UA, ADR, HA1/1, petty cash entry 25 May 1887.
79. Regrettably the burial records do not separate body parts from whole cadavers once the new burial plot was allocated; see ORO, C/ENG/1/A1/2, Botley Cemetery Records, 1894–1938. When the anatomy department was redesigned there was a clinical waste bin used to dispose of some parts and so any attempt to reconstruct data from the available records would be inaccurate.
80. Debates summarised in A. H. T. Robb-Smith (1968), 'The Development and Future of Oxford Medical School', *Transactions of the Society of Occupational Medicine*, XXVIII, pp. 13–21, and J. Howarth (1987), 'Science education in late-Victorian Oxford: a curious case of failure', *English Historical Review*, CII, pp. 334–71.
81. Bodl. Lib. UA, ADR, HA89, A. Thomson, 'Private Notes to Ministry of Health', 26 November 1919.
82. Ibid., HA89, large file of loose-leaf material including private notes made at meetings, confidential letters, and minutes of private conversations.
83. Richardson, *Death, Dissection*, pp. 256–7.
84. Bodl. Lib. UA, ADR, HA89, sequence of letters between Thomson and J. Pickering Pick, dated January–February 1907. Pick lived at 'The Hook, Great Bookham, Surrey' and he also went up and down to Oxford to discuss matters in person.
85. Ibid., HA1/1, see account books, held with the Westminster Bank, 1885–1931 and the payments made to the Leicester branch for guardians. HA89 details the Leicester contracts. In 1910, some Leicester guardians tried to stop the body trade on the basis that 'poor people never benefited from medical science' but the resolution was lost by '24 votes to 13'. It continued under the New Poor Law until its official demise in 1929. The opposite was reported inaccurately at the time to avoid any bad publicity.
86. Bodl. Lib. UA, ADR, HA89, J. Pickering Pick to Thomson, letter, 6 February 1907.
87. See, notably, C. Bellamy (1988), *Administering Central-Local Relations: The Local Government Board in its Fiscal and Cultural Context* (Manchester: Manchester University Press); E. T. Hurren (2005), 'Poor Law versus Public Health: diphtheria and the challenge of the crusade against outdoor relief to public health improvements in late-Victorian England, 1870–1900', *Social History of Medicine*, XXVIII, pp. 399–414
88. Bodl. Lib. UA, ADR, HA89, file notes – in his final report he crossed out the reference to the drop in mortality rates because it was emotive revealing anatomy's reliance on poor public health standards.
89. Ibid.
90. Refer, Hurren, *Protesting about Pauperism*, pp. 214–47.
91. Bodl. Lib. UA, ADR, HA89, copy of confidential memos penned by Thomson.
92. Ibid., HA89, opinions summarised in a final report 12 May 1921 sent to Sir George Newman (Ministry of Health) and Alexander McPhial (Anatomy Inspectorate).

93. Ibid.
94. Bodl. Lib. UA, ADR, HA51 gives handwritten statistics and HA64 file contains Thomson's student calculations and profiles in loose-leaf notes and letters to the University Chest, after he was asked to account for '1. How much he lectures, 2. The roll call of his classes and 3. Student attendance figures'.
95. Ibid., HA89, notes on Thomson's cadaver redistribution scheme.
96. Ibid.
97. Bodl. Lib. UA, ADR, HA89, Thomson's notes on his cadaver redistribution scheme and memorandum summarising the situation at the end of World War One.
98. Ibid., HA89, Thomson's private notes made as a record of the secret meeting.
99. Refer rest of career history in E. T. Hurren (2010), 'Remaking the Medico-Legal Scene: A Social History of the Coroner in Victorian Oxford, c. 1877 to 1894', *Journal of the History of Medicine and Allied Sciences*, CXV, pp. 207–52.
100. See ORO, MOR./LXXV/3, 5–6, personal file by his solicitor, Morell, Peel & Gamon, containing marriage settlement Edward Law Hussey, 1874–1876.
101. E. L. Hussey (1882), *Miscellanea Medico-Chirurgica: Cases in Practice, Reports, Letters & Occasional Papers* (Oxford: Pickard, Hall and Stacy), p. 256, entry 26 May 1875.
102. Oxford Library Local Studies Room (hereafter OLLSR), Anon. (1850), *Slatter's Oxford Trade Directory* (Oxford: Parker & Co), p. 56.
103. Refer E. L. Hussey (1883), *Extracts from various authors and fragments of table-talk* (Oxford: Parker & Co.), pp. 61–2.
104. See E. L. Hussey (1887), *Miscellanea Medico-Chirugica, 2nd part, Occasional Papers and Remarks* (Oxford: Parker & Co.), pp. 11–13.
105. See E. L. Hussey (1856), *Analysis of Cases of Amputations of the Limbs in the Radcliffe Infirmary, Oxford, 2nd edition*, (London: John Murray & Co.), daily workload.
106. See Hurren, 'Remaking the Medico-Legal Scene', pp. 207–52, which elaborates his background and professional problems.
107. See OXHA, Radcliffe Infirmary, RI/II/100(16) Coroner's accounts and inquests, 1830–37, which gives a background summary and caseload details. In the mid-Victorian period, for instance, John Marriott Davenport (1809–1882) was a University Coroner, a practicing solicitor, clerk to the Peace for the County of Oxford and Secretary to the Bishop of Oxford, as well as District Registrar of the Court of Probate for Oxfordshire.
108. Refer, notably, I. Burney (2000), *Bodies of Evidence: Medicine and the Politics of the English Inquest, 1830–1926* (Baltimore, MD: Johns Hopkins University Press).
109. Medical Opinion, 'After Death', *Pall Mall Gazette*, 9 July 1870, issue 1686, pp. 1–2. The article was written in response to a number of complaints about more extensive post-mortems at Guy's hospital in London (refer also Chapter 1).
110. Elaborated in Hurren, 'Remaking the Medico-Legal Scene', pp. 207–252.
111. E. L. Hussey (1894), *Miscellanea Medico-Chirugica: 3rd part, Occasional Papers and Remarks* (Oxford: Parker & Co.), p. 74.

112. Hussey, *Miscellanea Medico-Chirurgica* (1887), pp. 14–15.
113. On coroner's election, *Jackson's Oxford Journal*, 30 October 1877, p. 1, and reprinted in Hussey, *Miscellanea Medico-Chirurgica* (1887), p. 32.
114. See *Oxford Chronicle*, 27 October 1878, p. 5, for a lengthy summary of the case.
115. Ibid.
116. Hussey recounted this history numerous times in pamphlets, and published books. See, in detail, Hurren, 'Remaking the Medico-Legal Scene', pp. 207–52.
117. The legal case and medical facts were documented in a 'Coroner's Memorandum to the Governors of the Radcliffe Infirmary, 30 August 1888', reprinted in E. L. Hussey, *Miscellanea Medico-Chirurgica* (1894), p. 56.
118. Ibid.
119. Events are described in OHA, Hussey to the Governors 28 January 1879 in response to Quarterly Reports of the Governors at the Radcliffe Infirmary in 1878.
120. Anatomy records show five bodies before the controversy and none thereafter.
121. Bodl. Lib. UA, ADR, HA105, volumes 1 & 2, letters books, recount Thomson's frustrations and Hussey's resentment.
122. See forthright opinions weekly in *Jackson's Oxford Journal* & *Oxford Chronicle*.
123. The National Archives (hereafter TNA), MH12/9721, 'Coroner's Report dated 1888' sent by Hussey to the LGB. He subsequently published the data in a report for public circulation in Oxford.
124. Bodl. Lib. UA, ADR, HA105, volume 2, letter 16, September 1909, sent in to Thomson by the Relieving Officer for Oxford Union, complains that problems with the city-coroner's office have not stopped even though Hussey had retired from office in 1894.
125. Paupers' cases taken for dissection reconstructed from record linkage work on OHA, RI/I/50, Admission Registers of Medical Inpatients, 1890–1917; RI/I/58, Admission Registers of Surgical In-Patients, 1898–1924; RI/9/B1/3–6, Registers of Death, 1878–May 1908; RI9/b1/7–9, Registers of Death, March 1916–April 1920; RI/I/23, RI Letter books, 1893–1901 and 1897–1909.
126. Reconstructed from Bodl. Lib. UA, ADR, HA/1/1–3, 1885–1929 and ORO, C/ENG/1/A1/2, Botley Cemetery Records, 1894–1938, cadavers buried by Oxford.
127. OHA, Radcliffe Infirmary, RI/I/49, Medical Inpatients Register, 1888–1904, entry 4 July 1894, patient case history Martha Gates.
128. Bodl. Lib. UA, ADR, HA1/1, petty cash book entry by Thomson, 12 March 1886.
129. Ibid., entry 21 February 1887.
130. Bodl. Lib. UA, ADR, HA1/1–3; Thomson employed J. Stroud in central Oxford, and J. Burrough Esq., 40 Gale Street, Oxford. Both men charged a set fee of '£2 10s 7d'.
131. ORO, City of Oxford Cemetery Committee Minute Book, Thomson letters 1905.
132. Ibid., 'Table of Fees Payable at the Cemeteries of the Mayor, Aldermen and Citizens of Oxford', leaflet dated 7 November 1898.

348 *Notes*

133. All case histories cited can be located by date entries, in OHA, RI/I/49, Radcliffe Infirmary Medical In-Patient Records, 1888–1904 cross-checked to Bodl. Lib. UA, ADR, HA1/1–3 and cross-checked to ORO, C/ENG/1/A1/2, Botley cemetery records, 1894–1938, whole cadavers buried.
134. TNA, MH12/9720, Oxford, Case No. 348, notes, letters, and witness statements sent by individuals involved to LGB.
135. Ibid., Memorandum by Alan Thomson and Harold Thomson, Public Vaccinator & Surgeon respectively, written from No. 31 Beaumont Street, Oxford, 28 December 1885.
136. TNA, MH12/9720, LGB Memorandum no. 1638, copy witnesses statement & letter of Caroline Coolling, wife of Thomas Coolling, shoemaker, 16 English's Row, Oxford.
137. Medical Doctor and Fellow of the Royal College of Surgeons of Ireland.
138. A. H. T. Robb-Smith (1970), *A Short History of the Radcliffe Infirmary* (Oxford: The Church Army Press, for the United Hospitals), p. 175.
139. Bodl. Lib. UA, ADR, HA1/1, body sales cross-checked to ORO, ENG/1/A1/2, Botley cemetery records, 1894–1938. In the latter, often a brief case history survives in marginalia of the burial register copied by an anatomist.
140. Ibid., HA89, miscellaneous correspondence, Thomson to Macalister and various railway companies where rates and using a van are compared.
141. Bodl. Lib. UA, ADR, HA105-2, letter books, entry 9 February 1909, Thomson to Parker.
142. Ibid., HA89, draft report, 17 February 1920, written by Thomson and letter back from Alexander McPhial, Inspector of Anatomy.
143. Bodl. Lib. UA, ADR, HA89, Thomson to McPhial, private letter 'February 1920'.
144. He repeated this phrase often in his correspondence to and from central government.
145. Hussey and his predecessors had after all co-operated in the past.
146. Burney, *Bodies of Evidence*, p. 3.

7 'Better a third of a loaf than no bread': Manchester's Human Material

1. T. Turner (1840), 'Introductory Address: To the Students at the Royal School of Medicine and Surgery, Pine Street, Manchester for the Winter Session of 1840–41', *Provincial Medical & Surgical Journal*, Vol. 1, III (17 October), pp. 33–8, q. at p. 34.
2. Ibid., p. 35.
3. Turner, 'Introductory Address', p. 35; Turner bought the Museum of Gregory Smith of Windmill Street Anatomy School in London, founded by William Hunter. John Hunter taught there too. This was how he managed to acquire his famous dissection table, see J. Howson (ed.) (1902), *British Medical Association – Manchester Meeting 1902 – Handbook and Guide to Manchester* (Manchester: F. Ireland), p. 141.
4. On the social history of professional expansion, see A. J. Kidd and K. W. Roberts (eds) (1985), *City, Class and Culture: Studies of Cultural Production & Social Policy in Victorian Manchester* (Manchester: Manchester University Press).

5. F. Hutton, 'Medicine and Mutilation: Oxford, Manchester and the Impact of the Anatomy Act 1832 to 1870 (Oxford Brookes University, unpublished Ph.D., 2007), p. 99.
6. See T. Turner (1824), *Outlines of a System of Medico-Chirurgical Education Containing Illustrations of the Application of Anatomy, Physiology, and Other Sciences to the Principal Practical Points in Medicine and Surgery* (Manchester: Underwood).
7. See, notably, on smaller schools J. Pickstone (1985), *Medicine in Industrial Society: A History of Hospital Development in Manchester and its Regions, 1752–1946* (Manchester: Manchester University Press), p. 186.
8. Hutton, 'Medicine and Mutilation', p. 99.
9. Refer S. V. F. Butler (1986), 'A Transformation in Training: the Formation of University Faculties in Manchester, Leeds and Liverpool, 1820–1884', *Medical History*, XXX, pp. 115–32.
10. See context in Pickstone, *Medicine in Industrial Society*.
11. The National Archives (hereafter TNA), MH 74/10, Anatomy Inspectorate correspondence, 12 January 1849.
12. Refer J. Reinarz (2005), 'The Age of Museum Medicine: The Rise and Fall of the Medical Museum at Birmingham's School of Medicine', *Social History of Medicine*, XXVIII, pp. 419–37.
13. Anon. (5 October 1882), 'Report Of the Means Employed in Medical Schools of Great Britain For the Preservation of Subjects for Dissection', *British Medical Journal*, 614, II, pp. 382–4, q. at p. 383.
14. Ibid., p. 383.
15. See, throughout, TNA, MH12/6042 Local Government Board files for Manchester Poor Law Union and Crumpsall Workhouse, the sources of supply, in the Victorian era.
16. Hutton, 'Medicine and Mutilation', pp. 215–17.
17. TNA, MH74/10, Anatomy Inspectorate, private letter dated 1 November 1850 in reply to notification that the first body had been purchased at Chatham Street School.
18. Ibid., MH74/10 letters 26 January 1852, and 16 February 1852, outline complaints.
19. 'Chatham Street School of Medicine: Distribution of Prizes', *Manchester Guardian*, 16 October 1852, column 1, p. 8.
20. TNA, Anatomy Inspectorate, Pine Street re Chatham Street School, 25 October 1852.
21. Confirmed in Home Office papers, see, for example, HO144/148/A38382, Prison Commissions authorised to transfer and remove 'unclaimed' bodies dying in Leeds and Wakefield prisons to the Medical Department of Yorkshire College in Leeds.
22. A. J. Kidd (1985), 'Outcast Manchester: Voluntary Charity, Poor Relief and the Casual Poor, 1860–1905', in A. J. Kidd and K. W. Roberts (eds), *Class, City and Culture: Studies of Cultural Production and Social Policy in Victorian Manchester* (Manchester: Manchester University Press), chapter 3, pp. 48–73, q. at p. 54.
23. TNA, MH74/10, 13 January 1850, H. Waddington to Anatomy Inspectorate, private memorandum.

24. Ibid., MH74/36, 31 January 1890, Queen's College School Birmingham return and Witton Cemetery.
25. TNA, MH74/10, 13 January 1850, H. Waddington to Anatomy Inspectorate, private memorandum.
26. 'Manchester, from our own correspondent', *British Medical Journal*, (2 December 1871), p. 651.
27. Hutton, 'Medicine and Mutilation', pp. 268–71.
28. J. Thompson (1886), *The Owens College, Its Foundation and Growth* (Manchester: Cornish Press).
29. E. M. Brockbank (1936), *The Foundation of Provincial Medical Education in England and of the Manchester School in Particular* (Manchester: Manchester University Press).
30. R. H. Kargon (1977), *Science in Victorian Manchester: Enterprise and Expertise* (Manchester: Manchester University Press).
31. P. J. Hartog (ed.) (1901), *The Owens College Manchester (founded 1851): A Brief History of the Colleges and Descriptions of Its Various Departments* (Manchester: JF Cornish Publishers), p. 94.
32. Ibid., p. 94.
33. This arrangement with UCL was originally agreed in 1859.
34. Forerunner of Manchester University; see H. B. Charlton (1952), *Portrait of a University 1851–1951: To Commemorate the Centenary of Manchester University* (Manchester: Manchester University Press).
35. See also S. Oleesky (1971), 'Julius Dreschfeld and Late Nineteenth-Century Medicine in Manchester', *Manchester Medical Gazette*, CI, pp. 14–17.
36. Anon. (1908), *The Victoria University of Manchester Medical School* (Manchester: Manchester University Press), p. 16.
37. Manchester Record Office (hereafter MRO), M10/808, Hulme Pauper Letter Collection, series of short entries about burial.
38. See, notably, J- Marie Strange (2005), *Death, Grief and Poverty in Britain, 1870–1914* (Cambridge: Cambridge University Press).
39. Refer S. A. King (2000), *Poverty and Welfare in England, 1700–1850: A Regional Perspective* (Manchester: Manchester University Press).
40. Figure accepted and debated in contemporary sources; see, for example, 'Conflicting Mortality Statistics for Manchester', *British Medical Journal* (31 August 1878), p. 329.
41. Resurrection men had targeted Hulme graveyards before the Anatomy Act. Bodies were given pauper funerals and therefore easy to locate to exhume; see, for example, 'Resurrection Men', *Manchester Guardian*, 1 January 1825, p. 3, col. 3.
42. E. T. Hurren (2007), *Protesting about Pauperism: Poverty, Politics and Poor Relief in late-Victorian England* (Woodbridge: Boydell and Brewer), pp. 21–2, pauper settlement.
43. Northamptonshire Record Office (hereafter NRO), 251p/98, Great Oxendon Overseers of the Poor's letters to and from Manchester, 251p/98, dated in sequence 19 August 1832, 1 December 1833, and 22 January 1834.
44. Ibid., 1 December 1833.
45. Manchester Local Studies Room (hereafter MLSR), census for Manchester 1841–1891.
46. NRO, 251p/98, letter, 22 January 1834.

Notes 351

47. See also H. Gaulter (1833), *The Origins and Progress of the Malignant Cholera in Manchester* (London: Longman, Rees, Orme, Brown, Green & Longman), p. 35.
48. See also S. Burrell and S. Gill (2005), 'The Liverpool Cholera Epidemic of 1832 and Anatomical Dissection – Medical Mistrust and Civil Unrest', *Journal of the History of Medicine and Allied Sciences*, CX, IV, pp. 478–98. It was dangerous dissection work because corpses were so contagious.
49. Yakub Quereshi, 'Bones found at Victoria Station', *Manchester Evening News*, 22 January 2010, p. 1. Human remains were also found here in 1993 and 1995.
50. Four mail trains that ran daily into Manchester, connecting it with the rest of the country. They often carried the dead in the rear carriages; see A. Freeling (1838), *The Railway Companion from London to Birmingham, Liverpool and Manchester, etc.* (London: Green).
51. 'Bills, Bulls and Blunders', *The Age*, penny journal, printed in Manchester and London, 8 January 1832, p. 10.
52. R. Richardson (2001 edn), *Death, Dissection and the Destitute* (London: Phoenix Press).
53. See, for context, H. L. Platt (2005), *Shock Cities: the Environmental Transformation of Manchester and Chicago* (Chicago: Chicago University Press), pp. 24–7.
54. F. Engels (1887 edn), *Conditions of the Working Class of England (written September 1844 to March 1845) published first in Leipzig 1845, authorised English edition* (London, 1887 and New York, 1891), quote cited at the start of Chapter 2, 'The Great Towns'.
55. Anon. (1969 edition), *Manchester in 1844: Its Present Condition and Future Prospects by Leon Faucher translated from the French with copious notes appended by a Member of the Manchester Athenaeum* (Manchester: Franck Cass and Co. Ltd) [first published: Manchester: Abel Heywood, 1884]), pp. 15–16.
56. Ibid., p. 63.
57. *Manchester in 1844*, p. 109.
58. See religious geography in K. D. M. Snell and P. S. Ell (2000), *Rival Jerusalems: The Geography of Victorian Religion* (Cambridge: Cambridge University Press). Non-conformists were a broad spectrum of Jews, Methodists, Plymouth Brethren, Presbyterians, Quakers and Unitarians.
59. See, for context, P. A. Johnson (1986), *Saving and Spending: Working Class Economy in Britain, 1870–1939* (Oxford: Oxford University Press); T. Alborn (2001), 'Senses of Belonging: The Politics of Working-Class Insurance in Britain, 1880–1914', *Journal of Modern History*, CXXIII, IX, pp. 561–602.
60. Refer L. Hollen Lees (1998), *The Solidarities of Strangers: The English Poor Laws and the People, 1700–1948* (Cambridge: Cambridge University Press), pp. 217–30, on why 'No Irish Need Apply' for poor relief.
61. See, for example, TNA, FS/15/1402, Bennett Street Manchester Sunday School Sick Funeral Society, 1839–1951.
62. Commonly found in TNA, MH13/58 Crumpsall workhouse in-correspondence (1851–1871).
63. T. Hunt (2004), *Building Jerusalem: The Rise and Fall of the Victorian City* (London: Weidenfeld and Nicolson), p. 28.
64. TNA, MH12/6402/115, Abel Heywood (Manchester) to Poor Law Commissioners (London), letter dated 26 July 1845 and internal civil service

discussions item reference 221/4. Heywood was a member of Manchester Borough Council and active Radical (c. 1810–1893).
65. Ibid.
66. TNA, MH12/6042/180, letter, 30 May 1845, from Poor Law Commission to Abel Heywood.
67. Ibid., MH12/6042/124, folios 184–5, LGB internal memo No. 10631B45 and reply, 28 August 1845.
68. Ibid.
69. TNA, MH12/6042/579, 19 August 1846 resolution to employ William Gleaves.
70. Ibid., MH12/6042/124, folios 184–7, Poor Law Commissioners to Abel Heywood, medical school, and Ner Gardiner, Clerk to Manchester Board of Guardians confirming burial procedures after dissection.
71. TNA, MH12/6042/127, Draft letter to Abel Heywood, 12 August 1845.
72. Ibid., MH74/36, Sheffield School of Anatomy body scandals.
73. M. Douglas (1966), *Purity and Danger: An Analysis of Concepts of Pollution and Taboo* (London: Routledge), p. 36.
74. 'Intramural Interments', *Manchester Times*, 20 March 1850, 144, 2nd edition, p. 1, col. 2.
75. 'The Parish Burial Ground', letter to the editor, *Manchester Guardian*, 22 February 1851, p. 11, col. 1, signed 'NECROPOLIS'.
76. 'Public Cemeteries for the City of Manchester: To the editor of the Manchester Guardian written by a Visitor', *Manchester Guardian*, 27 June 1855, p. 7, cols. 1–2.
77. 'Manchester Guardians', *Manchester Guardian*, 20 July 1855, p. 4, col. 2.
78. Ibid.
79. 'Manchester Board of Guardians, Delayed Burials', *Manchester Guardian*, 11 February 1859, p. 3, col. 1.
80. Letter to the editor, 'Pauperism in Manchester', *Manchester Guardian*, 14 February 1859, p. 4, col. 1.
81. Context, in R. G. Kirby and A. E. Musson (1975), *The Voice of the People: John Doherty, 1798–1854 – Trade Unionist, Radical and Factory Reformer* (Manchester: Manchester University Press).
82. See also J. Roberts (1979), *Working Class Housing in Nineteenth Century Manchester: The Example of John Street, Irk Town* (Manchester: Richardson).
83. Corroborated in I. Loudon (6 September 1996), 'Obstretric care, social class and maternal mortality', *British Medical Journal*, volume 293, No. 6457, pp. 606–7.
84. 'Special Correspondence: Manchester – Medical Summer Session – Towns and Child Life', *British Medical Journal* (2 May 1903), p. 1058.
85. Refer H. M. Boot (1990), 'Unemployment and Poor Relief in Manchester, 1845–50', *Social History*, 15, II, pp. 217–28.
86. See, M. A. Busteed and R. I. Hodgson (1994), 'Irish migration and settlement in early nineteenth-century Manchester, with special reference to the Angel Meadow district in 1851', *Irish Geography*, XXVII, pp. 1–13, and (1996) 'Irish migrant responses to urban life in early nineteenth century Manchester', *Geographical Journal*, 162, II, 139–53.
87. This quotation was how a clerk at Crumpsall Workhouse referred to pauper sales that he arranged. That context will be elaborated later in this section.

Notes 353

88. The townships incorporated on 29 March 1841 under the New Poor Law were Blackley Bradford, Cheetham, Crumpsall, Failsworth, Harpurhey, Great Heaton, Little Heaton, Moston, Newton, Prestwich, and Manchester township. Children at this stage were separated by gender, boys sent to Blackley workhouse, girls to Prestwich workhouse; see A. Redford (1940), *The History of Local Government in Manchester*, Vol. II: *Borough and City* (London: Longman and Green & Co.), pp. 124–6.
89. See, for example, G. B. Hindle (1875), *Provision for the Relief of the Poor in Manchester 1754–1826* (Manchester: Manchester University Press); M. E. Rose (ed.) (1985), *The Poor and the City: The English Poor Law in Its Urban Context* (Leicester: Leicester University Press).
90. See *Manchester Archives and Local Studies: A Guide to Poor Law and Workhouse Records*, available online at www.manchester.gov.uk/libraries/arts.
91. Refer *British History Online*, 'Crumpsall', www.british-history.ac.uk.
92. 'The New Workhouse Hospital at Crumpsall', *Manchester Guardian*, 12 September 1878, p. 8, col. 3.
93. 'Proposed New Workhouse for Manchester', *Manchester Guardian*, 1 July 1854, p. 8, col. 1.
94. See central government memorandums at Bodleian Library (hereafter Bodl. Lib.), University Archives (UA), Anatomy Department Records (ADR), HA89.
95. These figures were also repeated in the medical press; see footnote 26 above.
96. See, for example, 'Manchester Board of Guardians, Delayed Burials', and 'Manchester Board of Guardians, The Burial of Paupers', *Manchester Guardian*, 11 February 1859, p. 3, and 19 July 1867, p. 3, respectively.
97. 'Manchester Burial Scheme', *Manchester Guardian*, 11 November 1856, p. 3, col. 2. There was to be one on the North and South side of the city.
98. 'Manchester Improvement and Water Bill', *Manchester Guardian*, 22 March 1872, p. 8, col. 1.
99. See end quote, 'Manchester Burial Scheme', *Manchester Guardian*, 11 November 1856, p. 3. col. 2. That report was accepted and eventually its recommendations became the responsibility of the city authorities.
100. See central government memorandums retained by Professor Arthur Thomson at the Bodl. Lib, UA, ADR, HA89.
101. The first workhouse census recorded 745 able-bodied unemployed men and women; 152 pregnant women; 248 idiots, imbeciles and epileptics; 255 children aged below sixteen; 60 probationers on good-behaviour orders; 200 of the sick poor; and 76 new-born babies.
102. Redford, *History of Local Government in Manchester*, p. 126.
103. Hutton, 'Medicine and Mutilation', appendix 2, pp. 296–325.
104. See C. Hawkins, 'Anatomical Subjects to the Editor', *Lancet*, 23 February 1867, p. 9; 'Conflicting Mortality Statistics for Manchester', *British Medical Journal*, 31 August 1878, p. 329.
105. History recounted in 'Local Gleanings XXVI: The Churchwardens Accounts, Parish of Manchester in the 19th century', *Manchester Guardian*, 24 May 1851, p. 9.
106. MLSR, MFRR/627, St. Michael's Angel Meadow Burials, 1813–1854, July 1849.

107. TNA, MH12/6080, Crumpsall workhouse report dated 1888 sent to LGB.
108. Ibid., Inspector's reply dated 1888.
109. TNA, MH12/6080, Crumpsall workhouse report and inspector's reply dated 1888.
110. Ibid.
111. TNA, MH12/6080, Crumpsall workhouse report and inspector's reply dated 1888.
112. Refer Inspector's Report (1900), 'Burial of Paupers', *Annual Local Government Board Report, 1899–1900* (London: George Edward Eyre and William Spottiswoode), p. 93.
113. British Parliamentary Papers (1893), *Royal Commission of the Aged Poor*, Minutes of evidence, Vol. II, Question 3180, p. 179.
114. Bodl. Lib. UA, ADR, HA89, 'Private and Confidential Copy of an Anatomy Report sent to Sir. George Newman by Professor Arthur Thomson', 12 May 1921, pp. 1–18, q. at p. 18.
115. MLSR, MFPR/706, Phipps Park Cemetery, Burial Registers, Roman Catholic part, 1888–1891, entries 20, 21, and 22 May, 1884.
116. Cambridge Record Office (hereafter CRO), St. Benedict's parish burial book, 25/1/22, 1893–1906, Crumpsall workhouse, Manchester bodies buried together.
117. Ibid.
118. Refer, recently, on lives of workhouse long-term residents, N. Goose (2005), 'Poverty, Old Age and Gender in 19th Century England: the Case of Hertfordshire', *Continuity and Change*, XX, pp. 43–76.
119. Cross reference to source(s) details in Graph 7.3 for three keys ways that the economy of supply has been painstakingly reconstituted to ensure accuracy.
120. TNA, MH12/6080, Manchester Union including Crumpsall workhouse to LGB memorandum no. 75119/88, calculations made by Narne Son & Pollitt, Chartered Accountants, 10 Marsden Street, Manchester, 10 August 1888, re 'Burial'.
121. A. McDougall (1885), *Further inquiries into the causes of pauperism in the township of Manchester: a paper read before guardians* (Manchester: Charles Sever Printer), pp. 1–24, Crumpsall mortality statistics at pp. 13–18.
122. Ibid., p. 16.
123. Ibid.
124. Ibid., p. 15–16.
125. Ibid., p. 13.
126. Oxfordshire Record Office (hereafter ORO), C/ENG/1/a1/2, Botley Cemetery Records, 1894–1938, entries for 1908 from Crumpsall workhouse sold to Oxford medical school and buried by anatomists.
127. 'Manchester Health Profile 2009 produced annually by the Association of Public Health Observatories', funded by the Department of Health, source National Statistics website, www.statistics.gov.uk.

Conclusion

1. Extract from Lord Byron (1866 English edn), "Aurora Leigh", p. 34, cited in the 'Statistical Review of Ten Years of Disease in Manchester and Salford: being an analysis of the week', *British Medical Journal*, 518, II, pp. 597–99, q. at p. 598.

2. See, notably, M. J. Cherry (2005), *Kidneys for Sale by Owner: Human Organs, Transplantation and the Market* (Washington, DC: Georgetown University Press); M. Goodwin (2006), *Black Markets: The Supply and Demand of Body Parts* (Cambridge: Cambridge University Press); C. Walby and R. Mitchell (2006), *Tissue Economies: Blood, Organs and Cell Lines in Late Capitalism* (New York: Duke University Press); D. Dickenson (2009), *Body Shopping: Converting Body Parts to Profit – The Economy Fuelled by Flesh and Blood* (London: One World Publications).
3. A. Cheney (2006), *Body Brokers: Inside America's Underground Trade in Human Remains* (New York: Broadway Books Random House), cited in *The Indian Journal for Medical Ethics*, 4 March 2010, I, pp. 1–2, q. at p. 1, book reviews. The American Uniform Anatomical Gift Act (1986, 1987) outlaws trading bodies but brokers are allowed to charge storage and transportation costs, and this is how body profits are made legal.
4. 'Body Shop with a Human Head on Sale', *Daily Mail*, 29 May 2010, p. 54.
5. Refer M. Z. Bookman and K. R. Bookman (2007), *Medical Tourism in Developing Countries* (Basingstoke and New York: Palgrave Macmillan).
6. See S. Wilkinson (2003), *Bodies for Sale: Ethics and Exploitation in the Human Body Trade* (London: Routledge).
7. C. Green (2008), 'Body Parts for Sale: Return of the BodySnatcher', *BBC Focus Magazine*, II, pp. 63–5.
8. See bio-ethics debates in A. Buchanan, D. W. Brook, N. Daniels, D. Wilker (2001 edn), *From Chance to Choice: Genetics and Justice* (Cambridge: Cambridge University Press).
9. R. Richardson (2001 edn), *Death, Dissection and the Destitute* (Chicago: Chicago University Press).
10. Bodleian Library, University Archives, Anatomy Department Records, HA 105–3, loose leaf private letter from Alexander McPhial to Professor Arthur Thomson, 2 March 1928.
11. See series of recent articles in the *Evening Standard* by David Cohen, 'My baby boy was buried with 13 others in a pit left open for months', 18 March 2010; 'Fox digs up baby in paupers' grave in London', 18 March 2010; 'Policeman told us: A fox has taken your dead baby', 29 March 2010; 'The Dispossessed: Couple make first visit to son's mass grave after fox dug up body', 30 March 2010.
12. Quoted in V. Glendinning (1983 edn), *Edith Sitwell, A Unicorn Among Lions*, (Oxford: Oxford University Press), p. 132.
13. A. Macalister (1908), 'Fifty Years of Medical Education', *British Medical Journal*, Vol. 2, no. 2492, X, pp. 957–60, quote at p. 960.

Select Bibliography

Primary Research

Berkshire Record Office
Bradfield Union, Correspondence, 1845–1841, Tilehurst parish, D/P 132/19/7

Bodleian Library, Oxford
University Archives, Anatomy Department Records, HA/1/1–3, 1885–1929, HA50, HA51, HA64, HA89, HA105/1–2, HA107

British Parliamentary Papers
www.parliament.uk/parliamentary.../parliamentary.../archives_electronic.cfm – 1831–32 (419), 2 William IV (Session, 1831–2) – *A Bill [as amended on the second re-commitment] For Regulating the Schools of Anatomy, ordered by the House of Commons to be printed 8 May, 1832*, ref: 419, pp. 1–6

An Act to Amend the Regulating Schools of Anatomy [34 & 35 Vict. C. 16, 1871]

(1893), *Royal Commission of the Aged Poor*, Minutes of evidence, Vol. II, Question 3180, p. 179

Inspector's Report (1900), 'Burial of Paupers', *Annual Local Government Board Report, 1899–1900* (London: George Edward Eyre and William Spottisswoode), p. 93

www.publications.parliament.uk/pa/.../g/cmhuman.htm, Hansard online, Standing Committee G, 7th sitting, 5 February 2004, column 221

www.parliament.uk/parliamentary.../parliamentary.../archives_electronic.cfm, House of Commons parliamentary papers online (2006), Proquest Information and Learning Company

Cambridge Central Library Local Studies Collection
Popular ballads and poster collection, nineteenth century

Cambridgeshire Record Office
St. Benedict Parish Records (burial), P25/1/21–23
City-centre, Census, 1881

Downing College, Cambridge
Macalister Papers, 1885–1919

Journals and Newspapers
Age
Baner Cymru
Birmingham Daily Post

Select Bibliography 357

Bristol Mercury
British and Foreign Medico-Surgical Review
British Medical Journal
Bury and Norwich Post
Cambridge Review
Chamber's Journal
Cheshire Observer
Contemporary Review
Daily Mail
Daily News
Essex Standard, West Suffolk Gazette and Eastern Counties' Advertiser
Evening Standard
Examiner
Figaro in London
Financial Times Magazine
Fortnightly Review
Funny Folks
Guardian
Household Words
Hull Packet and East Riding Times
Illustrated Police News
Indian Journal for Medical Ethics
Jackson's Oxford Journal
Journal of Anatomy
Lancet
Leeds Mercury
Leicester Chronicle and Mercury
Liverpool Mercury
London Review
Lloyds Weekly Newspaper
Man (published by the Royal Anthropological Institute of Great Britain)
Manchester Evening News
Manchester Guardian
Manchester Times
Medical and Philosophical Commentaries
Medical Society Magazine
Northamptonshire Herald
Northern Star and National Trades Journal of Leeds
Notes of the Proceedings of the Royal Society
Nottingham Evening Post
Observer
Once a Week
Oxford Chronicle
Oxford Magazine
Oxford University Gazette
Pall Mall Gazette
Penny Satirist
Provincial Medical and Surgical Journal
Reynolds Miscellany

Sheffield and Rotherham Independent
Sunday Times
Times
Trewman's Exeter Flying Post

London Metropolitan Archives

Holborn Union, Guardian Minute Books, 1838–1930, and financial records, 1848–1903, HO107/670
St. Mary's Newington, P92/MRY and census returns, 1861
Undertaking records, ACC/1416/47 & BRA898/B/186

Manchester Central Library Local Studies Collection

City-centre Census for Manchester Township, 1841–1891
www.1901censusonline.com/, Manchester, 1901
www.visionofbritain.org.uk/data, Total deaths 1871–1911, Manchester Poor Law Registration District
www.manchester.gov.uk/libraries/arts, *Manchester Archives and Local Studies: A Guide to Poor Law and Workhouse Records*
www.british-history.ac.uk, *British History Online*, 'Crumpsall'
Community History Report (2004) 'St. Michael's Flags and Angel Meadow, Then and Now', produced by the Friends of Angel Meadow
St. Michael's Angel Meadow Burials, 1813–1854, July 1849, MFRR/627
Phipps Park Cemetery, Burial Registers, RC Part, 1883–1891, MFPR/706

Manchester Record Office

Hulme Pauper Letter Collection, M10/808
www.statistics.gov.uk, "Manchester Health Profile 2009 produced annually by the Association of Public Health Observatories" funded by the Department of Health

Northampton Central Library Local Studies Collection

Popular and theatre posters collection, nineteenth century

Northamptonshire Record Office

Bills of supply 4 March 1848 from 'Mary March haberdasher and hatter and funerals neatly furnished Market Place Kettering'
Brackley Union, pasted scraps of pauper letters (most undated), PL/1/37
Brackley Union, letter book, PL/1/47
Brackley Union, correspondence and parish books, 42p/17/3
Brackley Union, Letter Book, 1835–1843
Brixworth Union, ZA2037/I, 'Evidence of Mr Rosebrook Morris, surgeon' (1844)
Great Oxenden, Overseers of the Poor's letters to and from Manchester, 251p/98
Inquest Returns, Dean and Chapter of Peterborough, Box X.88, 1813–1842, The Case of Lucy Ladds, 9 March 1841
Roade Burial Board, Minute Books, 281p/102, recently re-catalogued as 281p/517

Select Bibliography 359

Old Bailey Court Records (www.oldbaileyonline.org)
Rex versus Feist [1858], *The Proceedings of the Old Bailey, London's Central Criminal Court, 1674–1913*, OB case number: t18580222-354, 22 February 1858

Oxfordshire Health Archives
Admission Registers of Medical Inpatients, 1890–1917, RI/I/50
Admission Registers of Surgical In-Patients, 1898–1924, RI/I/58
Registers of Death, 1878–May 1908, RI/9/B1/3–6,
Registers of Death, March 1916–April 1920, RI9/B1/7–9
RI Letter books, 1893–1901, RI/I/23, & 1897–1909, Mary Coolling 14 July 1886
Coroner's accounts and inquests, 1830–1837, RI/I/100(16)
Medical Inpatients Register, 1888–1904, RI/I/49
Admission Registers of Medical Inpatients, 1890–1917, RI/I/50
Admission Registers of Surgical In-Patients, 1898–1924, RI/I/58
Registers of Death, 1878–May 1908, RI/9/B1/3–6
Registers of Death, March 1916–April 1920, RI9/b1/7–9
RI Letter books, 1893–1901 & 1897–1909, RI/I/23

Oxfordshire Record Office
Botley Cemetery Records, ENG/1/A1/2, 1894–1938
City-centre, Census returns, 1871, 1881, 1891
Solicitor's records, MOR./LXXV/3, 5–6, personal file held by Morell, Peel & Gamon, containing marriage settlement of Edward Law Hussey, 1874–1876

Shropshire Local History Society
Popular and political poster, nineteenth century

Southwark Local History Library
Parish of St Mary Newington, 'Legal Papers about the Indictment of Alfred Feist, Workhouse Master, for the Removal of Paupers' Bodies to Guy's Hospital School of Anatomy, 1858'
Lambeth Police Court, Monday, 20 January 1858 and 'Copy of evidence taken by Mr Farnall, Poor Law Inspector, 22 January 1857'

St. Bartholomew's Hospital Trust
Dissection registers, MS/81–6
Order in Council, HAL/20
Hospital Reports (1886)
Hospital Accounts, HA1/23, 1866–1872
Treasurer's Letter Book, 1867–1874, HA10/1
Treasurer and Almoner's Book, 30 July 1874–27 December 1877
Correspondence from Poor Law Unions, MS13/19, 1871
Medical School Cash Payments, MS 42/1, cash book, 1879–1891

The National Archives
Anatomy Inspectorate TNA, MH74/10, MH74/12, MH74/15, MH74/16, MH74/36

Brixworth Union correspondence, MH12/8701, 1878
Oxford Poor Law Union, MH12/9720, Oxford, Case No. 348, LGB Memorandum no. 1638, MH12/9721, 1888
Manchester Poor Law Union (including Crumpsall workhouse), MH12/6042/115 April 1845, MH12/6042/125, fols. 184–7 August 1845, MH12/6042/127 August 1845, MH12/6042/180 May 1845, MH12/6042/579 August 1845
Bennett Street Manchester Sunday School Sick Funeral Society, 1839–1951, FS/15/1402,
Crumpsall workhouse in-correspondence (1851–71), MH13/58
Crumpsall workhouse report dated 1888 sent to LGB, MH12/6080
Poor Law Union records for England and Wales, MH12 series
St. Mary's Newington (burial registers) RG8/42–51
Anatomy Inspectorate Reports to the Home Office, HO45/10062/B2694, 1905
Anatomy Inspectorate Reports to the Home Office, HO144/148/A38382, Prison Commission, Leeds and Wakefield prisons and Medical Department of Yorkshire College, Leeds

Wellcome Trust Collection

Slide Number, L0013321, 'The dissection of a young beautiful woman directed by J. CH. G. Lucas (1814–1885) in order to determine the ideal female proportions', chalk drawing by J. H. Hasselhorst, (1864)
Slide L0002687, 'Photograph of the interior of the Department of Anatomy at Cambridge University, 1888'
Slide L0013441, 'Photograph of Edinburgh University dissection room, 1889'
Slide L0014980, 'Photograph of The Dissection Room, Medical School, Newcastle Upon Tyne, 1897'
Slide L0039195, 'Photograph of The Interior of a dissecting room: five students and/or teachers dissect a corpse at University College, London'

Pre-1900 published sources

Anon. (1818, 2nd edn), *The Hospital Pupil's Guide* (London: E. Cox and Sons)
Anon. (1851), *The Overseers Handbook* (London: John Murray)
Anon. (1850), *Slatter's Oxford Trade Directory* (Oxford: Parker & Co.)
Anon. (1858 edn), *Oxford English Dictionary* (Oxford: Oxford University Press)
Anon (1858 edn), *Webster's Medical Dictionary* (London: John Murray)
Anon. (1884, 1969 edition), *Manchester in 1844: Its Present Condition and Future Prospects by Leon Faucher, Translated from the French with Copious Notes Appended by a Member of the Manchester Anthenaeum* (Manchester: Franck Cass and Co. Ltd) [first published by Abel Heywood: Manchester, 1884]
Anon. (1886), *St. Bartholomew's Hospital Reports*, Vol. XXII (London: Hospital)
Anon. (1882), *London Winter Temperatures from 1858 to 1881* (London: John Murray)
Bellars, J. (1714), *Essay Towards the Improvement of Physick in Twelve Proposals by which the Lives of Many Thousands of the Rich as well as the Poor may be SAVED Yearly* (London: J. Sowle)
Cunningham, G. W. (1881, 4th edn), *The Law Relating to Burial of the Dead: Including the Burial Acts, 1852 to 1871* (London: Shaw and Sons)

Darwin, C. [1758–1778] (1780), *Experiments Establishing a Criterion between Mucaginous and Purulent matter. And an Account of the Retrograde Motion from the Absorbent Vessels of Animal Bodies in Some Diseases* (Edinburgh: Green)

Dearsly, H. R., and T. Bell (1858), *Crown Cases Reserved for Consideration, and Decided by the Judges of England* (London: Stevens & Norton, H. Sweet & W. Maxwell)

Engels, F. (1887 edn), *Conditions of the Working Class of England (written September 1844 to March 1845) Published First in Leipzig 1845, Authorised English Editions* (London, 1887 and New York, 1891)

Freeling, A. (1838), *The Railway Companion from London to Birmingham, Liverpool and Manchester, etc.* (London: Green)

Gaskell, E. (1840), *Mary Barton: A Tale of Manchester Life 1840* (London: Chapman and Hall)

Gaulter, H. (1833), *The Origins and Progress of the Malignant Cholera in Manchester* (London: Longman, Rees, Orme, Brown, Green & Longman)

George, H. (1879), *Poverty and Progress: An Inquiry into the Cause of Industrial Depressions and of Increase of Want with Increase of Wealth – The Remedy* (London: Doubleday)

Hearne, T. (1884), *Remarks and Collections, 1705–1735*, 11 vols. (Oxford: Oxford Historical Society Publication), Vol. 1

Hussey, E. L. (1856), *Analysis of Cases of Amputations of the Limbs in the Radcliffe Infirmary, Oxford*, 2nd edn, (London: John Murray & Co)

Hussey, E. L. (1882), *Miscellanea Medico-Chirurgica: Cases in Practice, Reports, Letters & Occasional Papers* (Oxford: Pickard, Hall and Stacy)

Hussey, E. L. (1883), *Extracts from Various Authors and Fragments of Table-Talk* (Oxford: Parker & Co.)

Hussey, E. L. (1887), *Miscellanea Medico-Chirugica*, 2nd part, *Occasional Papers and Remarks* (Oxford: Parker & Co.)

Hussey, E. L. (1894) *Miscellanea Medico-Chirurgica:* 3rd part, *Occasional Papers and Remarks* (Oxford: Horace Hart)

Kahn, J. (1855), *The Evangel of Human Nature; Being Fourteen Lectures, on the Various Organs of the Human Frame, in Health and Disease* (London: James Gilbert)

Lee, S. (ed.) (1860), *The Dictionary of National Biography*, Vol. 60 (London: John Murray)

S. Lee (ed.) (1901), *Dictionary of National Biography*, Supplement 3 (London: Smith, Elder & Co.)

Lewis, S. (ed.) (1848), *A Topographical Dictionary of England* (London: Institute of Historical Research Publication)

Macalister, A. (1891), *The History of the Study of Anatomy in Cambridge* (Cambridge: Cambridge University Press)

Macalister, A. (1891) *A History of the Study of Anatomy at Cambridge: A Lecture* (Cambridge: Cambridge University Press)

McDougall, A. (1885), *Further Inquiries into the Causes of Pauperism in the Township of Manchester: A Paper Read Before Guardians* (Manchester: Charles Sever Printer)

Nettleship, E. (FRCS) (1875), 'The State of the Eyelids in the Orphan and Deserted Children as Compared to Those Who Have Parents', *4th Annual Report of the Local Government Board Report, 1874–5* (London: George Edward Eyre and William Spottiswoode), pp. 113–131

Storey, J. (1895), *Historical Sketches of Some Principal Works and Undertakings of the Council of the Borough of Leicester by the Town Clerk, September 1874 to October 1894* (Leicester: Leicester Publishers)
Tennyson, A. (1859 8th edn), *In Memoriam* (London: Edward Moxon & Co.)
Thomson, A. (1899), *Handbook of Anatomy for Arts Students* (Oxford: Clarendon Press)
A. Thomson (1899), *Anatomy for Art Students* (Oxford: Clarendon Press)
Thompson, J. (1886), *The Owens College, Its Foundation and Growth* (Manchester: Cornish Press)
Turner, T. (1824), *Outlines of a System of Medico-Chirurgical Education Containing Illustrations of the Application of Anatomy, Physiology, and Other Sciences to the Principal Practical Points in Medicine and Surgery* (Manchester: Underwood)
Walker, G. A. (1839), *Gatherings from Graveyards: Particularly those of London. With a Concise History of the Modes of Interment Among Different Nations and a Detail of the Results Produced by the Custom of Exhuming the Dead Amongst the Living* (London: Nottingham Publishers)

Secondary Research

Alberti, S. J. M. M. (2008), 'Wax Bodies: Art and Anatomy in Victorian Medical Museums', *Museum History Journal*, 2, I, pp. 7–36
Alborn, T. (2001), 'Senses of Belonging: The Politics of Working-Class Insurance in Britain, 1880–1914', *Journal of Modern History*, CXXIII, IX, pp. 561–602
Anderson, T. P. (2006), *Performing Early Modern Trauma from Shakespeare to Milton* (Aldershot: Ashgate)
Anning, S. T., and W. J. K. Walls (1982), *A History of the Leeds School of Medicine: One and a Half Centuries, 1831–1981* (Leeds: Leeds University Press)
Anon. (1908), *The Victoria University of Manchester Medical School* (Manchester: Manchester University Press)
Archer, A. (2008), 'Mysterious and Suspicious Deaths: Missing Homicides in North-West England, 1850–1900', *Crime, Historie et Societies/Crime, History and Societies*, XII, pp. 45–63
Avery, G., and K. Reynolds (eds) (2000), *Representations of Childhood Death* (Basingstoke and New York: Palgrave Macmillan)
Bailey, V. V. (2000), *This Rash Act: Suicide Across the Life Cycle of the Victorian City* (Stanford: Stanford University Press)
Barclay-Smith, E. (1919), 'In Memoriam: Professor Alexander Macalister M.D. F.R.S., etc., 1844–1919', *Journal of Anatomy*, 54, I, pp. 96–9
Barclay-Smith, E. (1937), *The First Fifty Years of the Anatomical Society of Great Britain and Ireland: A Retrospect* (London: John Murray)
Barnard, S. V. (1990), *To Prove I'm Not Forgot: Living and Dying in the Victorian City* (Manchester: Manchester University Press)
Bartlett, E. T. (1995), 'Differences between Death and Dying', *Journal of Medical Ethics* XXI, pp. 270–6
Bashford, A. (1998), *Purity and Pollution: Gender, Embodiment and Victorian Medicine* (Basingstoke: Macmillan Press Ltd)
Bates, A. W. (2008), '"Indecent and Demoralising Representations": Public Anatomy Museums in Mid-Victorian England', *Medical History*, 52, I, pp. 1–22

Begg, P. (2009), *Jack the Ripper: The Facts* (London: Portico Publishers)
Bellamy, C. (1988), *Administering Central-Local Relations: The Local Government Board in Its Fiscal and Cultural Context* (Manchester: Manchester University Press)
Bellamy, J. (1988), 'Barriers of Silence: Women in Victorian Fiction', in E. M. Sigsworth (ed.), *In Search of Victorian Values: Aspects of Nineteenth Century Thought and Society* (Manchester: Manchester University Press), pp. 131–46
Bellamy, L., and T. Williamson (eds) (1999), *Life in the Victorian Village: The Daily News Survey of 1891*, Vol. 1 (London: Caliban Books)
Benson, J. (1989), *The Working Class in Britain, 1850–1939* (London: Longman)
Bonner, T. (both 1995 and 2000 edns cited), *Becoming a Physician: Medical Education in Britain, France, Germany and the United States, 1750–1945* (Oxford and Baltimore: Oxford University Press and Johns Hopkins Press)
Bookman, M. Z., and K. R. Bookman (2007), *Medical Tourism in Developing Countries* (Basingstoke and New York: Palgrave Macmillan)
Boot, H. M. (1990), 'Unemployment and Poor Relief in Manchester, 1845–50', *Social History*, 15, II, pp. 217–28
Borgstedt, T. (2009), *Topology of the Sonnet: Theory and History of its Genre* (Verlag Germany: Max Niemeyer, English translation)
Bowler, P. J. (1996), *Charles Darwin* (Cambridge: Cambridge University Press)
Brock, M. G., and M. C. Curthoys (eds) (1997), *The History of the University of Oxford*, Vol. VI, *19th Century, Part One* (Oxford: Clarendon Press)
Brockbank, E. M. (1936), *The Foundation of Provincial Medical Education in England and of the Manchester School in Particular* (Manchester: Manchester University Press)
Brodie, M. (2003), 'Free Trade and Cheap Theatre: Sources of Politics for the 19th-Century London Poor', *Social History*, 28, III, pp. 346–60
Brodie, M. (2004), *The Politics of the Poor: The East End of London, 1885–1914*, (Oxford: Oxford University Press)
Browne, J. (2003 edn) *Charles Darwin: The Power of Place* (New York and London: Pimlico)
Buchanan, A., D. W. Brook, N. Daniels, D. Wilker (2001 edn), *From Chance to Choice: Genetics and Justice* (Cambridge: Cambridge University Press)
Buklijas, T. (2008), 'Cultures of Death and Politics of Corpse Supply: Anatomy and Vienna, 1848–1914', *Bulletin of the History of Medicine*, 82, III, pp. 570–607
Burgh, D. (2007), *Digging Up the Dead: Uncovering the Life and Times of an Extraordinary Surgeon* (London: Random House)
Burnby, J. G. L. (1983), *A Study of the English Apothecary from 1660–1760* (London: Medical History, Supplement III, Wellcome Inst. Hist. Medicine)
Burney, I. (1996), 'Making Room at the Public Bar: Coroners' Inquests, Medical Knowledge and the Politics of the Constitution in Early Nineteenth-Century England', in James Vernon (ed.), *Re-reading the Constitution: New Narratives in the Political History of England's Long Nineteenth Century* (Cambridge: Cambridge University Press), pp. 123–53
Burney, I. (2000), *Bodies of Evidence: Medicine and the Politics of the English Inquest, 1830–1926* (Baltimore: Johns Hopkins University Press)
Burrell, S., and G. V. Gill (2005), 'The Liverpool Cholera Epidemic of 1832 and Anatomical Dissection – Medical Mistrust and Civil Unrest', *Journal of the History of Medicine and Allied Sciences*, 60, IV, pp. 478–98

Bury, P. (1952), *The Colleges of Corpus Christi and the Blessed Virgin Mary: A History, 1922–1952*, (Cambridge: Cambridge University Press), pp. 278–89

Busteed, M. A., and R. I. Hodgson (1994), 'Irish Migration and Settlement in Early Nineteenth Century Manchester, with Special Reference to the Angel Meadow District in 1851', *Irish Geography*, XXVII, pp. 1–13

Busteed, M. A. and R. I. Hodgson (1996) 'Irish Migrant Responses to Urban Life in Early Nineteenth Century Manchester', *Geographical Journal*, 162, II, pp. 139–53

Cannadine, D. (1981), 'War and Death, Grief and Mourning in Modern Britain' in J. Whalley (ed.), *Mirrors of Mortality: Studies in the Social History of Death* (London: Europa), pp. 187–241

Charlton, H. B. (1952), *Portrait of a University 1851–1951: To Commemorate the Centenary of Manchester University* (Manchester: Manchester University Press)

Cheney, A. (2006), *Body Brokers: Inside America's Underground Trade in Human Remains* (New York: Broadway Books Random House)

Cherry, M. J. (2005), *Kidneys for Sale by Owner: Human Organs. Transplantation and the Market* (Washington, DC: Georgetown University Press)

Childs, J. (2006), *Henry VIII's Last Victim: The Life and Times of Henry Howard, Earl of Surrey* (London: Vintage)

Clarke, J. M. (1995), *The Brookwood Necropolis Railway, Locomotion Papers No. 143* (London: Oakwood Press)

Condrau, F., and M. Worboys (2009), 'Second Opinion: Final Response – Epidemics and Infections in Nineteenth-Century Britain', *Social History of Medicine*, 22, I, pp. 165–171

Cooter, R. (1983), *Surgery and Society in Peace and War: Orthopaedics and the Organisation of Modern Medicine, 1880–1948* (Basingstoke and New York: Palgrave Macmillan)

Crook, T. (2008), 'Accommodating the Outcast: Common Lodging Houses and the Limits of Urban Governance in Victorian and Edwardian London', *Urban History*, 35, III, pp. 414–36

Curl, J. (2000 edn), *The Victorian Celebration of Death* (London: History Press Ltd).

Deech, R., and A. Smaidor (2007), *From IVF to Immortality: Controversy in the Era of Reproductive Technology* (Oxford: Oxford University Press)

Desmond, A., and J. Moore (1992), *Darwin* (Chicago: Chicago University Press)

Dickenson, D. (2009), *Body Shopping: Converting Body Parts to Profit – The Economy Fuelled by Flesh and Blood* (London: One World Publications)

Digby, A. (1994), *Making a Medical Living: Doctors and Patients in the English Market of Medicine, 1720–1911* (Cambridge: Cambridge University Press)

Dijkstra, B. (1986), *Idols of Perversity: Fantasies of Feminine Evil in Fin-de-Siècle Culture* (Oxford: Oxford University Press)

Dollimore. J. (2000), *Death, Desire and Loss in Western Culture* (New York and London: Routledge)

Douglas, M. (1966), *Purity and Danger: An Analysis of the Concepts of Pollution and Taboo* (London: Ark Paperbacks)

Dunbabin, J. P. D. (1978), 'Oxford College Finances: A Reply', *Economic History Review*, 2nd series, XXII, I, p. 439 [a one page rejoinder]

Durey, M. J. (1976), 'Body Snatchers and Benthamites: The Implications of the Dead Body Bill for the London Schools of Anatomy, 1820–42', *London Journal*, II, pp. 200–25

Eliot, S. (2006), 'What Price Poetry?: Selling Wordsworth, Tennyson, and Longfellow in Nineteenth and Early Twentieth-Century Britain', *Papers of the Bibliographical Society of America*, 100, IV, pp. 425–45
Emmons, R. A., and M. E. McCullogh (eds) (2004), *The Psychology of Gratitude*, (Oxford: Oxford University Press)
Englander, D. (1998), *Poverty and Poor Law Reform in 19th Century Britain, 1834–1914: From Chadwick to Booth* (New York: Longman)
Englander, D. (2000), 'From the Abyss: Pauper Petitions and Correspondence in Victorian London', *London Journal*, 25, I, pp. 71–83
Engle, A. J. (1978), 'Oxford College Finances, 1871–1913: A Comment', and J. P. D. Dunbabin, 'Oxford College Finances: A Reply', *Economic History Review*, 2nd series, XXI, I, p. 439 [one page rejoinder]
Eyler, J. M. (1997), *Sir Arthur Newsholme and State Medicine, 1885–1935* (Cambridge: Cambridge University Press)
Flanders, J. (2003), *The Victorian House: Domestic Life from Childbirth to Deathbed* (London: Harper Collins)
Fortley, R. (2008), *Dry Store Room No. 1: The Secret Life of the Natural History Museum* (London: Harper Perennial)
Franklin, K. J. (1936), 'A Short Sketch of the History of the Oxford Medical School', *Annals of Science* I, pp. 431–46
French, S. (1978), *The History of Downing College, Cambridge*, Vol. Two, *1888–1894*, (Cambridge: Cambridge University Press)
Fumerton, P. (2006), *Unsettled: The Culture of Mobility and the Working Poor in Early Modern England* (Chicago: Chicago University Press)
Garrett, E., A. Reid, K. Schurer and S. Szreter (eds) (2001), *Changing Family Size in England and Wales: Place, Class and Demography, 1891–1911* (Cambridge: Cambridge University Press)
Gattrell, V. A. C., B. Lenman and G. Parker (eds) (1980), *Crime and the Law: The Social History of Crime in Western Europe since 1500* (London: Europa)
Gestrich, A., L. Raphael, and S. A. King (eds) (2006), *Being Poor in Modern Europe: Historical Perspectives 1800–1940* (Oxford: Peter Lang)
Gilbert, S. M. (2006), *Death's Door: Modern Dying and the Ways We Grieve* (New York: Norton and Co)
Gittings, C., and P. C. Jupp (eds) (1999), *Death in England: An Illustrated History*, (Manchester: Manchester University Press)
Glendinning, V. (1983 edn), *Edith Sitwell, A Unicorn Among Lions* (Oxford: Oxford University Press)
Goodwin, M. (2006), *Black Markets: The Supply and Demand of Body Parts* (Cambridge: Cambridge University Press)
Goose, N. (2005), 'Poverty, Old Age and Gender in 19th Century England: The Case of Hertfordshire', *Continuity and Change*, XX, pp. 43–76
Gordon, D. I. (1968), *A Regional History of the Railways of Great Britain*, Vol. 5: *The Eastern Counties* (Newton Abbott, Devon: David and Charles Publishers)
Gray, L. (2006), 'The Early Days of the Anatomical Society of Great Britain and Ireland', *Anastomosis*, I, p. 1–8
Green, C. (2008), 'Body Parts for Sale: Return of the Bodysnatcher', *BBC Focus Magazine*, II, pp. 63–5
Green, D. R. (1995), *From Artisans to Paupers: Economic Change and Poverty in London, 1790–1870* (Aldershot: Ashgate)

Green, D. R. (2006), 'Pauper Protests: Power and Resistance in Early Nineteenth-Century London Workhouses', *Social History*, 31, II, pp. 137–59

Gregory, D. E. (2006), *Victorian Song-Hunters: The Recovery and Editing of English Vernacular Ballads and Folk Lyrics, 1820–1883* (Lanham, MD: Scarecrow Publications)

Hallam, E., J. Hockney and G. Howarth (eds) (1999), *Beyond the Body: Death and Social Identity* (London: Routledge)

Hamrighaus, R. E. (2001), 'Wolves in Women's Clothing: Baby Farming and the British Medical Journal, 1860–72', *Journal of Family History*, XXVI, pp. 350–72

Hardy, A. (1993), *The Epidemic Streets: Infectious Disease and the Rise of Preventative Medicine, 1856–1900* (Oxford: Clarendon Press)

Hartog, P. J. (1901), *The Owens College Manchester (founded 1851): A Brief History of the Colleges and Descriptions of Its Various Departments* (Manchester: JF Cornish Publishers)

Hayes, B. (2009), *The Anatomist: A True Story of Gray's Anatomy* (New York: Bellevue Literary Press)

Higgs, E. (2004), *The Information State in England: The Central Collection of Information on Citizens, 1500–2000* (Basingstoke and New York: Palgrave Macmillan)

Hindle, G. B. (1975), *Provision for the Relief of the Poor in Manchester 1754–1826* (Manchester: Manchester University Press)

Hollen Lees, L. (1979), *Exiles of Erin: Irish Migrants in Victorian London*, (Manchester: Manchester University Press)

Hollen Lees (1998), *The Solidarities of Strangers: The English Poor Laws and the People, 1700–1948* (Cambridge: Cambridge University Press)

Hopwood, N. (1999), '"Giving Body" to Embryos: Modelling, Mechanism and Microtone in Late-Nineteenth Century Anatomy', *Isis*, CXXXX, I, pp. 462–96

Hotz, M. E. (2008), *Literary Remains: Representations of Death and Burial in Victorian England* (New York: State University of New York Press)

Houlbrooke, R. (ed.) (1989), *Death, Ritual and Bereavement* (London: Routledge)

Howarth, J. (1987), 'Science Education in Late-Victorian Oxford: A Curious Case of Failure', *English Historical Review*, CII, pp. 334–71

Howkins, A. (1991), *Reshaping Rural England: A Social History 1850–1929* (London: Routledge)

Howson, J. (ed.) (1902), *British Medical Association – Manchester Meeting 1902 – Handbook and Guide to Manchester* (Manchester: F. Ireland)

Humphreys, R. (1995), *Sin, Organized Charity and the Poor Law in Victorian England* (Basingstoke and New York: Palgrave Macmillan)

Hunt, T. (2004), *Building Jerusalem: The Rise and Fall of the Victorian City* (London: Wedenfield and Nicolson)

Hurren, E. T. (2004), 'A Pauper Dead-House: The Expansion of Cambridge Anatomical Teaching School under the late-Victorian Poor Law, 1870–1914', *Medical History*, 48, I, pp. 69–94

Hurren, E. T. (2005), 'Poor Law versus Public Health: Diphtheria and the Challenge of the "Crusade Against Outdoor Relief" to Public Health Improvements in late-Victorian England, 1870–1900', *Social History of Medicine*, XXVIII, pp. 399–414

Hurren, E. T. and S. A. King (2005), 'Begging for a Burial: Form, Function and Conflict in Nineteenth Century Burial', *Social History*, XXX, 321–41

Hurren, E. T. (2007), *Protesting about Pauperism: Poverty, Politics, and Poor Relief in Late-Victorian England* (Woodbridge: Boydell & Brewer)

Hurren, E. T. (2008), 'Whose Body Is It Anyway? Trading the Dead Poor, Coroner's Disputes and the Business of Anatomy at Oxford University, 1885–1929', *Bulletin of the History of Medicine*, 82, pp. 775–819

Hurren, E. T. (2010), 'Remaking the Medico-Legal Scene: A Social History of the late-Victorian Coroner in Oxford', *Journal of the History of Medicine and Allied Sciences*, 65, IV, pp. 207–252

Hurren, E. T., and I. Scherder (2011), 'Dignity in Death? The Dead Body as an Anatomical Object in England and Ireland c. 1832 to 1929', in A. Gestrich and S. A. King (eds), *Dignity of the Poor* (Oxford: Oxford University Press), pp. 1–30

Jalland, P. (1996), *Death and the Victorian Family* (Oxford: Oxford University Press)

Johnson, P. A. (1986), *Saving and Spending: Working Class Economy in Britain, 1870–1939* (Oxford: Oxford University Press)

Jones, G. D., and M. I. Whitaker (2009), *Speaking for the Dead: Cadavers in Biology and Medicine* (Aldershot: Ashgate)

Jones, N., and P. Jones (1976), *The Rise of the Medical Profession: A Study of Collective Mobility* (London: Croom Helm)

Jupp, P., and G. Howarth (eds) (1997), *The Changing Face of Death: Historical Accounts of Death and Disposal* (Basingstoke and New York: Palgrave Macmillan)

Kargon, R. H. (1977), *Science in Victorian Manchester: Enterprise and Expertise* (Manchester: Manchester University Press)

Kidd, A. J., and K. W. Roberts (eds) (1985), *City, Class and Culture: Studies of Cultural Production and Social Policy in Victorian Manchester* (Manchester: Manchester University Press)

Kidd, A. J. (1985), 'Outcast Manchester: Voluntary Charity, Poor Relief and the Casual Poor, 1860–1905', in A. J. Kidd and K. W. Roberts (eds), *Class, City and Culture: Studies of Cultural Production and Social Policy in Victorian Manchester* (Manchester: Manchester University Press), pp. 48–73

King, P. J. R. (2006), *Crime and Law in England, 1750–1840: Remaking Justice from the Margins* (Cambridge: Cambridge University Press)

King, S. A. (2000), *Poverty and Welfare in England, 1700–1850: a Regional Perspective* (Manchester: Manchester University Press)

King, S. A. (2005, reissued in paperback 2010), *"We Might Be Trusted": Women, Welfare and Local Politics c. 1880–1920* (Brighton: Sussex Academic Press)

King, S. A. (2007), 'Pauper Letters as a Source', *Family & Community History*, 10, II, pp. 167–70

King, S. A. (2008), 'Friendship, Kinship and Belonging in the Letters of Urban Paupers 1800–1840', *Historical Social Research*, 33, III, pp. 249–77

King, S. A. (2011), *The Sick Poor in England, 1750–1850* (Basingstoke and New York: Palgrave Macmillan)

King, S. A., and G. Timmins (2001), *Making Sense of the Industrial Revolution: English Economy and Society 1700–1850* (Manchester: Manchester University Press)

King, S. A., and A. Tomkins (eds) (2003), *The Poor in England 1700–1850: An Economy of Makeshifts* (Manchester: Manchester University Press)

Kirby, R. G., and A. E. Musson (1975), *The Voice of the People: John Doherty, 1798–1854 – Trade Unionist, Radical and Factory Reformer* (Manchester: Manchester University Press)

Jordanova, L. (1998), *Sexual Visions: Images of Gender in Science and Medicine between the Eighteenth and Twentieth Centuries* (Madison, WI: University of Wisconsin Press)

Laqueur, T. (1983), 'Bodies, Death, and Pauper Funerals', *Representations*, I, pp. 109-31

Laqueur, T. (1994), 'Cemeteries, Religion and the Culture of Capitalism', in J. A. James and M. Thomas (eds), *Capitalism in Context: Essays on Economic Development and Cultural Change in Honour of R. M. Hartwell* (Chicago: Chicago University Press), pp. 138-55

Lawrence, C. (1994), *Medicine in the Making of Modern Britain, 1700-1920*, (London: Routledge)

Lawrence, J. (1998), *Speaking for the People: Party, Language and Popular Politics in England, 1867-1914* (Cambridge: Cambridge University Press)

Lawrence, S. (1991), 'Private Enterprise and Public Interest: Medical Education and the Apothecaries' Act, 1780-1825', in R. K. French and A. Wear (eds), *British Medicine in the Age of Reform* (London: Routledge), pp. 45-73

Lawrence, S. (1993), 'Educating the Senses: Students, Teachers and Medical Rhetoric in Eighteenth Century London', in W. F. Bynum and R. Porter (eds), *Medicine and the Five Senses* (Cambridge: Cambridge University Press), pp. 154-78

Lawrence, S. (1996), *Charitable Knowledge: Hospital Pupils and Practitioners in Eighteenth Century London* (Cambridge: Cambridge University Press)

Litton, J. (1998), 'The English Funeral, 1700-1850', in M. Cox (ed.) *Grave Concerns: Death and Burial in England, 1700-1850* (York: Council for British Archaeology), pp. 3-16

Litton, J. (2002), *The English Way of Death: The Common Funeral since 1450* (London: Robert Hale Ltd)

Livingstone, D. (2007), 'Science, Site and Speech: Scientific Knowledge and the Spaces of Rhetoric', *History of Human Sciences*, XX, I, pp. 71-98

Loudon, I. (1986), *Medical Care and the General Practitioner, 1750-1850* (Oxford: Oxford University Press)

Loudon, I. (1992), 'Medical Practitioners and Medical Reform in Britain 1750-1850', in A. Wear (ed.), *Medicine in Society: Historical Essays* (Cambridge: Cambridge University Press), pp. 219-37

Loudon, I. (6 September 1996), 'Obstetric Care, Social Class and Maternal Mortality', *British Medical Journal*, Vol. 293, No. 6457, pp. 606-7

Macalister, A. (1908), 'Fifty Years of Medical Education', *British Medical Journal*, 2492, II, pp. 957-60

MacDonald, H. (2006), *Human Remains: Dissection and Its Histories* (New Haven and London: Yale University Press)

MacDonald, H. (2009), 'Procuring Corpses: The English Anatomy Inspectorate, 1842-1858', *Medical History*, 53, III, pp. 379-96

Mason, M. (1994), *The Making of Victorian Sexuality* (Oxford: Oxford University Press)

Maude, W. C. (1903), *The Poor Law Handbook* (London: Poor Law Officers Journal)

Maulitz, R. (1987), *Morbid Appearances: The Anatomy of Pathology in the Early Nineteenth Century* (Cambridge: Cambridge University Press)

McGregor, A. (2007), *Curiosity and Enlightenment: Collectors and Collections from the Sixteenth to the Nineteenth Century* (New Haven and London: Yale University Press)

Medvei, V., and J. Thornton (1974), *The Royal Hospital of St. Bartholomew's, 1123–1973* (London: Hospital Publications)

Meyers, D. W. (2006 edn), *The Human Body and the Law: A Medico-Legal Study* (Chicago: Aldine Transactions)

Mooney, G. (2007), 'Infectious Diseases and Epidemiologic Transition in Victorian Britain? Definitely', *Social History of Medicine*, XX, pp. 595–606

Morrell, J., and A. Thackray (1981), *Gentlemen of Science: The Early Years of the British Association for the Advancement of Science* (Oxford: Oxford University Press)

Moscucci, O. (1999), *The Science of Women: Gynaecology and Gender in England, 1800–1929* (Cambridge: Cambridge University Press)

Nead, L. (1988), *Myths of Sexuality: Representations of Women in Victorian Britain* (London: Willey Blackwell)

Nutton, V., and R. Porter (eds) (1995), *The History of Medical Education in Britain* (Amsterdam: Rodopi)

Oleesky, S. (1971), 'Julius Dreschfeld and late-19th Century Medicine in Manchester', *Manchester Medical Gazette*, CI, pp. 14–17

Paget, S. (ed.) (1901), *Memoirs and Letters of Sir James Paget* (London: Longman, Green & Co.)

Parsons, M. (2002), *Yorkshire and the History of Medicine* (York: Sessions of York)

Pelling, M. (1998), *The Common Lot: Sickness, Medical Occupations and the Urban Poor in Early Modern England* (London: Longman)

Pelling, M. (2006), 'Corporatism or Individualism: Parliament, the Navy, and the Splitting of the London Barber-Surgeons' Company in 1745', in I. A. Gadd and P. Wallis (eds), *Guilds and Association in Europe, 900–1900* (London: Centre for Metropolitan History, Institute of Historical Research), pp. 57–82

Pelling, M., and F. White (2003), *Medical Conflicts in Early Modern London: Patronage, Physicians and Irregular Practitioners, 1550–1640* (Oxford: Oxford University Press)

Peterson, M. J. (1978), *The Medical Profession in Mid-Victorian London* (Berkley: University of California Press)

Picard, L. (2005), *Victorian London: The Life of a City, 1840–1870* (London: Orion)

Pickstone, J. (1985), *Medicine in Industrial Society: A History of Hospital Development in Manchester and its Regions, 1752–1946* (Manchester: Manchester University Press)

Platt, H. L. (2005), *Shock Cities: The Environmental Transformation of Manchester and Chicago* (Chicago: Chicago University Press)

Porter, R. (1987), 'The Patient's View: Doing Medical History from Below', *Theory and Society*, XIV, pp. 167–74

Porter, R. (1989), *Health for Sale: Quackery in England, 1660–1850* (Manchester: Manchester University Press)

Porter, R. (1997), *The Greatest Benefit to Mankind: A Medical History of Humanity from Antiquity to the Present* (London: Harper Collins)

Porter, R. (2001), *Bodies Politic: Disease, Death and Doctors in Britain, 1650–1900* (London: Reaktion)

Porter, R., and M. Teich (eds) (1994), *Sexual Knowledge, Sexual Science* (Cambridge: Cambridge University Press)
Redford, A. (1940), *The History of Local Government in Manchester*, Vol. II: *Borough and City* (London: Longman and Green & Co.)
Reinarz, J. (2005), 'The Age of Museum Medicine: The Rise and Fall of the Medical Museum at Birmingham's School of Medicine', *Social History of Medicine*, 18, III, pp. 419–37
Richardson, R. (1991), ' "Trading Assassins" and the Licensing of Anatomy', in R. French and A. Wear (eds), *British Medicine in the Age of Reform* (Cambridge: Cambridge University Press), pp. 74–91
Richardson, R. (2001 edn) *Death, Dissection and the Destitute* (London: Phoenix Press)
Richardson, R. (2006), 'Human Dissection and Organ Donation: a Historical and Social Background', *Mortality*, 11, II, pp. 151–65
Richardson, R. (2008), *The Making of Mr. Gray's Anatomy: Bodies, Books, Fortune and Fame* (Oxford: Oxford University Press)
Rivière, P. (ed.) (2007), *A History of Oxford Anthropology* (Oxford: Oxford University Press)
Robb-Smith, A. H. T. (1968), 'The Development and Future of Oxford Medical School', *Transactions of the Society of Occupational Medicine*, XXVIII, pp. 13–21
Robb-Smith, A. H. T. (1970), *A Short History of the Radcliffe Infirmary* (Oxford: The Church Army Press, for the United Hospitals)
Roberts, J. (1979), *Working Class Housing in Nineteenth Century Manchester: The Example of John Street, Irk Town* (Manchester: Richardson Press)
Rolleston, H. D. (1932), *The Cambridge Medical School: A Biographical History*, (Cambridge: Heffer)
Rose, M. E. (ed.) (1985), *The Poor and the City: The English Poor Law in its Urban Context* (Leicester: Leicester University Press)
Roter, D. L., and J. A. Hall (2006), *Doctors Talking with Patients/Patients Talking with Doctors: Improving Communication in Medical Visits* (Westport, CT: Praeger Publishers)
Rowell, G. (1974), *Hell and the Victorians* (Oxford: Oxford University Press)
Rugg, J. (1998), 'A New Burial and Its Meanings: Cemetery Establishment in the First Half of the 19th Century', in M. Cox (ed.), *Grave Concerns: Death and Burial in England 1700–1850*, (London: Council for British Archaeology), pp. 44–54
Sandel, M. J. (2007), *The Case Against Perfection: Ethics in the Age of Genetic Engineering* (Cambridge, MA: Harvard University Press)
Sappol, M. (2002), *A Traffic of Dead Bodies: Anatomy and Embodied Social Identity in 19th Century America* (Princeton: Princeton University Press)
Schoenfeldt, M. (ed.) (2007), *A Companion to Shakespeare's Sonnets* (Oxford: Blackwell Publishers Ltd)
Sen, A. (1992), *Inequality Reexamined* (Oxford: Oxford University Press)
Sinclair, H. M., and A. H. T. Robb-Smith (1950), *A History of Anatomy in Oxford* (Oxford: Clarendon Press)
Snell, K. D. M. (2006), *Parish and Belonging: Community, Identity and Welfare in England and Wales, c. 1700–1950* (Cambridge: Cambridge University Press)

Snell, K. D. M., and P. S. Ell (2000), *Rival Jerusalems: The Geography of Victorian Religion* (Cambridge: Cambridge University Press)

Sokoll, T. (2001 edn), *Essex Pauper Letters, 1731–1837* (Oxford: Oxford University Press)

Sokoll, T. (2006), 'Writing for Relief: Rhetoric in English Pauper Letters, 1800–1834', in A. Gestrich, S. A. King, and L. Raphael (eds), *Being Poor in Modern Europe: Historical Perspectives, c. 1840–1900* (Oxford: Peter Lang), pp. 91–112

Sponza, L. (1988), *Italian Immigrants in Nineteenth Century Britain: Realities and Images* (Leicester: Leicester University Press)

Stephenson, J. S. (1985), *Death, Grief and Mourning: Individual and Social Realities*, (New York: Free Press)

Strange, J-M. (2000), 'Menstrual Fictions: Languages of Medicine and Menstruation, 1850–1930', *Women's History Review*, 9, III, pp. 607–28

Strange, J-M. (2000), 'Review Article – Death and Dying: Old Themes and New Directions', *Journal of Contemporary History*, 35, III, pp. 491–9

Strange, J-M. (2003), 'Only A Pauper Whom Nobody Owns: Reassessing the Pauper Grave c. 1880–1914', *Past & Present*, CLXXVIII, pp. 148–75

Strange, J-M. (2005), *Death, Grief and Poverty in Britain, 1870–1914* (Cambridge: Cambridge University Press)

Syfret, R. H. (1950), 'Some Early Reactions to the Royal Society', *Notes Rec. Royal Society of London*, VII, pp. 207–58

Szreter, S. (1994), 'Mortality in England in the 18th and 19th Centuries: A Reply to Sumit Guha', *Social History of Medicine*, 7, II, pp. 269–82

Tarlow, S. (2007), *The Archaeology of Improvement in Britain, 1750–1850* (Cambridge: Cambridge University Press)

Taylor, H. (1998), 'Rationing Crime: The Political Economy of Criminal Statistics since the 1850s', *English History Review*, 3rd series, IX, pp. 569–90

Thomson, A. (1912), *The Anatomy of the Human Eye, as Illustrated by Enlarged Stereoscopic Photographs* (Oxford: Clarendon Press)

Vaidya, J. H. (1921), 'A Point in Favour of Professor Arthur Thomson's Theory of the Production of Glaucoma', *The British Journal of Opthalmology*, 5, IV, pp. 172–5

Vincent, D. (1980), 'Love and Death in the 19th Century Working Class', *Social History*, V, pp. 223–47

Waddington, K. (2003), *Medical Education at St. Bartholomew's Hospital 1123–1995* (Woodbridge: Boydell and Brewer)

Walby, C., and R. Mitchell (2006), *Tissue Economies: Blood, Organs and Cell Lines in Late Capitalism* (New York: Duke University Press)

Walkowitz J. R. (1992), *City of Dreadful Delight: Narratives of Sexual Danger in Late-Victorian London* (Chicago and London: Chicago University Press)

Walter [pseudonym] (1966), *My Secret Life, 11 Vols., 1888–1894* (New York: Grove Press edn, 2 Vols)

Walter, S. H. L. (2006), *John Milton, Radical Politics, and Biblical Republicanism* (Newark: University of Delaware Press)

Waters, C. (2003), ' "Trading Death": Contested Commodities in "Household Words" ', *Victorian Periodicals Review*, 36, IV, pp. 313–30

Watt, T. (1991), *Cheap Print and Popular Piety, 1550–1640* (Cambridge: Cambridge University Press)

Weatherall, M. W. (1996), 'Making Medicine Scientific: Empiricism, Rationality and Quackery in Mid-Victorian Britain', *Social History of Medicine*, IX, pp. 175–94

Weatherall, M. W. (2000), *Gentlemen, Scientists and Doctors: Medicine at Cambridge, 1800–1940* (Cambridge: Cambridge University Press)

Whaley, J. (ed.) (1981), *Mirrors of Mortality: Studies in the Social History of Death* (New York: St. Martins Press)

White, J. (2008), *London in the 19th Century: A Human Awful Wonder of God* (London: Vintage Books Ltd)

Wilkinson, S. (2003), *Bodies for Sale: Ethics and Exploitation in the Human Body Trade* (London: Routledge)

Williams, K. (1981), *From Pauperism to Poverty* (Manchester: Manchester University Press)

Wilson, D. S. (2003 edn), *Darwin's Cathedral: Evolution, Religion and the Nature of Society* (Chicago: Chicago University Press)

Wise, S. (2005), *The Italian Boy: Murder and Grave-Robbery in 1830s London* (London: Jonathan Cape)

Wise, S. (2008), *The Blackest Street: The Life and Death of a Victorian Slum* (London: Random House)

Wohl, A. S. (1977), *The Eternal Slum: Housing and Social Policy in Victorian London* (London: Transaction Publishers)

Wolffe, J. (2000), *Great Deaths: Grieving, Religion and Nationhood in Victorian and Edwardian Britain* (Oxford: Oxford University Press)

Wrigley, E. A. (2004), 'British Population During the "Long" Eighteenth century 1680–1840', in R. Floud and P. Johnson (eds.), *The Cambridge Economic History of Modern Britain: Vol. 1, Industrialisation* (Cambridge: Cambridge University Press), chapter 3, pp. 57–95

Young, L. (2002), *The Book of the Heart* (London: Harper Collins Publishers)

Unpublished Thesis

Butler, S. V. F. (1981), "Science and the Education of Doctors during the Nineteenth Century: A Study of British Medical Schools with Particular Reference to the Development and Uses of Physiology" (unpublished Ph.D., UMIST)

Hutton, F. (2007), 'Medicine and Mutilation: Oxford, Manchester and the Impact of the Anatomy Act 1832 to 1870 (unpublished Ph.D., Oxford Brookes University) [co-supervised by Dr. E. T. Hurren]

Unpublished Dissertation

Buckley, V. (2007), 'Oxford Medical School in the 19th Century' (unpublished B.A. Diss., Oxford University) [supervised by Dr. E. T. Hurren]

Index

Abnormalities and Deformities register, 131, 200
abortion, 125–7
Addenbrookes Hospital, Cambridge, 102, 187, 198
Addison, Right Hon Dr Christopher MP, 244
alcoholism, 128–9, 248–9
amputation, 137, 140
anatomical burial, after dissection, 43, 51, 57, 67, 71, 190, 192–3, 195, 198, 199–200, 211–14, 237, 276–85, 290–1, 293–4
Anatomical Society of Great Britain and Ireland, 230
Anatomical Supply Committee, also called Anatomical League, and later styled National Anatomical Supply Committee System, 20–4, 239, 241, 243–4
Anatomy Act (1832)
   anatomy and forensic medicine, 44–9, 218–19
   anti-anatomy posters and slang, 24–7, 187, 193–4
   creationism, 95
   general clauses, xvi, 4, 5, 15, 17–18, 22–3, 24–5, 28–31, 77, 184–5, 266
   historiography, 32–7
   images, 104–113
   Inspectorate, 39, 187, 240, 266, 268, 289–90, 293, 298
   museums, 75
   Official Secrets Act (1889), 71, 97
   oral performance of terms, 22–7
   recent legislation, 20–1, 28
   re-reading, 20–32, 37–40
anatomy of human eye, 228–9, 274
Anson, Mr Charles, 46
anti-vaccination, 257–9
Apothecaries Act (1815), 80, 81; *see also* Worshipful Company of Apothecaries

Ashton, Sarah, dissected servant, London, 125–7
asylums, 136, 163, 213

Bashford, Alison, 104–5, 106
Bateson, Dr, medical officer of health, South London, 55
Bellars, Mr John, Quaker, 222
Benbow, Alice, police court witness, London, 98
Bethnal Green Union, London, 62
Biggleswade Poor Law Union, Bedfordshire, 198, 213
biology, 21
biomedical age and internet body business, xvi, 303–5
Birmingham Daily Post, 68–9
Birmingham Poor Law Union, 69, 231, 269
Birmingham, Queen's College anatomy school, 68, 69, 269
Birmingham, Witton Cemetry, pauper burial site, 70–1
Births, Deaths and Marriages Registration Act (1836–7), xvi, 35
Blackfriars, London, 98–101
Brackley Poor Law Union, Northamptonshire, 52–3, 56, 64, 66–7
Bradfield Poor Law Union, Berkshire *see also* Lee, John, 52
Briers, Ms, female body dealer, St. James, London, 172–3
Brighton Poor Law Union, Sussex, 212
British Empire Mutual Life Assurance Company, 99
British Medical Journal, 216, 303, 310
Brixworth Poor Law Union, Northamptonshire
   anti-poor law poster, 27
   sick poor mistreatment, 178–180

Brown, Dr James Johnston, anatomy school, Stamford Street, London, 100
Bull, William, poor law clerk, 12
Bury and Norwich Post, 19

calculative reciprocity, 160–1
Cambridge anatomy school
  basic design and size, 23, 102
  body business, 175–217, 306
  body parts, 185–6, 214
  pauper riots, 34
  photographs, 109–110
  surplus bodies traded, 297–8
Camden Square churchyard, 3
cancer, 129, 179
Carter, Henry Vandyke, illustrator, *Gray's Anatomy*, 1858, 139
Chamber's Journal, 93
Charity Organisation Society, 134
Cheapside, London, pauper cemetery, 92
Chelmsford Poor Law Union, East Anglia, 184
cholera, 273–95
Chorlton Poor Law Union, Lancashire, 288
Christ Church College, anatomy school, Oxford, 219, 223
Church Of England Sunday School Institute, 99
Church, Mary, undertaker, Kettering Union, Northamptonshire, 64
Clarke family of Walsall, West Midlands
  James, father, dissected pauper, 69–70
  Anne, widow, 69
  Mrs Margaret Longmine, estranged daughter, 70
  see also Walsall Poor Law Union
Clarke, Phoebe, 15
  see also Matthews sisters
Clements, Mr Charles, Poor Law Inspector, 280
Colchester Poor Law Union, East Anglia, 207–8
Colney Hatch Asylum, 199
Colney Hatch Cemetery, 149

Common Metropolitan Act (1867), 9
Cooke, Mr Thomas, anatomist
  neighbours prosecute in police court, 98–101
  private anatomy school, Blackfriars, London, 98–101
  surgeon, Westminster Hospital, London, 97–9, 111
Cooling family of Oxford, paupers
  Caroline, grandmother, 258
  Alice, mother, 259
  Mary, cousin, dissected, 259–60, 263
  James, dead infant, 258–9
Cooter, Roger, 82
coroner's cases, 44–8, 55–6, 88–93, 182–4, 205–6, 218–19, 245–52
courtship, failed, 125–7
Croydon Poor Law Union, 207
Crumpsall workhouse, near Manchester, 276, 278, 285, 286–99, 306
Curley, Margaret, brothel girl and coroner's witness, London, 46

Daily News, 97, 100, 155
Darwin family
  Charles Darwin senior, death from dissection, 95
  Charles Darwin junior, student dissections, 109
dead house, 10–11, 13, 15, 45, 88, 100, 102, 127–9, 138–9, 142, 184, 223, 236, 248, 250, 280, 290
dead trains, railway transportation, 189–90, 237–8, 261, 306–7
death and dying
  childbirth cases, 128, 182–4
  children, xv, 58
  historiography, 49–60
  infants, 258–60, 284–5
Deaths and Births Registration Act (1926), 130
Dentwart, Mr William, undertaker, St. Bartholomew's hospital, 92
Department for Work and Pensions, 42
Dickens, Charles, on dissection, 38–9, 63, 96–7
Dijkstrra, Bram, 106
discovery of DNA, 310

diseases of the destitute, 165–8
dissection
  animals, 98
  death certificates, 16
  domestic violence, 249–50
  ethnicity profiles, Irish and Italian, 151–3
  images and dramatic performance, 104–113
  preservation methods, 267
  private schools, 98–101
  room design, 84–7, 90, 190
  suicide victims, 88–93
  techniques, 189–90
Doncaster Poor Law Union, 212
Doughty, William, medical student, London, 100
Dumvill, Arthur, Manchester anatomist, 267–8
Durmer, Edward Cormer, churchwarden and guardian, London, 9–10

Engels, Frederick, 299
Essex Standard, West Suffolk Gazette and Eastern Counties, 207–8

faceless corpse, 44–9
Falmouth, for good health, 3
Fiest, Albert, *see under Rex versus Feist (1858)*
Figaro in London, 60
Fleming Directive (1871), 134
forensic medicine, 44–9, 89, 90–1
Fortnightly Review, 113
Foster, E. M, *Howard's End*, 1910, 168
Franall, Mr Harry Burrell, Poor Law Board inspector, 10–11
Freeman, Mr B, artist, anatomist and London journalist, 103
Fulbourn asylum, Cambridge, 213

Gabriel, Sir Thomas, police court judge, London, 100
Gannon, John, medical officer, Newington workhouse, 15
Goschen Minute (1869), 133
gratuities, for corpses, 204–5

grave-robbing, crime of Burking and sack men, 5–6
Gray's Anatomy (1858), 75, 254
Great Stink of London, 3
Greenland family, Newington, London, dissection case, 12
Gunell family, Manchester, dissection case, 273–5
Guy's hospital, London
  accountant, 14, 16
  dissection cases, 7–20
  gratuities for bodies, 16
  principal anatomy lecturer, 14, 15

Hardingstone Poor Law Union, Northamptonshire, 54
Harpurphey cemetery, Manchester, later renamed Phipps Park, 282–4, 287, 288, 290–1
Harris, Mr Albert, police court officer, London, 98
Harris, Right Hon Mr Evan, MP for Oxford, West and Abingdon, 20
Hastingdene, Mr John Milne, anatomist and body–dealer, Oxford, 236
Hawkins, Mr Charles, Anatomy Inspector, London, 100
Hewett, Mr B, undertaker, St. Bartholomew's hospital, 63
Heywood, Mr Abel, Manchester radical, 279–81
Hogg, Mr, body dealer, London undertaker, 11–19, 142–4, 173, 305
Holborn Union, London
  murders, 172
  paupers, 127, 140
  population, 147
  St. Giles parish, Little Ireland, 148–9
Holman Hunt, William, 106
Home Department, later the Home Office, 240
  see also Ministry of Health
Hookham family, Oxford, pauper protests about dissection,
  William, elderly pauper, protestor, 223–4
  Daniel, son, protestor, unemployed, 220–5, 239, 251

Hospital Pupil's Guide (1832), 93
houses of correction, 157–8, 163
Howard, Henry, Earl of Surrey, 24
    see also sonnets
Howlett, Mr Christopher Wright, police court witness, London, 99
Hull Packet and East Riding Times, 197
Hull Poor Law Union, 188, 215
Hulme township, near Manchester, pauper letters, 271–5
Human Tissue Act (2004), 20–1, 28, 38, 39
Humphrey, Professor G. M, anatomist and Chair of Surgery, Cambridge, 1886–1883, 192
Humphreys, Mr, city–coroner, Marylebone, London, 45
Hunt, Tristam, 279
Hussey, Dr Edward Law, Oxford city–coroner, 245–52
Hutton, Fiona, 265, 270, 276, 289–90

Jackson, Emma, murdered prostitute, London, 44
Jackson's Oxford Journal, 218–25, 233
Jack-the-Ripper, and body trade, 169–73
Jalland, Patricia, 49
Jenkins, Mr, master, Hull Union, 188–9
Johnson, Paul, 278
Jones, Rev. Pryse, Northamptonshire, 177–8
Jordan, Joseph, Manchester anatomist, 265–6
Jordanova, Ludmilla, 105

Kettering Poor Law Union, Northamptonshire, 65–6
    see also Martin, Mr B
Kidd, Alan, 269
King, Steven, 78

Ladds family, Peterborough, death in childbirth and dissection, Mary, midwife, 183
    Lucy, dead pauper, 182–4
    Joseph, widower, 182–4
Ladyman, Right Hon Steven, MP and Parliamentary Under Secretary of Health, 20
Lambeth workhouse, London, 173
Lancet, 150, 171
Lane, John, anatomist, body dealer, Cambridge, 213–14
Lang, Rev J. T, senior tutor, Corpus Christi College, Cambridge, anatomy burials, 199
Lawrence, Susan, 80
Lee, John, from Burghfield, Berkshire, 51–2
Leeds anatomy school, 269
Leicester Poor Law Union, 201, 231, 239–40
License of the Society of Apothecaries (LSA), 81
    see also Apothecaries Act (1815)
Litton, Julian, 63
Local Government Act, (1888) and (1894/5), 135, 200, 223
Local Government Board, 71, 157, 240
London, overview of poor law unions, 9
London Review, 44, 45, 47
Longley Strategy (1873), 134
Luton Poor Law Union, Bedfordshire, 198

Macalister, Professor Alexander, Chair of Anatomy, Cambridge, 1883–1919, 110–11, 140, 175–217, 261, 310, 311
Manchester anatomy school
    background, 264–8
    body business 264–302, 306
    dissection room, 270
    medical marketplace, 266
    surplus sold, 276, 297–8
Manchester Poor Law Union, Lancashire, 212, 280, 283, 285, 286–99
Martin, Mr B, relieving officer, Kettering Poor Law Union, Northamptonshire, 66
Marylebone workhouse infirmary, London, 150–1
Matthews sisters, paupers, Newington, London, 15
    see also Phoebe Clarke
McDougall, Alexander, Chair of board of guardians, Manchester, 299–300

McDougall, James, police court witness, London, 99
McPhial, Alexander, Anatomy Inspector, 261, 308
measles, 163
Medical Act (1858), 3–4, 34, 36, 77, 80, 93, 100, 133, 174, 193, 223
Medical Act (Extension) (1885), 34, 77, 134, 225, 231, 242
Medical and Philosophical Commentaries, 95
Medical education
  brief history, 77–80
  basic training, 80–7
  marketplace, 70, 242
  recruitment images 104–113
  numbers of recruits, 114, 138–9, 191, 242
  student experiences, 77, 80–8, 90, 94, 100, 104–113, 225–6, 242
  wax models, 232–3
Medical Register (1859), 106
Medical Relief (Disqualification Removal) Act, 134–5, 196–8, 206, 228
Medical Society Magazine, 213
Medical Society of London, 45
Mental hospitals, 157–8
Midland Free Press, 62
Milton, John, 24
  see also sonnets
Ministry of Health, 231, 244, 262
miscarriage, xv–vi
  see also still births
Mixer, Louisa, pauper, 11
Morning Chronicle, 61
Morris, Dr Rosebrook, surgeon, Northamptonshire, 178–81
Morton family, Leeds, Arthur and Mary, grieving paupers, 57–60
Murrison, Right Hon Andrew, MP for Westbury, 20

Newcastle–Upon–Tyne anatomy school
  body business, 269
  dissection room photographs, 111–12
Newington, St. Mary's workhouse, South London

clerk, 12
funerals, 7–9
historical lessons, 56
medical officer, 15
sick ward, 142; *see also Rex versus Feist* (1858)
Nicholson, Mr H. J, undertaker, St. Bartholomew's hospital, 92
Northampton Herald, 181
Northamptonshire, anti–anatomy poster, 24
Northern Star and National Trades' Journal, 57
Nottingham Poor Law Union, dissection cases and undertaker dispute, 135, 201, 202–3

Oakes, Thomas, lime pit death, East London, 62
Official Secrets Act (1889), on dissection, 71, 97
Old Age Pension Act (1908), 136, 241
Old Bailey, *see under Rex versus Feist* case
overseer of the poor, 51–2, 65–6
Oxford anatomy school
  background trends, 34, 102, 306
  body business, 218–63
  body parts, 228, 234–6, 256
  burial board, 233, 255
  coronial bodies, 245–52
  financial problems, 229–31
  medical marketplace, 246–7
  surplus sales, 301
  transportation costs, 256, 261
  undertaking costs, 235–6, 254–5, 256, 262
Oxford City local government
  cemetery committee, 255
  police court, 251
  workhouse, 249

Paget, Sir James, as medical student, St. Bartholomew's hospital, 90, 94
Paris
  anatomical method and popularity, 5, 39, 82
  public morgue, 86–7
Parker, Mr C. L, apothecary, St. Bartholomew's hospital, 90

pathology, 165
pauper funerals
  begging for burials, 50, 206, 224
  common burials, 43, 64, 70–1, 92, 149, 233, 271–6
  death customs, 49–60, 65, 181, 205–6, 219, 259–60, 276–85
  ethnicity, 151–2, 307 (Irish and Italians)
  non-conformist ceremonies, 69, 278–9
  parish coffins, 10, 63
  paupers interviewed by newspapers, 207–9
  poor law interments, 51, 52, 54, 64, 70–1, 149
  popular discourse, 68, 70–1
  pub raffles, 6, 51
  religious belief, 226, 278–9
  shrouds, 10
  state-funded today, 41–2
pauper stories, Northamptonshire, 54
pawnbroker, 88–90
Payne, Sergeant William, City of London coroner, 88–90, 125–7
Pegg, John, police court witness, London, 99
Pender, Right Hon James, Conservative MP, Mid–Northamptonshire 1895–1906, 135
Penny Satirist, 83, 86
Pickering Pick, Mr J, Inspector of Anatomy, 239–40
poison, 88–93
Poland, Mr Alfred, principal teacher in anatomy, Guy's hospital, 14, 15
Poor Law Amendment Act 1834 (New Poor Law), 4, 8, 19, 34–5, 43, 53, 64, 75–6, 154, 174, 195, 197, 209, 215, 240–2, 269, 276, 305, 307
Poor Law Commissioners, London, 279
Poor Law Handbook (1903), 53
post-mortems
  general caseload, 131, 247–8
  photography, 216–17; *see also* skiagraph
prisons, 157

prostitution, females and males, 127–9
Provincial and Medical Association, Oxford, 223

Radcliffe Infirmary, Oxford, 102, 222–3, 228, 234–5, 237, 247–8, 250–1, 254, 256–7, 260
railway fees, journeys and stations, 36, 189–190, 237–8, 256, 261, 275, 306–7
  see also dead trains
Reading Poor Law Union, Berkshire, 212, 231, 234–5, 237
Reed, Mr Alexander, physician, Westminster Hospital, 100
Reeves family, Catherine, Whitechapel burial, 156
Rex versus Feist (1858), Old Bailey case, 7–20, 37, 38, 56, 142, 144, 305
Reynolds' Magazine, 103
Richardson, Dr Benjamin Ward, anatomist, Manchester Square, London, 45, 46, 47
Richardson, Ruth, 20, 21, 32, 33, 34, 37, 75, 239, 277, 304, 307
riots, anti–Anatomy Act and its extension
  Cambridge, 6, 133, 193
  Liverpool, 133
  Manchester, 97
  Oxford, 222
  Sheffield, 97, 197, 269
  York, 6, 133
Robinson, Mr, QC prosecution counsel, Old Bailey, 10, 17
Royal College of Physicians, 79, 80–1
Royal College of Surgeons, 79, 81
Royal Commission on the Aged Poor (1894–5), 295

St. Bartholomew's hospital, London
  anatomy museum, 131
  body business, 18, 33–4, 76, 87, 119–74, 289
  body dealers, 127–30, 142–4
  body parts, 137–41
  brief history 121–5
  burial fees, 123–4

St. Bartholomew's hospital, London – *continued*
  burial ground, 128
  dissection rooms, 74, 102
  size, 33–4
  street geography, 145–54
  training, 245
  turnover of corpses, 141–2
  undertakers, 63–4, 92, 128, 130
St. George's hospital, London, 45
St. George's parish, Southwark, London, 55
St. Luke's Capital Dispensary, London, 140
St. Michael's and Angel Meadow, pauper burial ground, Manchester, 285, 290–1
Salford Poor Law Union, Lancashire, 290–1
Sappol, Michael, 74–5, 76
Saunders, Dr Sedgwick, medical officer of health, London, 100
Second Report of the Select Committee of Metropolitan Improvements (1838), 148
Select Committee on Medical Poor Relief (1844), 178
Shattock, Mr Mark, accountant, Guy's hospital, 14, 16
Sheffield anatomy school, 269
Shoreditch workhouse, London, 163
Shropshire, anti-anatomy poster, 24
Sims, Mr T. H, anatomist, body dealer, Cambridge, 192
Sitwell, Edith, poet and social commentator, 310
skiagraph, early X-rays, of dissections, 216, 310
sonnets, 23–4
  see also Anatomy Act posters and performance
Southam, Mr George, Manchester anatomist, 268
Southampton Poor Law Union, Hampshire, 212
Spaldling Poor Law Union, Lincolnshire, 206

Spencer, Right Hon Charles Robert, Liberal MP, Mid-Northamptonshire, 1892–5, 135
Statistical Society of Manchester, 278
stillbirths, xvi, 130–1, 182–4, 308–10
  see also miscarriage
Stokes, Mr, boot maker, coroner's witness, London, 46
Strange, Julie-Marie, 43, 50, 55–6
suicide, dissections of, 88–93, 124–7
surgeons
  dissection cases, 119, 178–9, 182–3, 202–3, 218
  general workload, 247
  qualifications, 79, 80–2, 270–1
syphilis, 128, 129

Taylor, Right Hon Richard, MP for Wyre Forest, 20
Teale, Mr William, undertaker, St. Bartholomew's hospital, 92
Tennyson, Lord Alfred, 42–3
Thomas, Dr, sick poor, Northamptonshire, 66
Thompson, Dr Alan, public vaccinator, Oxford, 257
Thompson family, Newington, London, dissection case, 14
Thomson, Professor Arthur (Chair of Anatomy, Oxford, 1885–1929), 203–4, 218–63, 287, 294–5, 310
Times, dissected paupers, 55, 91, 132, 138
Turby, Martha, asylum patient, pauper letter, Northamptonshire, 56
Turner, Dr Thomas, Manchester anatomist, 264–8
Turner, Mr J, London undertaker, 63

United Company of Barber Surgeons, 79
  see also Royal College of Surgeons
University College London, photographs, 111, 113
Unsworth, John, suicide, dissected St. Bartholomew's hospital, 91

Victoria cemetery, Hackney, London, 11
Victorian society, general discussion of
  contagious diseases, 42
  death and dying cultures, 43
  nature of information state, 7, 34–5
  population figures, 78
  railway trains for the dead, 36, 306–7
  rates of pauperism, 8–9
  teaching hospitals, 94–5
  underworld, 7

Waddington, Keir, 77, 90, 154
Walker, Dr George Alfred, *Gatherings from Graveyards*, 1839, 61–2
Walsall Poor Law Union, West Midlands, 69–70
wax models, 232–3
Weatherall, Mark, 191
Wellington, Duke of, state funeral, 42
Welsh, Mr John, meteorologist, Kew Gardens, London, 3
West Bromwich Poor Law Union, West Midlands, 69
Weston, Mr Robert, poor law clerk, Northamptonshire, 53

Wheeler, Mr Thomas Martin, Secretary to the National Charter Association and National Land Company, critic of pauper funerals, 57–8
Whitechapel Poor Law Union, London, 156, 172
  see also Jack-the-Ripper
Whiteman, Mr Justice, Old Bailey judge, 17–18
Whitehead, Mary, Newington, London, dissected pauper, 7–8, 11, 17–18
Wiggins, Mr John, grieving father, Northamptonshire pauper, 65, 67
Windle, Professor Bertram C. A, Chair of Anatomy, Birmingham Medical School, 69
Wisbech Poor Law Union, Norfolk, 213
Wolverhampton Poor Law Union, West Midlands, 231
Wordsworth, William, 24
  see also sonnets
Worshipful Company of Apothecaries, 80
  see also Apothecaries Act (1815)
Wyatt, Sir Thomas, 24
  see also sonnets

X-ray, *see skiagraph*

CPI Antony Rowe
Chippenham, UK
2016-12-23 13:18